AF355966

Advances in Neuroinflammation

Advances in Neuroinflammation

Junhui Wang
Hongxing Wang
Jing Sun

Basel • Beijing • Wuhan • Barcelona • Belgrade • Novi Sad • Cluj • Manchester

Editors

Junhui Wang
Lunenfeld Tanenbaum
Research Institute
Mount Sinai Hospital
Toronto
Canada

Hongxing Wang
Neurology
Xuanwu Hospital, Capital
Medical University
Beijing
China

Jing Sun
Pathology
Capital Medical University
Beijing
China

Editorial Office
MDPI AG
Grosspeteranlage 5
4052 Basel, Switzerland

This is a reprint of articles from the Special Issue published online in the open access journal *Brain Sciences* (ISSN 2076-3425) (available at: www.mdpi.com/journal/brainsci/special_issues/IDOF63A8VR).

For citation purposes, cite each article independently as indicated on the article page online and as indicated below:

Lastname, A.A.; Lastname, B.B. Article Title. *Journal Name* **Year**, *Volume Number*, Page Range.

ISBN 978-3-7258-2346-8 (Hbk)
ISBN 978-3-7258-2345-1 (PDF)
doi.org/10.3390/books978-3-7258-2345-1

© 2024 by the authors. Articles in this book are Open Access and distributed under the Creative Commons Attribution (CC BY) license. The book as a whole is distributed by MDPI under the terms and conditions of the Creative Commons Attribution-NonCommercial-NoDerivs (CC BY-NC-ND) license.

Contents

About the Editors

Junhui Wang

Dr. Junhui Wang received his Ph.D in neurobiology with specialty in astrocytes at Peking University, China. He then trained in behavioral studies during his postdoc period at the University of Manitoba, Canada. Afterwards, he continued his training in neuroscience research at the University of Alberta and University of Toronto, Canada, as a postdoc or research associate. His current research focuses on the interactions between neurons and astrocytes, and he is especially interested in the role of astrocytes in neurodegenerative diseases, especially tauotopathy, lewy body, and AD.

Hongxing Wang

Dr. Hongxing Wang graduated from Shan'xi Medical University, China, with his M.D. and obtained his Ph.D. from Shanghai Jiaotong University, China. He completed his postdoc training at the University of Manitoba, Canada, and Johns Hopkins University, US. He is a neurologist and has conducted extensive research, including clinical studies on neuropsychiatric diseases and neurodegenerative diseases. His current research focuses on how gut microbiota affect Alzheimer's disease at the molecular level, especially utilizing the translational medicine route to further understand brain dysfunction.

Jing Sun

Dr. Jing Sun obtained her M.D at Shandong University, China, and her Ph.D. at Capital Medical University, China. She is a pathologist, and her research interests mainly include clinical neuropathology, neuroglial cells, and neuroinflammation.

Preface

This Special Issue, organized by Dr. Junhui Wang (Mount Sinai Hospital, Toronto, Canada), Dr. Jing Sun (Capital Medical University, Beijing China), and Dr. Hongxing Wang (Xuan Wu Hospital), aims to present the recent advances in neuroinflammation research on preclinical and clinical areas, especially focusing on, but not limited to, neurodegenerative diseases. In vivo and in vitro studies related to advances in conception, technology renovation, novel models, etc., which could be applied to study the role of neuroinflammation in neurodegenerative diseases, are especially welcome. In addition to research articles, reviews or mini reviews covering the recent progress of the neuroinflammatory role in neurodegenerative diseases are encouraged too. Case reports presenting some rare glia-related conditions (neuroinflammation) from clinical settings with imaging or neuropathology patterns are also welcome to be submitted.

This Special Issue collated 10 articles from multifaced subjects of neuroscience and broadened the deep understanding of the role of neuroinflammation in multiple neurological diseases. We have no doubt that this Special Issue will benefit both researchers and clinicians in this field.

Finally, we sincerely express our gratitude to the Editorial Office of *Brain Sciences* for their generous support, especially Mr. Eric Yu. We would also like to thank all the authors of the submitted articles for their willingness to submit their research to our Special Issue.

Junhui Wang, Hongxing Wang, and Jing Sun
Editors

Editorial

Advances in Neuroinflammation

Junhui Wang [1,2,*], **Jing Sun** [3] **and Hongxing Wang** [4]

1. Lunenfeld-Tanenbaum Research Institute, Mount Sinai Hospital, Toronto, ON M5G 1X5, Canada
2. Thyropathy Hospital, Sun Simiao Hospital, Beijing University of Chinese Medicine, Tongchuan 727000, China
3. Department of Pathology, Capital Medical University, Beijing 100069, China; sunjing@ccmu.edu.cn
4. Department of Neurology, Xuanwu Hospital, Capital Medical University, Beijing 100053, China; wanghongxing@xwh.ccmu.edu.cn
* Correspondence: junhui@lunenfeld.ca

Citation: Wang, J.; Sun, J.; Wang, H. Advances in Neuroinflammation. *Brain Sci.* **2024**, *14*, 965. https://doi.org/10.3390/brainsci14100965

Received: 20 August 2024
Revised: 24 September 2024
Accepted: 25 September 2024
Published: 26 September 2024

Copyright: © 2024 by the authors. Licensee MDPI, Basel, Switzerland. This article is an open access article distributed under the terms and conditions of the Creative Commons Attribution (CC BY) license (https://creativecommons.org/licenses/by/4.0/).

Recent research in neuroscience has shown significant advancements in relation to neuroinflammation, especially its role in neurological diseases, including neurodegenerative diseases. Over-activated central nerve system (CNS) "immune cells"—astrocytes and microglial cells with their subsequent cytokines and chemokines—are, *sine qua non*, fundamental mechanisms of pathogenesis in many neurological diseases [1]. Neuroinflammation acts as a double-edged sword in the CNS. While working as a defense response against brain insults by removing toxic agents and minimizing the detrimental effects, a prolonged battle in the process of inflammation will result in the development of chronic neuroinflammatory conditions, such as neurodegenerative diseases [2]. Therefore, we are facing a dilemma in pinpointing the role of neuroinflammation in the pathological or even physiological processes of the CNS. The primary goal of this Special Issue, titled "Advances in Neuroinflammation", is to investigate the recent advances in neuroinflammation research in both preclinical and clinical areas, in particular, but not limited to, neurodegenerative diseases. We collected and published five research and five review papers from distinguished scientists in the field; each publication addresses the role of neuroinflammation in neurological diseases from the novel perspective of the authors, presenting possible new direction for future study in the field.

In the review papers, the authors aim to decipher the possible role and importance of neuroinflammation in CNS diseases, especially neurodegenerative diseases. For years, contradictory findings have existed concerning the association between both aberrant apolipoprotein E (APOE)-ε4 and serum lipids and the occurrence of Alzheimer's disease (AD) [3,4]. In the first paper, Xu et al. performed a meta-analysis to investigate the relationship of apolipoprotein E alleles and serum lipids with; they found that the elevated total cholesterol (TC) and low-density lipoprotein (LDL) levels showed considerable heterogeneity between patients with AD and healthy controls. Higher TC and LDL levels were found in APOEε4 allele carriers compared with non-carriers, and the difference was more significant in patients with AD compared to healthy controls. Their results supported the hypothesis that the APOEε4 allele might lead to the development of AD through influencing lipid metabolism (Contribution 1).

As mentioned above, neuroinflammation is closely related to the onset of many neurological diseases. However, the underlying mechanisms have not yet been fully deciphered. Balistreri et al. presented a review paper to discuss the recent progress in this field, focusing on the important roles of peripheral and infiltrated monocytes and clonotypic cells, the gut–brain axis, the apelinergic system, the endothelial glycocalyx of the endothelial component of neuronal vascular units, non-coding RNA and other types of gene expression, etc., in the development of neurological diseases. This review illustrated the complex neuroinflammatory reactions and novel mechanisms proposed by the authors, significantly adding valuable information to the comprehension of the complex etiological link between neuroinflammation and neurological diseases (Contribution 2).

AD is a detrimental CNS disease with immense complexity in terms of its mechanisms, which is a major obstacle in understanding its pathogenesis. The third paper, authorized by prestigious scientist Weaver Donald, proposed that neuroinflammation was the central player in unifying 30 different risk factors of AD; this review identified 30 risk factors for AD and extended the analysis to further identify neuroinflammation as a unifying player presented in all of these risk factors. In this review, the author claimed that the dysfunction of the neuroimmune–neuroinflammation axis was central to all 30 identified risk factors. Though the nature of the neuroinflammatory involvement varies in different conditions, the activation of glial cells and the release of pro-inflammatory proteins are common pathways shared by all these risk factors. While discussing a very novel point of view, this review article provided further evidence of the importance of neuroinflammatory mechanisms in the etiology of AD (Contribution 3).

In another review paper, Yang et al. detailed the role of astrocytes in Amyotrophic Lateral Sclerosis (ALS) and emphasized the importance of the non-cell-autonomous role of astrocytes in this detrimental CNS disease. According to this paper, astrocytes can play a pivotal role in ALS development via participating in calcium homeostasis imbalance, mitochondrial dysfunction, abnormal lipid and lactate metabolism, glutamate excitotoxicity, etc. This review systemically outlined the possible contributions of reactive astrocytes in the pathogenesis of ALS. More importantly, comprehensive evidence is provided stating that astrocytes could be the potential therapeutic target for the treatment of ALS (Contribution 4).

Glioblastoma, as the most common and malignant brain tumor, usually presents with high morbidity and mortality [5]. Extreme complication of the tumor microenvironment is a formidable challenge in advancing glioblastoma therapy for the medical community, and neuroinflammation is characterized by a variety of resident or infiltrating inflammatory cells—key players in creating this complexity. In their review paper, Li et al. pointed out that neuroinflammation not only builds a unique tumor environment for glioblastoma cells to develop and grow, but also played important roles in regulating tumor aggressiveness and treatment resistance; they also emphasized that the anti-tumor microenvironment interventions, such as anti-inflammation, could be used as potential therapeutic tools against glioblastoma in the future (Contribution 5).

Chronic idiopathic demyelinating polyneuropathy (CIDP) is an acquired, immune-mediated neuropathy with very limited treatment options thus far [6]. Recently, subcutaneous immunoglobulins (SCIgs) have been employed in clinical setting as a maintenance therapy for CIDP. In the first research article of this Special Issue, Alonge et al. retrospectively explored electrophysiological and efficacy data from 15 patients who received the SCIg treatment. They reported that SCIg maintenance therapy could preserve nerve function in CIDP with good efficacy and safety properties. Electronystagmography can be used to evaluate treatment effectiveness and was a useful instrument for the follow-up and prognostic assessment of CIDP. This study further strengthened evidence in relation to the efficacy of SCIg maintenance therapy in CIDP (Contribution 6).

Sleep deprivation could adversely impact immune function, cognitive memory, learning ability, etc. Studies have revealed that sleep deprivation can lead to inflammatory responses in the CNS. Li et al. investigated the protective role of dexmedetomidine, an anesthetic compound, in sleep deprivation by focusing on its possible anti-inflammatory role in the CNS. In a sleep deprivation mouse model, the authors claimed that dexmedetomidine could significantly improve anxiety-like behaviors in the sleep-deprived mice and could attenuate inflammatory responses and oxidative stress in the CNS by inhibiting the activation of the p38/MSK1/NFκB pathway (Contribution 7).

Epidermal growth factor receptor (EGFR) gene deficits and the subsequent activation of the EGFR signaling pathway promote the genesis of gliomas [7]. However, whether factors exist within the microenvironment that can lead to EGFR activation is currently unknown. In a clinical study on glioma, Zhou et al. probed the association between the EGFR and IFN-γ pathways and their possible synergistic effects on survival prediction and immune escape in glioma patients. Their study concluded that cytokine IFN-γ might be an upstream trigger of EGFR signaling activation, and EGFR-related and IFN-γ-related signatures could be jointly used to stratify patients into well-defined risk groups. High-risk patients tended to have a poorer prognosis and a more inhibitory microenvironment. They claimed that these patients might be more suitable for immune checkpoint blockade (ICB) therapy or other immunotherapeutic approaches (Contribution 8).

In a clinical study of insomnia, Wang et al. investigate the clinical efficacy of biofeedback on insomnia and its potential neural mechanisms. They found that a biofeedback treatment based on alpha power and prefrontal EMG could relieve insomnia and ameliorate anxiety and depression. The underlying mechanism may be attribute to increased alpha power, decreased beta and theta power, and decreased EMG power (Contribution 9).

The last study in this Special Issue, authored by Li et al., used a mendelian randomization (MR) method to investigate the causal effect of circulating inflammatory proteins on multiple sclerosis (MS) by digging into the data from a large-scale genome-wide association study (GWAS). They reported that 91 circulating inflammatory proteins were closely associated with the onset and progression of MS. They claimed that this study provided new insights into the relationship between circulating inflammatory proteins and MS, and discussed the possibility of using these circulating inflammatory proteins as potential biomarkers and therapeutic targets for MS in the future (Contribution 10).

In a nutshell, the above studies in our Special Issue pinpoint the importance and complexity of neuroinflammation from multiple perspectives, and strengthen the idea that neuroinflammation plays a very important role in a variety of neurological diseases, especially neurodegenerative diseases. In general, our Special Issue highlights the urgency to further investigate the inflammatory role in disease genesis with continuous research by using diverse new tools and technologies.

Author Contributions: Conceptualization, J.W.; writing—original draft preparation, J.W.; writing—review and editing, J.S. and H.W. All authors have read and agreed to the published version of the manuscript.

Conflicts of Interest: The authors declare no conflicts of interest.

List of Contributions

1. Xu, H.; Fu, J.; Mohammed Nazar, R.B.; Yang, J.; Chen, S.; Huang, Y.; Bao, T.; Chen, X. Investigation of the Relationship between Apolipo protein E Alleles and Serum Lipids in Alzheimer's Disease: A Meta-Analysis. *Brain Sci.* **2023**, *13*, 1554.
2. Balistreri, C.R.; Monastero, R. Neuroinflammation and Neurodegenerative Diseases: How Much Do We Still Not Know? *Brain Sci.* **2024**, *14*, 19.
3. Weaver, D.F. Thirty Risk Factors for Alzheimer's Disease Unified by a Common Neuroimmune–Neuroinflammation Mechanism. *Brain Sci.* **2024**, *14*, 41.
4. Yang, K.; Liu, Y.; Zhang, M. The Diverse Roles of Reactive Astrocytes in the Pathogenesis of Amyotrophic Lateral Sclerosis. *Brain Sci.* **2024**, *14*, 158.
5. Li, X.; Gou, W.; Zhang, X. Neuroinflammation in Glioblastoma: Progress and Perspectives. *Brain Sci.* **2024**, *14*, 687.
6. Alonge, P.; Di Stefano, V.; Lupica, A.; Gangitano, M.; Torrente, A.; Pignolo, A.; Maggio, B.; Iacono, S.; Gentile, F.; Brighina, F. Clinical and Neurophysiological Follow-Up of Chronic Inflammatory Demyelinating Polyneuropathy Patients Treated with Subcutaneous Immunoglobulins: A Real-Life Single Center Study. *Brain Sci.* **2023**, *13*, 10.
7. Li, J.; Zhang, H.; Deng, B.; Wang, X.; Liang, P.; Xu, S.; Jing, Z.; Xiao, Z.; Sun, L.; Gao, C.; et al. Dexmedetomidine Improves Anxiety-like Behaviors in Sleep-Deprived Mice by Inhibiting the p38/MSK1/NFκB Pathway and Reducing Inflammation and Oxidative Stress. *Brain Sci* **2023**, *13*, 1058.
8. Zhou, X.; Liang, T.; Ge, Y.; Wang, Y.; Ma, W. The Crosstalk between the EGFR and IFN-γ Pathways and Synergistic Roles in Survival Prediction and Immune Escape in Gliomas. *Brain Sci.* **2023**, *13*, 1349.
9. Wang, H.; Hou, Y.; Zhan, S.; Li, N.; Liu, J.; Song, P.; Wang, Y.; Wang, H. EEG Biofeedback Decreases Theta and Beta Power While Increasing Alpha Power in Insomniacs: An Open-Label Study. *Brain Sci.* **2023**, *13*, 1542.

10. Li, X.; Ding, Z.; Qi, S.; Wang, P.; Wang, J.; Zhou, J. Genetically Predicted Association of 91 Circulating Inflammatory Proteins with Multiple Sclerosis: A Mendelian Randomization Study. *Brain Sci.* **2024**, *14*, 833.

References

1.   Li, T.; Lu, L.; Pember, E.; Xinuo, L.; Zhang, B.; Zhu, Z. New Insights into Neuroinflammation Involved in Pathogenic Mechanism of Alzheimer's Disease and Its Potential for Therapeutic Intervention. *Cells* **2022**, *11*, 1925. [CrossRef] [PubMed]
2.   Wyss-Coray, T.; Mucke, L. Inflammation in Neurodegenerative Disease—A Double-Edged Sword. *Neuron* **2002**, *35*, 419–432. [CrossRef] [PubMed]
3.   Wang, P.; Zhang, H.; Wang, Y.; Zhang, M.; Zhou, Y. Plasma cholesterol in Alzheimer's disease and frontotemporal dementia. *Transl. Neurosci.* **2020**, *11*, 116–123. [CrossRef] [PubMed]
4.   Raygani, A.V.; Rahimi, Z.; Kharazi, H.; Tavilani, H.; Pourmotabbed, T. Association between apolipoprotein E polymorphism and serum lipid and apolipoprotein levels with Alzheimer's disease. *Neurosci. Lett.* **2006**, *408*, 68–72. [CrossRef] [PubMed]
5.   Yu, M.W.; Quail, D.F. Immunotherapy for Glioblastoma: Current Progress and Challenges. *Front. Immunol.* **2021**, *12*, 676301. [CrossRef] [PubMed]
6.   Dziadkowiak, E.; Nowakowska-Kotas, M.; Budrewicz, S.; Koszewicz, M. Pathology of Initial Axon Segments in Chronic Inflammatory Demyelinating Polyradiculoneuropathy and Related Disorders. *Int. J. Mol. Sci.* **2022**, *23*, 13621. [CrossRef] [PubMed]
7.   Saadeh, F.S.; Mahfouz, R.; Assi, H.I. EGFR as a clinical marker in glioblastomas and other gliomas. *Int. J. Biol. Markers* **2018**, *33*, 22–32. [CrossRef] [PubMed]

Disclaimer/Publisher's Note: The statements, opinions and data contained in all publications are solely those of the individual author(s) and contributor(s) and not of MDPI and/or the editor(s). MDPI and/or the editor(s) disclaim responsibility for any injury to people or property resulting from any ideas, methods, instructions or products referred to in the content.

Article

Clinical and Neurophysiological Follow-Up of Chronic Inflammatory Demyelinating Polyneuropathy Patients Treated with Subcutaneous Immunoglobulins: A Real-Life Single Center Study

Paolo Alonge, Vincenzo Di Stefano , Antonino Lupica , Massimo Gangitano *, Angelo Torrente , Antonia Pignolo , Bruna Maggio, Salvatore Iacono , Francesca Gentile and Filippo Brighina

Neurology Unit, Department of Biomedicine, Neuroscience, and Advanced Diagnostics (BiND), University of Palermo, 90129 Palermo, Italy
* Correspondence: massimo.gangitano@unipa.it

Abstract: Background: chronic idiopathic demyelinating polyneuropathy (CIDP) is an acquired, immune-mediated neuropathy characterized by weakness, sensory symptoms and significant reduction or loss of deep tendon reflexes evolving over 2 months at least, associated with electrophysiological evidence of peripheral nerve demyelination. Recently, subcutaneous immunoglobulins (SCIg) have been introduced in clinical practice as a maintenance therapy for CIDP; nevertheless, electrophysiological and efficacy data are limited. Methods: to evaluate SCIg treatment efficacy, we retrospectively reviewed data from 15 CIDP patients referring to our clinic, receiving SCIg treatment and who performed electrophysiological studies (NCS) and clinical scores (MRC sumscore, INCAT disability score and ISS) before starting the treatment and at least one year after. Results: NCS showed no significant changes before and during treatment for all the nerves explored. Clinical scores did not significantly change between evaluations. Correlation analysis evidenced a positive correlation of cMAPs distal amplitude with MRC sumscore and a trend of negative correlation with the INCAT disability score. Conclusions: SCIg maintenance therapy preserves nerve function in CIDP with a good efficacy and safety. Treatment effectiveness can be assessed with ENG, which represents a useful instrument in the follow-up and prognostic assessment of CIDP.

Keywords: CIDP; SCIg; cMAP; SNAP; ISS; INCAT; MRC; subcutaneous immunoglobulin

Citation: Alonge, P.; Di Stefano, V.; Lupica, A.; Gangitano, M.; Torrente, A.; Pignolo, A.; Maggio, B.; Iacono, S.; Gentile, F.; Brighina, F. Clinical and Neurophysiological Follow-Up of Chronic Inflammatory Demyelinating Polyneuropathy Patients Treated with Subcutaneous Immunoglobulins: A Real-Life Single Center Study. *Brain Sci.* **2023**, *13*, 10. https://doi.org/10.3390/brainsci13010010

Academic Editor: Marco Luigetti

Received: 25 November 2022
Revised: 17 December 2022
Accepted: 18 December 2022
Published: 21 December 2022

Copyright: © 2022 by the authors. Licensee MDPI, Basel, Switzerland. This article is an open access article distributed under the terms and conditions of the Creative Commons Attribution (CC BY) license (https://creativecommons.org/licenses/by/4.0/).

1. Introduction

Chronic idiopathic demyelinating polyneuropathy (CIDP) is an acquired, immune-mediated polyradiculoneuropathy evolving over 2 months at least [1]. The typical form is characterized by sensory symptoms (e.g., paresthesia, sensory loss), distal muscle weakness and reduced or absent deep tendon reflexes, with a distal and symmetrical involvement that progresses proximally [2]. Cranial nerves and the autonomic system are usually spared in CIDP. However, there are also atypical forms in which some of the classical symptoms are absent (e.g., motor CIDP or sensory CIDP) or an asymmetrical or focal involvement is observed [3]. Electrophysiological findings play a key role in the diagnosis and monitoring of CIDP: according to the EAN/PNS (European Academy of Neurology/Peripheral Nerve Society) diagnostic criteria, the demonstration of peripheral nerve demyelination in two or more nerves is required for a defined diagnosis [4]. Electrophysiological variables have been also extensively used in clinical trials to evaluate the response to treatment and the progression of the disease [5,6].

There are several therapeutic options for CIDP, which include intravenous immunoglobulin (IVIG), plasma exchange (PEX) and glucocorticoids. After an induction therapy, most patients require a maintenance therapy with periodic IVIG administration, PEX procedures or immunodepressants (steroids, rituximab) to prevent relapses and progression [4,7,8]. Subcutaneous immunoglobulin therapy (SCIg) has been used as an alternative to IVIG in primary immunodeficiencies for over thirty years. Compared to IVIG, which are administered every 3–4 weeks, SCIg are administered in smaller doses; hence, the frequency of administration is higher (once or twice weekly). Evidence shows that, while the SCIg efficacy is similar to IVIG, patients usually report a lower incidence of side effects (e.g., headache, local reactions in injection site, renal and cardiac impairment) and a better quality of life; this is attributed to the lower peak serum dose reached by SCIg compared to IVIG (61%); another advantage is that SCIg therapy does not require an intravenous access. Hence, SCIg is commonly administered at home [9,10]. Recently, SCIg has been introduced in clinical practice as a maintenance therapy even for CIDP; nevertheless, data on the efficacy of SCIg and electrophysiological data are limited.

2. Materials and Methods

2.1. Study Procedures

The study was conducted in accordance with the Declaration of Helsinki and approved by the Ethics Committee Palermo I, University of Palermo (Protocol code 07/2020; 13 July 2020). In this study, we present a retrospective evaluation of the efficacy of SCIg treatment in a population of CIDP patients using electroneurography (ENG) and clinical scores.

2.2. Patient Demographics

We reviewed data from patients referring to our clinic ("Policlinico Paolo Giaccone di Palermo" –"Centro per la Diagnosi e Cura della Malattie Neuromuscolari Rare") from January 2014 to September 2022.

Inclusion criteria:

Patients accessing our clinic were enrolled in the presence of the following criteria:

- *Age >18 years;*
- *Diagnosis of definite CIDP according to the EAN/PNS 2021 criteria;*
- *Treatment with SCIg;*
- *Evaluation with apposite clinical scales (INCAT, ISS, MRC) and nerve conduction studies.*
- *Exclusion criteria:*
- *Lack of infomed consent to participation;*
- *Diagnosis of probable or possible CIDP according to the EAN/PNS 2021 criteria;*
- *Lack of response to IVIg.*

2.3. Clinical Assessment

The inflammatory neuropathy cause and treatment (INCAT) disability score is calculated by summing a score measuring arm impairment (0 = no upper limb problems; 5 = inability to use either arms for purposeful movements) and another measuring leg impairment (0 = walking unaffected; 5 = restricted to wheelchair). The INCAT score can range from 0 (no disability) to 10 (maximum disability) [11].

Similarly, the INCAT sensory sumscore (ISS) evaluates sensory impairment by measuring pinprick sensation, vibration sensation and two-point discrimination at arms and legs. It is calculated by adding scores obtained by all four limbs and it ranges from 0 (normal sensation) to 20 (severe sensory deficit) [12].

The Medical Research Council (MRC) sumscore measures strength by applying the MRC 5-point system (0 = no movement; 5 = movement completed against full resistance) to six muscle groups (abduction of the arm, flexion of the forearm, extension of the wrist, flexion of the leg, extension of the knee, dorsal flexion of the foot) of both sides. It ranges from 0 (minimum strength) to 60 (maximum strength) [13].

The INCAT disability score, the ISS and the MRC sumscore were developed specifically to evaluate patients affected by inflammatory polyneuropathies and have been used in several clinical trials to estimate the efficacy of treatments and the clinical progression of the disease over time.

2.4. Nerve Conduction Studies (NCS)

Sensory nerve action potentials (SNAPs) and compound muscle action potentials (CMAPs) were recorded, analyzing distal latencies (dL), negative peak amplitudes (dA) and conduction velocities (CV). We investigated median and ulnar nerves for the upper limbs and peroneal, tibial and sural nerves for the lower limbs, according to standard procedures (i.e., bipolar surface stimulating electrodes delivering rectangular pulses 0.1–0.5 ms in duration with recording electrodes placed over the recording site, with a ground electrode placed between the recording electrodes and stimulation site). In particular, the study protocol was defined as follows:

For upper-limb SNAPs: stimulation at wrist and registration from II digit (medial nerve) and V digit (ulnar nerve);

For upper-limb CMAPs: stimulation at the wrist and elbow and recording from abductor pollicis brevis (APB) for median nerve; stimulation at the wrist and elbow 4 cm distal from the medial epicondyle of the humerus and recording from the abductor digiti minimi (ADM) muscle for the ulnar nerve;

For lower-limb SNAPs: stimulation at the posterior-lateral calf, recording from the lateral malleolus for the sural nerve;

For lower-limb CMAPs: stimulation at the medial malleolus and popliteal fossa, recording from the abductor hallucis brevis (AHB) muscle for the tibial nerve; stimulation at the anterior ankle and popliteal fossa, recording from the extensor digitorum brevis (EDB) muscle for the peroneal nerve.

2.5. Statistical Analysis

Neurophysiological variables (continuous) and clinical scores (discrete) were compared using the Mann–Whitney test to detect changes between the evaluations before and during therapy. Correlations between clinical scores and neurophysiological variables (cMAPs and SAPs distal amplitude) were evaluated using Pearson's r value. Significance was set at 0.05 for all the analyzed variables. Data are presented as a median $\pm$ interquartile range, except for correlations, which are presented as a mean value $\pm$ standard deviation. All the analyses were performed with JASP (version 0.16.2; computer software).

3. Results

Out of 19 patients, 3 were lacking a baseline electrophysiological evaluation and 1 did not reach a 1-year follow-up. Fifteen patients were included in the final analysis (9 males, 56 $\pm$ 13 years; see Table 1 for population characteristics). Thirteen patients (87%) showed a typical CIDP phenotype, while two showed an atypical pattern (motor CIDP). Five patients out of fifteen did not have an evaluation with the mentioned clinical scales before and/or after treatment start; therefore, analysis on clinical scores and correlations were performed on the remaining 10 patients. Evaluations during SCIg treatment (both clinical and neurophysiological) were performed after a median interval of 16 months (IQ 13–19) from the start. Median time between ENG evaluations was 37 months (IQ range 29–42). The median time between clinical evaluations was 35 months (IQ range 10–20; see Figure 1 for histogram of follow-up time). Before starting SCIg as maintenance therapy, nine patients (60%) were treated with prednisone; nine patients (60%) received IVIg as a maintenance therapy; one patient (6%) was treated with cyclophosphamide; and one (6%) with azathioprine.

At baseline, the median nerve (both motor and sensitive) was tested in 60% of patients, the ulnar (both motor and sensitive) nerve in 60%, the peroneal motor nerve in 60%, the tibial nerve in 73.3% and the sural nerve in 20%. Repeat testing was conducted in 100% of patients for all the nerves, except for the sural nerve, which was tested in 66% of the patients at follow-up.

NCS showed no significant changes before and during treatment for all the nerves explored (Table 2); a trend of worsening was observed for dA (11.9 vs. 4.3 uV; $p = 0.078$) and CV (44.46 vs. 32.13 m/sec; $p = 0.17$) registered from the right sensitive median nerve, while

CV registered from the right peroneal nerve improved at follow-up (29.82 vs. 44.78 m/sec; $p= 0.15$).

Table 1. Clinical data in our cohort of CIDP patients treated with SCIg. Chronic inflammatory demyelinating polyneuropathy, CIDP; intravenous immunoglobuline, IVIg; subcutaneous immunoglobulins, SCIg.

Patient Code	Age (Years)	Sex	Duration of Disease (Months)	CIDP Phenotype	Previous Maintenance Therapy	Duration of SCIg Treatment (Months)	SCIg Dosage (Monthly)	Other Conditions
001	64	M	90	Motor CIDP	IVIg	84	60 g	Diabetes mellitus Osteopenia
002	59	F	21	Typical	Prednisone	19	80 g	-
003	52	F	204	Motor CIDP	Prednisone IVIg	33	80 g	-
004	74	M	31	Typical	Prednisone IVIg	13	120 g	Atrial fibrillation
005	80	M	336	Typical	IVIg cyclophosphamide	16	30 g	-
006	48	F	20	Typical	Prednisone	15	60 g	-
007	48	M	21	Typical	Prednisone	18	60 g	-
008	44	M	72	Typical	Prednisone	13	80 g	-
009	44	F	36	Typical	Azathioprine	14	80 g	-
010	61	M	48	Typical	IVIg	13	80 g	-
011	48	M	20	Typical	Prednisone	17	120 g	-
012	73	M	96	Typical	IVIg	30	100 g	Peripheral arterial disease, diabetes mellitus
013	60	M	84	Typical	Prednisone IVIg	13	80 g	Dyslipidemia
014	73	F	76	Typical	IVIg	15	60 g	HCV-related hepatopathy, osteoporosis
015	55	F	50	Typical	Prednisone IVIg	21	60 g	-

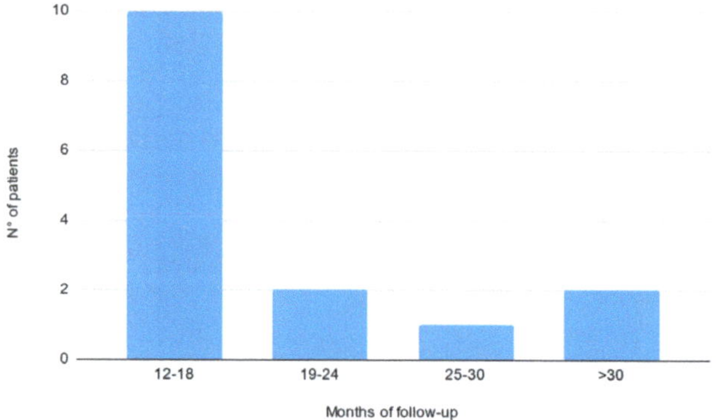

Figure 1. Histogram of follow-up time reported in months in our cohort of CIDP patients treated with SCIg.

Table 2. Electrophysiological data. Time 1: pre-SCIg; time 2: during SCIg; N: number of patients; Dx: right nerve; Sn: left nerve; dL: distal latency; dA: distal amplitude; SAP: sensory action potential; cMAP: compound motor action potential; CV: conduction velocity; SD: standard deviation.

	Time	N	Mean	SD	*p*
SAPs median Dx dL (ms)	1	9	2.319	1.122	
	2	10	2.340	1.845	0.774
SAPs median Dx dA (uV)	1	9	11.867	15.890	
	2	10	4.370	7.170	0.078
SAPs median Dx CV (m/s)	1	9	44.467	20.205	
	2	10	32.130	23.677	0.175
cMAPs median Dx dL (ms)	1	10	4.640	2.118	
	2	12	5.342	1.718	0.291
cMAPs median Dx dA (mV)	1	10	4.430	2.113	
	2	12	3.692	2.190	0.339
cMAPs median Dx CV (m/s)	1	10	39.540	15.752	
	2	12	39.942	12.487	0.947
SAPs median Sn dL (ms)	1	6	2.000	1.698	
	2	9	2.367	2.170	0.810
SAPs median Sn dA (uV)	1	6	2.133	1.900	
	2	9	7.022	11.782	0.904
SAPs median Sn CV (m/s)	1	6	29.583	25.443	
	2	9	24.211	25.225	0.626
cMAPs median Sn dL (ms)	1	9	5.011	2.913	
	2	10	4.650	1.828	0.837
cMAPs median Sn dA (mV)	1	9	4.078	2.100	
	2	10	5.260	3.364	0.513
cMAPs median Sn CV (m/s)	1	9	37.189	15.430	
	2	10	40.250	14.481	0.842
SAPs ulnar Dx dL (ms)	1	6	1.797	1.059	
	2	10	2.367	1.603	0.301
SAPs ulnar Dx dA (uV)	1	6	16.933	20.476	
	2	10	9.370	11.741	0.703
SAPs ulnar Dx CV (m/s)	1	6	44.150	23.869	
	2	10	32.360	25.265	0.444
cMAPs ulnar Dx dL (ms)	1	9	3.452	1.896	
	2	9	4.100	1.429	0.250
cMAPs ulnar Dx dA (mV)	1	9	4.956	2.220	
	2	9	4.511	2.723	0.690
cMAPs ulnar Dx CV (m/s)	1	9	45.267	20.194	
	2	9	40.011	14.217	0.354
SAPs ulnar Sn dL (ms)	1	4	2.100	1.490	
	2	8	2.357	1.221	0.864
SAPs ulnar Sn dA (uV)	1	4	3.250	2.575	
	2	8	8.588	14.304	0.865

Table 2. *Cont.*

	Time	N	Mean	SD	p
SAPs ulnar Sn CV (m/s)	1	4	33.025	23.085	1.000
	2	8	34.112	23.129	
cMAPs ulnar Sn dL (ms)	1	5	5.320	3.214	0.224
	2	9	3.644	1.696	
cMAPs ulnar Sn dA (mV)	1	5	2.800	2.623	0.317
	2	9	4.456	2.706	
cMAPs ulnar Sn CV (m/s)	1	5	32.160	17.373	0.257
	2	9	42.889	14.089	
cMAPs peroneal Dx dL (ms)	1	10	7.530	5.707	0.967
	2	9	6.078	2.621	
cMAPs peroneal Dx dA (mV)	1	10	1.990	1.959	0.870
	2	9	1.756	1.490	
cMAPs peroneal Dx CV (m/s)	1	10	29.820	14.193	0.156
	2	9	44.789	21.443	
cMAPs peroneal Sn dL (ms)	1	5	4.160	3.010	0.881
	2	3	5.233	4.546	
cMAPs peroneal Sn dA (mV)	1	5	1.480	1.462	1.000
	2	3	1.567	1.845	
cMAPs peroneal Sn CV (m/s)	1	5	33.940	21.367	0.453
	2	3	24.533	21.804	
cMAPs tibial Dx dL (ms)	1	11	5.709	3.901	0.526
	2	10	6.160	3.382	
cMAPs tibial Dx dA (mV)	1	11	3.818	3.919	0.526
	2	10	4.630	8.823	
cMAPs tibial Dx CV (m/s)	1	11	33.318	16.021	0.549
	2	10	26.500	20.007	
cMAPs tibial Sn dL (ms)	1	6	5.717	2.927	0.470
	2	6	3.600	3.109	
cMAPs tibial Sn dA (mV)	1	6	3.133	3.840	0.378
	2	6	1.733	2.229	
cMAPs tibial Sn CV (m/s)	1	6	49.817	24.108	0.229
	2	6	27.183	24.222	
SAPs sural Dx dL (ms)	1	3	2.200	2.066	1.000
	2	2	1.450	2.051	
SAPs sural Dx dA (uV)	1	3	9.400	10.054	0.554
	2	2	2.650	3.748	
SAPs sural Dx CV (m/s)	1	3	29.767	30.003	1.000
	2	2	25.850	36.557	
SAPs sural Sn dL (ms)	1	2	1.350	1.909	1.000
	2	3	1.767	1.537	

Table 2. *Cont.*

	Time	N	Mean	SD	*p*
SAPs sural Sn dA (uV)	1	2	6.000	8.485	1.000
	2	3	7.333	6.351	
SAPs sural Sn CV (m/s)	1	2	22.200	31.396	0.554
	2	3	35.033	30.679	

The median scores before the start of the therapy were 3 (IQ range 3–4) for the INCAT disability score, 10 (IQ range 8–10) for ISS and 54 (IQ range 47–58) for the MRC sumscore. The scores did not significantly change at follow-up (Table 3).

Table 3. Clinical scores. Time 1: pre-SCIg; time 2: during SCIg; MRC: Medical Research Council sumscore; ISS: INCAT sensory scale; SD: standard deviation.

	Time	N	Mean	SD	*p*
MRC	1	9	53.111	6.660	0.85
	2	9	51.889	7.688	
ISS	1	9	8.444	3.609	0.32
	2	9	6.111	5.183	
INCAT disability score	1	9	2.667	1.803	0.36
	2	9	3.333	2.121	

Correlation analysis (Table 4) evidenced a positive correlation of cMAPs distal amplitude with an MRC sumscore ($r = 0.2$; $p = 0.05$) and a trend of negative correlation with the INCAT disability score ($r = -0.156$; $p = 0.15$).

Table 4. Pearson's correlations.

Variable		SAPs dA	cMAPs dA
MRC sumscore	Pearson's r	-	0.21
	p-value	-	0.05
ISS	Pearson's r	−0.05	-
	p-value	0.78	-
INCAT disability score	Pearson's r	−0.12	−0.15
	p-value	0.54	0.15

SAP: sensory nerve action potential; cMAP: compound muscle action potential; dA: distal amplitude.

4. Discussion

Only a few studies reported electrophysiological data of CIDP patients undergoing SCIg treatment. The PATH study, which is the largest trial to evaluate SCIg efficacy in CIDP, reported no significant changes in nerve conduction variables after six months in 115 patients, equally divided in two treatment regimens (0.2 g/kg/week vs. 0.4 g/kg/week), while the placebo group (57 patients) showed a slight worsening of proximal motor latencies and conduction velocities in median, ulnar and peroneal motor nerves. Cirillo et al. reported how SCIg therapy is effective in preserving nerve function in the long term in a population of 14 patients, which also showed an improvement of CMAP amplitude and CVs after 48 months of treatment [14–16].

Our data provide additional support that SCIg maintenance therapy is effective in preventing nerve function deterioration in CIDP patients, confirming the findings of the aforementioned studies. The absence of cMAP amplitude improvement in our population could be due to the shorter median follow-up time after the start of therapy compared to the study of Cirillo et al. However, an improvement was seen in single patients (see Figure 2; a block conduction on motor median nerve resolved after treatment in patient 001). Furthermore, we confirmed the presence of a positive correlation between cMAPs amplitude and MRC sumscore, suggesting that ENG variables could hold a role as prognostic factors to estimate treatment efficacy and duration time.

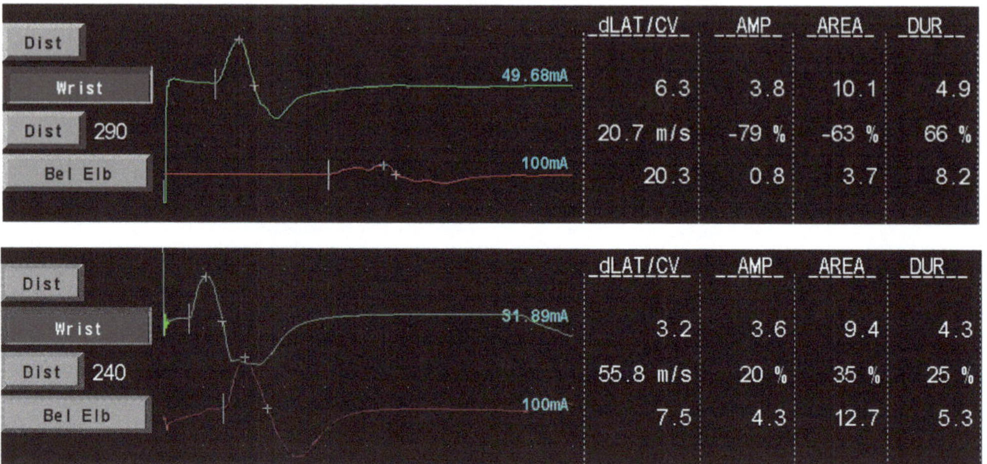

Figure 2. Patient 001 motor medial nerve conduction before (**upper image**) and during (**lower image**) SCIg therapy. The conduction block resolves with treatment. dLAT: distal latency; CV: conduction velocity; AMP: amplitude (mV); DUR: duration (msec).

There were no differences between patients with typical and atypical CIDP phenotype in our population. However, the number of patients with atypical characteristics in our population was too small to draw conclusions from our findings. Considering that atypical phenotypes are reported to respond poorly to immunoglobulin treatment [17], further studies are required to investigate whether the efficacy of SCIg treatment changes in atypical CIDP.

A relevant strength of this study is the long follow-up of our cohort of CIDP patients compared to previous studies. Indeed, prolonged therapy with SCIg was safe and provided stable disease burden and neurophysiological data. Moreover, an improvement in the MCV of peroneal nerves support the idea that the demyelinating process and inflammation in CIDP recede from this treatment; also, stable CMAP amplitude on motor nerves confirm that no more significant axonal loss happens during SCIg treatment in CIDP. Another relevant point is the use of neurophysiological variables to assess treatment efficacy. Despite being cheap, easily reproducible and easy to perform, ENG has been seldom used as a long-term follow-up technique in clinical studies. We suggest that neurophysiological examination could provide more detailed information on SCIg treatment efficacy compared to clinical scores.

Our study has some limitations; first, the small size of the population analyzed; second, the lack of a homogenous protocol of ENG testing and clinical follow-up among patients, which reduces the significance of our data; and third, the absence of a control group (i.e., with IVIg administration).

5. Conclusions

Our data strengthen the evidence on the efficacy of SCIg maintenance therapy in CIDP. Indeed, no patients presented a worsening of symptoms during maintenance treatment with SCIg and there was good safety. Nerve conduction studies are a useful instrument not only in the diagnostic process, but also in the follow-up and prognostic assessment of CIDP.

Author Contributions: Conceptualization, V.D.S. and F.B.; methodology, V.D.S. and F.B.; software, P.A.; formal analysis, P.A.; investigation, A.T., A.L., P.A. and S.I.; data curation, B.M., A.P. and F.G.; writing—original draft preparation, P.A.; writing—review and editing, V.D.S., A.T. and M.G.; supervision, F.B and M.G. All authors have read and agreed to the published version of the manuscript.

Funding: This research received no external funding.

Institutional Review Board Statement: The study was conducted in accordance with the Declaration of Helsinki and approved by the Ethics Committee Palermo I, University of Palermo (Protocol code 07/2020; 13 July 2020).

Informed Consent Statement: Informed consent was obtained from all the subjects involved in the study.

Data Availability Statement: Data are available from the corresponding author upon a reasonable request.

Conflicts of Interest: The authors declare no conflict of interest.

References

1. Kuwabara, S.; Misawa, S. *Chronic Inflammatory Demyelinating Polyneuropathy*; Springer: Singapore, 2019; pp. 333–343. [CrossRef]
2. Di Stefano, V.; Barbone, F.; Ferrante, C.; Telese, R.; Vitale, M.; Onofrj, M.; Di Muzio, A. Inflammatory polyradiculoneuropathies: Clinical and immunological aspects, current therapies, and future perspectives. *Eur. J. Inflamm.* **2020**, *18*, 2058739220942340. [CrossRef]
3. Nobile-Orazio, E. Chronic inflammatory demyelinating polyradiculoneuropathy and variants: Where we are and where we should go. *J. Peripher. Nerv. Syst.* **2014**, *19*, 2–13. [CrossRef] [PubMed]
4. Bergh, P.Y.K.V.D.; Doorn, P.A.; Hadden, R.D.M.; Avau, B.; Vankrunkelsven, P.; Allen, J.A.; Attarian, S.; Blomkwist-Markens, P.H.; Cornblath, D.R.; Eftimov, F.; et al. European Academy of Neurology/Peripheral Nerve Society guideline on diagnosis and treatment of chronic inflammatory demyelinating polyradiculoneuropathy: Report of a joint Task Force—Second revision. *Eur. J. Neurol.* **2021**, *28*, 3556–3583. [CrossRef] [PubMed]
5. Donofrio, P.D.; Bril, V.; Dalakas, M.C.; Deng, C.; Hanna, K.; Hartung, H.-P.; Hughes, R.; Latov, N.; Merkies, I.; Van Doorn, P. Safety and Tolerability of Immune Globulin Intravenous in Chronic Inflammatory Demyelinating Polyradiculoneuropathy. *Arch. Neurol.* **2010**, *67*, 1082–1088. [CrossRef] [PubMed]
6. Alcantara, M.; Hartung, H.; Lawo, J.; Durn, B.L.; Mielke, O.; Bril, V. Electrophysiological predictors of response to subcutaneous immunoglobulin therapy in chronic inflammatory demyelinating polyneuropathy. *Clin. Neurophysiol. Off. J. Int. Fed. Clin. Neurophysiol.* **2021**, *132*, 2184–2190. [CrossRef]
7. Doneddu, P.E.; Nobile-Orazio, E. Management of chronic inflammatory demyelinating polyradiculopathy. *Curr. Opin. Neurol.* **2018**, *31*, 511–516. [CrossRef] [PubMed]
8. Bunschoten, C.; Jacobs, B.C.; van den Bergh, P.Y.K.; Cornblath, D.R.; van Doorn, P.A. Progress in diagnosis and treatment of chronic inflammatory demyelinating polyradiculoneuropathy. *Lancet Neurol.* **2019**, *18*, 784–794. [CrossRef] [PubMed]
9. Berger, M.; Allen, J.A. Optimizing IgG therapy in chronic autoimmune neuropathies: A hypothesis driven approach. *Muscle Nerve* **2015**, *51*, 315–326. [CrossRef] [PubMed]
10. Gentile, L.; Mazzeo, A.; Russo, M.; Arimatea, I.; Vita, G.; Toscano, A. Long-term treatment with subcutaneous immunoglobulin in patients with chronic inflammatory demyelinating polyradiculoneuropathy: A follow-up period up to 7 years. *Sci. Rep.* **2020**, *10*, 7910. [CrossRef] [PubMed]
11. Hughes, R.; Bensa, S.; Willison, H.; van den Bergh, P.; Comi, G.; Illa, I.; Nobile-Orazio, E.; van Doorn, P.; Dalakas, M.; Bojar, M.; et al. Randomized controlled trial of intravenous immunoglobulin versus oral prednisolone in chronic inflammatory demyelinating polyradiculoneuropathy. *Ann. Neurol.* **2001**, *50*, 195–201. [CrossRef] [PubMed]
12. Merkies IS, J.; Schmitz PI, M.; Van Der, M.e.c.h.é.; FG, A.; Samijn JP, A.; Van Doorn, P.A. Psychometric evaluation of a new handicap scale in immune-mediated polyneuropathies. *Muscle Nerve* **2002**, *25*, 370–377. [CrossRef] [PubMed]
13. Syndrome, G.-B. Interobserver agreement in the assessment of muscle strength Guillain-Barre Syndrome. *Muscle Nerve* **1991**, *14*, 1103–1109.
14. Lamb, Y.N.; Syed, Y.Y.; Dhillon, S. Immune Globulin Subcutaneous (Human) 20% (Hizentra®): A Review in Chronic Inflammatory Demyelinating Polyneuropathy. *CNS Drugs* **2019**, *33*, 831–838. [CrossRef] [PubMed]
15. Cirillo, G.; Todisco, V.; Tedeschi, G. Long-term neurophysiological and clinical response in patients with chronic inflammatory demyelinating polyradiculoneuropathy treated with subcutaneous immunoglobulin. *Clin. Neurophysiol. Off. J. Int. Fed. Clin. Neurophysiol.* **2018**, *129*, 967–973. [CrossRef] [PubMed]

16. Cirillo, G.; Todisco, V.; Ricciardi, D.; Tedeschi, G. Clinical-neurophysiological correlations in chronic inflammatory demyelinating polyradiculoneuropathy patients treated with subcutaneous immunoglobulin. *Muscle Nerve* **2019**, *60*, 662–667. [CrossRef] [PubMed]
17. Dziadkowiak, E.; Waliszewska-Prosół, M.; Nowakowska-Kotas, M.; Budrewicz, S.; Koszewicz, Z.; Koszewicz, M. Pathophysiology of the Different Clinical Phenotypes of Chronic Inflammatory Demyelinating Polyradiculoneuropathy (CIDP). *Int. J. Mol. Sci.* **2021**, *23*, 179. [CrossRef] [PubMed]

Disclaimer/Publisher's Note: The statements, opinions and data contained in all publications are solely those of the individual author(s) and contributor(s) and not of MDPI and/or the editor(s). MDPI and/or the editor(s) disclaim responsibility for any injury to people or property resulting from any ideas, methods, instructions or products referred to in the content.

Article

Dexmedetomidine Improves Anxiety-like Behaviors in Sleep-Deprived Mice by Inhibiting the p38/MSK1/NFκB Pathway and Reducing Inflammation and Oxidative Stress

Jiangjing Li [1], Heming Zhang [1], Bin Deng [2], Xin Wang [3], Peng Liang [4], Shenglong Xu [5], Ziwei Jing [1], Zhibin Xiao [6], Li Sun [1], Changjun Gao [1], Jin Wang [5,*] and Xude Sun [1]

[1] Department of Anesthesiology, The Second Affiliated Hospital of Air Force Medical University, Xi'an 710038, China; lulu2790@163.com (J.L.); gaocj74@163.com (C.G.); sunxudes@163.com (X.S.)

[2] Department of Anesthesiology & Center for Brain Science, The First Affiliated Hospital of Xi'an Jiaotong University, Xi'an 710065, China

[3] Department of Otolaryngology Head and Neck Surgery, Shaanxi Provincial People's Hospital, Xi'an 710068, China

[4] Department of Rehabilitative Physioltherapy, The Second Affiliated Hospital of Air Force Medical University, Xi'an 710038, China

[5] Department of Radiation Medical Protection, Ministry of Education Key Lab of Hazard Assessment and Control in Special Operational Environment, School of Military Preventive Medicine, The Fourth Military Medical University, Xi'an 710068, China

[6] Department of Anesthesiology, The 986th Air Force Hospital, Xijing Hospital, The Fourth Military Medical University, Xi'an 710032, China

* Correspondence: wangjinn@fmmu.edu.cn

Citation: Li, J.; Zhang, H.; Deng, B.; Wang, X.; Liang, P.; Xu, S.; Jing, Z.; Xiao, Z.; Sun, L.; Gao, C.; et al. Dexmedetomidine Improves Anxiety-like Behaviors in Sleep-Deprived Mice by Inhibiting the p38/MSK1/NFκB Pathway and Reducing Inflammation and Oxidative Stress. *Brain Sci.* **2023**, *13*, 1058. https://doi.org/10.3390/brainsci13071058

Academic Editors: Junhui Wang and Mona Dehhaghi

Received: 6 June 2023
Revised: 26 June 2023
Accepted: 6 July 2023
Published: 11 July 2023

Copyright: © 2023 by the authors. Licensee MDPI, Basel, Switzerland. This article is an open access article distributed under the terms and conditions of the Creative Commons Attribution (CC BY) license (https://creativecommons.org/licenses/by/4.0/).

Abstract: (1) Background: Sleep deprivation (SD) triggers a range of neuroinflammatory responses. Dexmedetomidine can improve sleep deprivation-induced anxiety by reducing neuroinflammatory response but the mechanism is unclear; (2) Methods: The sleep deprivation model was established by using an interference rod device. An open field test and an elevated plus maze test were used to detect the emotional behavior of mice. Mouse cortical tissues were subjected to RNA sequence (RNA-seq) analysis. Western blotting and immunofluorescence were used to detect the expression of p38/p-p38, MSK1/p-MSK1, and NFκBp65/p- NFκBp65. Inflammatory cytokines were detected using enzyme-linked immunosorbent assay (ELISA); (3) Results: SD triggered anxiety-like behaviors in mice and was closely associated with inflammatory responses and the MAPK pathway (as demonstrated by transcriptome analysis). SD led to increased expression levels of p-p38, p-MSK1, and p-NFκB. P38 inhibitor SB203580 was used to confirm the important role of the p38/MSK1/NFκB pathway in SD-induced neuroinflammation. Dexmedetomidine (Dex) effectively improves emotional behavior in sleep-deprived mice by attenuating SD-induced inflammatory responses and oxidative stress in the cerebral cortex, mainly by inhibiting the activation of the p38/MSK1/NFκB pathway; (4) Conclusions: Dex inhibits the activation of the p38/MSK1/NFκB pathway, thus attenuating SD-induced inflammatory responses and oxidative stress in the cerebral cortex of mice.

Keywords: dexmedetomidine; sleep deprivation; neuroinflammation; p38 MAPK; oxidative stress

1. Introduction

Sleep is critical for health and normal brain function. Sleep deprivation (SD)—defined as inadequate sleep below baseline requirements—is known to affect overall health and wellness. SD reduces immune function, cognitive memory, and learning ability and disrupts emotional health, thereby affecting the daily life activities of individuals [1–4]. SD impairs the functioning of the sympathetic nerve system, leading to metabolic dysregulation [5]. In addition, SD triggers a range of neuroinflammatory responses that modulate immune function by increasing the release of pro-inflammatory cytokines such as interleukin-1β (IL-1β), interleukin-6 (IL-6), tumor necrosis factor (TNF-α), and C-reactive protein (CRP) [6,7].

Growing evidence suggests that SD-induced anxiety behaviors should be closely related to the activation of astrocytes and microglia in the central nervous system (CNS), leading to increased levels of proinflammatory markers and nerve damage [6,8–10]. Although neuroinflammatory responses and oxidative stress are both key factors related to the adverse effects of SD, little is known about the regulatory mechanisms that mitigate these factors [11].

Dexmedetomidine (Dex) is a potent and selective agonist of α2-adrenergic receptors that has seen widespread clinical use since its approval by the US Food and Drug Administration in 1999. Dex exerts protective effects on the nervous system, maintains anesthetic activity, and attenuates immune suppression without causing respiratory depression [12]. Dex has been shown to have neuroprotective effects both in vivo and in vitro [13–15] and these effects are increasingly considered to have clinical implications. Because Dex sedation is closer to the characteristics of natural sleep [16], Dex is often used to improve the sleep quality of perioperative patients and critically ill patients [17–21]. Previous studies have shown that dexmedetomidine can improve the emotional behavior [22] and cognitive dysfunction [23] caused by sleep deprivation but the mechanism is not clear. Recent studies have shown that mice deprived of acute rapid eye movement (REM) sleep for 3 days show increased expression of IL-17A and IL-17f and activation of the p38 MAPK pathway in the hippocampus [24]. In addition, the p38 signaling pathway is involved in SD-induced activation of the NLRP3/pyroptosis axis [25]. Therefore, we hypothesized that Dex might alleviate the anxiety behaviors of sleep-deprived mice by reducing the inflammatory response through the p38 MAPK pathway. In this study, we investigated the effects of Dex on the inhibition of inflammatory response pathways in sleep-deprived mice. We evaluated Dex-induced improvements in emotional behavior in sleep-deprived mice. Our results provide support for the use of Dex in the treatment of sleep disorder-related diseases.

2. Materials and Methods

2.1. Induction of SD

The animal model of SD was established (ZL-013, Anhui Yaokun Biotechnology Co., Ltd., Hefei, China) [6]. Male C57BL/6 J mice were placed in a transparent Plexiglas cylinder (400 mm × 390 mm) and allowed to move, feed, and drink water freely. The cylinder contained a horizontal bar at the bottom that rotated in a random direction at a speed of 5 rpm. The duration of SD was 7 days, with only 4 h of sleep per 24 h (during the last 4 h of the light period). Specifically, the sleep disruption bar was rotated for 20 h per day, and the rotation was halted during 15:00–19:00 each day to allow the mice to sleep. This was continued for 7 days. The procedures of Experiment 1, Experiment 2, and Experiment 3 are shown in Figure 1A. Experimental protocols were approved by the Medical Experimental Animal Administrative Committee of Air Force Medical University (No. IACUC-20210963) and strictly followed the Guidelines from the National Institute of Health (U.S.) regarding the care and use of animals for experimental procedures. Every effort was made to minimize the number of animals for experiments and any pain or discomfort they experienced.

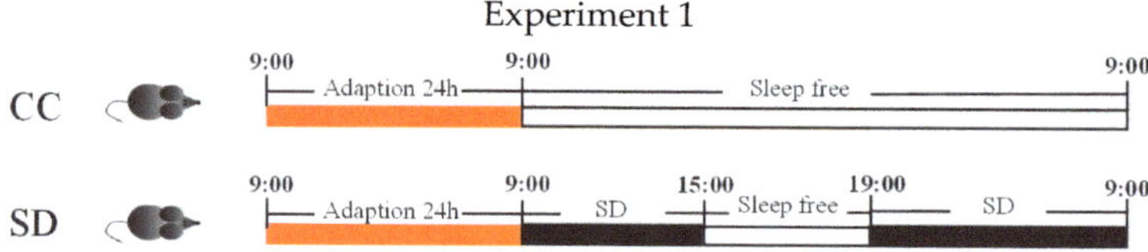

Figure 1. *Cont.*

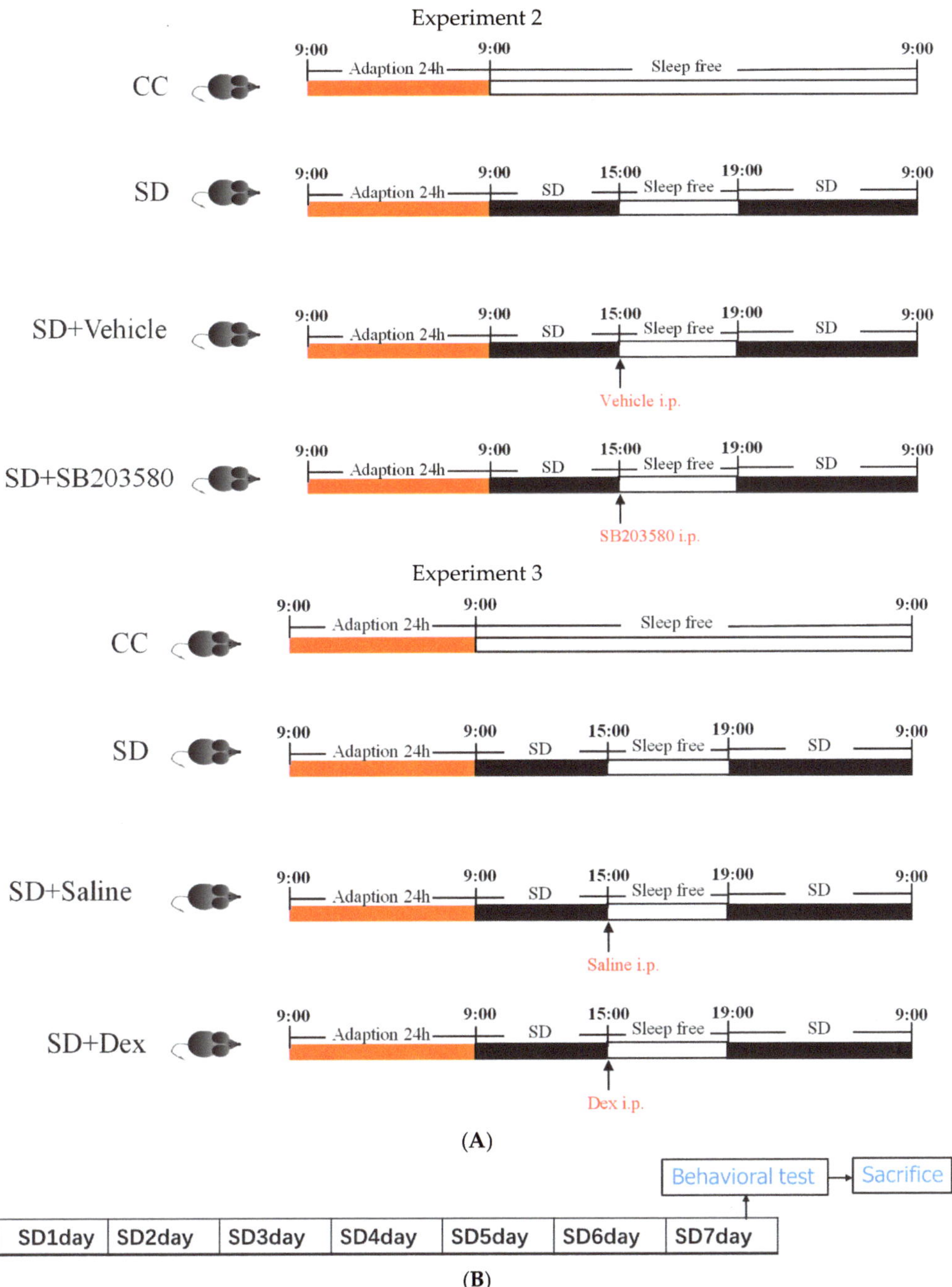

Figure 1. Establishment of the sleep deprivation models. (**A**) Single-day sleep deprivation model with 4 h of sleep every 24 h (during the last 4 h of the light phase) in Experiment 1, Experiment 2, and Experiment 3. (**B**) Seven-day sleep deprivation model.

2.2. Experimental Animals and Pharmacological Treatments

Male C57BL/6 J mice aged 8–10 weeks were used in the experiments. All animals were purchased from the Animal Research Center of the Air Force Medical University. The mice were reared under controlled conditions (ambient temperature, 23 °C; 12-h photoperiod with illumination during 7:00–19:00) and were provided with water and food ad libitum. In Experiment 1, mice were randomly divided into two groups: the control cage (CC) group and the sleep deprivation (SD) group. In Experiment 2, mice were randomly divided into four groups: the control cage (CC) group and the sleep deprivation (SD) group, the sleep deprivation and SB203580 (SD+SB203580; daily intraperitoneal injection of 0.5 mg/kg SB203580 [26] for 6 days) group, and the sleep deprivation and vehicle (SD + vehicle; intraperitoneal injection of vehicle equal in volume to SB203580) group. SB203580 (0.5 mg/kg) was purchased from MedChemExpress Co., Ltd. (Shanghai, China). The vehicle was 0.1% DMSO. In Experiment 3, the animals were divided into four groups: the control cage (CC) group, the sleep deprivation (SD) group, the sleep deprivation and dexmedetomidine (SD + Dex; daily intraperitoneal injection of 100 µg/kg Dex for 6 days) group, and the sleep deprivation and saline (SD + saline; intraperitoneal injection of saline equal in volume to Dex) group [22,23]. Dex was purchased from Yangtze River Pharmaceutical (Group) Co., Ltd. (Taizhou, China). Mice were anesthetized with an O_2 −2% isoflurane mask before specimen collection.

2.3. Open-Field Experiment

In a quiet environment, the mice were introduced into a 50 cm × 50 cm open field from a fixed position and allowed to move freely. Their activities and movements were recorded for 10 min and the data were extracted for analysis. After each test, the field was cleaned of feces and urine stains and wiped with alcohol before another mouse was introduced. The central area was set as the target area, we mainly analyzed the duration of mice in the target area. The average speed of movement of the mice was used to judge whether the model and treatment affected the movement ability of the mice. The behavioral experimental data were recorded using Supermaze software, which was provided by Shanghai XinRuan Information Technology Co., LTD (Shanghai, China).

2.4. Elevated plus Maze Experiments

In a quiet environment, the mice were gently placed in the center of an elevated plus maze (arm width, 5 cm; arm length, 35 cm; closed arm height, 15 cm; maze height, approximately 40–55 cm above the ground). The mice were oriented to face the open arm and were allowed to roam, and their activities were recorded for 5 min. After each test, the maze was cleaned of feces, urine stains, and other debris and wiped with alcohol before another mouse was introduced. The open arm was set as the target area, we mainly analyzed the duration of mice in the target area. The average speed of movement of the mice was used to judge whether the model and treatment affected the movement ability of the mice. The behavioral experimental data were recorded using Supermaze software, which was provided by Shanghai XinRuan Information Technology Co., Ltd. (Shanghai, China).

2.5. mRNA Sequencing

RNA sequencing was performed on mice cortical tissue. Total RNA was isolated from each sample using the standard TRIzol protocol (Invitrogen, Carlsbad, CA, USA). RNA quality was examined using gel electrophoresis and with a Nanodrop spectrophotometer (Thermo, Waltham, MA, USA). Strand-specific libraries were constructed using the TruSeq RNA sample preparation kit (Illumina, San Diego, CA, USA). The libraries were sequenced by Genergy Biotechnology Co., Ltd. (Shanghai, China) using the Illumina Novaseq 6000 instrument.

The raw data were processed in Perl and the data quality was checked with FastQC v0.11.2. The mapped genome data were annotated using the GFF3 file provided by

Huang et al. [27]. The expression level of transcripts was evaluated by calculating the fragments per kilobase of the exon model per million mapped reads (FPKM) in Perl. software was used to screen differentially expressed genes between different groups. The thresholds for determining DETs were $p < 0.05$ and absolute fold change ≥ 2. Then, the identified DETs were used for functional annotation and pathway enrichment analysis using the Gene Ontology (GO) and Kyoto Encyclopedia of Genes and Genomes (KEGG) databases, respectively. Significantly enriched pathways were determined at $p < 0.05$ and at least two related genes were included.

2.6. Immunohistochemical Assays

Mouse brain tissue was fixed with 4% formaldehyde for 24 h, transferred to a 30% sucrose solution until the tissue sank, was frozen and then sectioned into 10-μm-thick slices. The tissue sections were fixed on slides, which were washed three times with PBS solution for 5 min each time, blocked with blocking solution at 25 °C for 90 min, and incubated with primary antibody anti-p-p38, 1:200, CST, 4511; anti-p-NFκBp65, 1:200, CST, 3033T; anti-Iba1, 1:1000, Servicebio, GB12105) at 4 °C overnight. The slides were washed three times with PBS solution for 5 min each time and incubated with the secondary antibody for 1 h away from the light at room temperature. Next, the slides were washed three times with PBS solution for 5 min each time and stained with DAPI. After a final wash with PBS, an anti-quenching agent was added dropwise to mount the slide.

2.7. Western Blotting

After deep anesthesia, the mice were sacrificed by decapitation and the cerebral cortex tissue was collected. The cortical tissue was lysed using a high-throughput tissue grinder and lysis mixing buffer (RIPA+PMSF+ protease inhibitors), and the samples were left on crushed ice for 10 min. After centrifugation at 12,000 r/min for 15 min at 4 °C, the supernatant was separated according to the manufacturer's instructions, and the sample protein concentration was determined by the BCA Protein Assay kit (BOSTER, Wuhan, China). The same amount of protein was separated using electrophoresis on SDSPAGE gel (BOSTER, Wuhan, China) and transferred to a PVDF membrane at constant pressure. The membranes were blocked with 5% skim milk for 120 min at room temperature and washed with TBST. The membranes were cut according to different molecular weight proteins and the corresponding primary antibodies (anti-p38: CST, 8690; anti-p-p38: CST, 4511; anti-MSK: CST, 3489S; anti-p-MSK: CST, 9595S; anti-NFκBp65: CST, 8242; anti-p-NFκBp65: CST, 3033T; anti-GAPDH: BOSTER, A00227-1) and the membranes were placed in antibody diluent using primary antibody diluent (BOSTER, Wuhan, China) diluted at a ratio of 1:1000 and incubated at 4 °C overnight. The next day, the membranes were washed three times with TBST for 5 min each time and incubated with the corresponding secondary antibody (BOSTER, Wuhan, China) for 2 h at room temperature. Finally, all target membrane bands were imaged using a gel imaging system (Bio-Rad, San Francisco, CA, USA) and the gray value of the target band was analyzed using Quantity one software.

2.8. Enzyme-Linked Immunosorbent Assay (ELISA)

The concentrations of TNF-α, IL-1β, IL-6, and IL-10 in the cortical tissue of mice were measured using ELISA kits (BOSTER, Wuhan, China) according to the manufacturer's instructions. The concentrations of COX2 and iNOS in the cortical tissue were measured using ELISA kits (Elabscience Biotechnology, China) according to the manufacturer's instructions.

2.9. Superoxide Dismutase Activity

Superoxide dismutase (SOD) activity in the cortical brain tissue was measured using an SOD assay kit (Beyotime, Shanghai, China) per the manufacturer's instructions.

2.10. Data Analysis

All data are presented as the mean ± standard error of the mean (SEM). The data were analyzed in SPSS (version 26.0, IBM Corp., Armonk, NY, USA). All data were tested for normal distribution and homogeneity of variance. Between-group differences were analyzed with a Student's *t*-test. Multi-group comparisons were performed with one-way analysis of variance (ANOVA). The Least Significant Difference (LSD) was used for multiple comparisons of data that met the homogeneity of variance in the one-way analysis of variance. Statistical significance was indicated at $p < 0.05$. The data were visualized with GraphPad Prism 8.3.0 (GraphPad Software, Inc., San Diego, CA, USA).

3. Results

3.1. Effect of SD on Emotional Behavior in Mice

We evaluated the effect of SD on the emotional behavior of mice using the open-field and elevated plus maze experiments. In the open-field experiment, mice in the SD group spent significantly less time in the open central area than mice in the CC group ($p < 0.01$). Similarly, in the elevated plus maze experiment, mice in the SD group spent significantly less time in the open arm than mice in the CC group ($p < 0.01$). The movement speed of the mice did not change significantly, indicating that this SD model triggers anxiety-like emotional behavior in the mice without affecting their mobility (Figure 2). Results for non-target zones are presented in the Supplementary Material (Supplementary Figure S1).

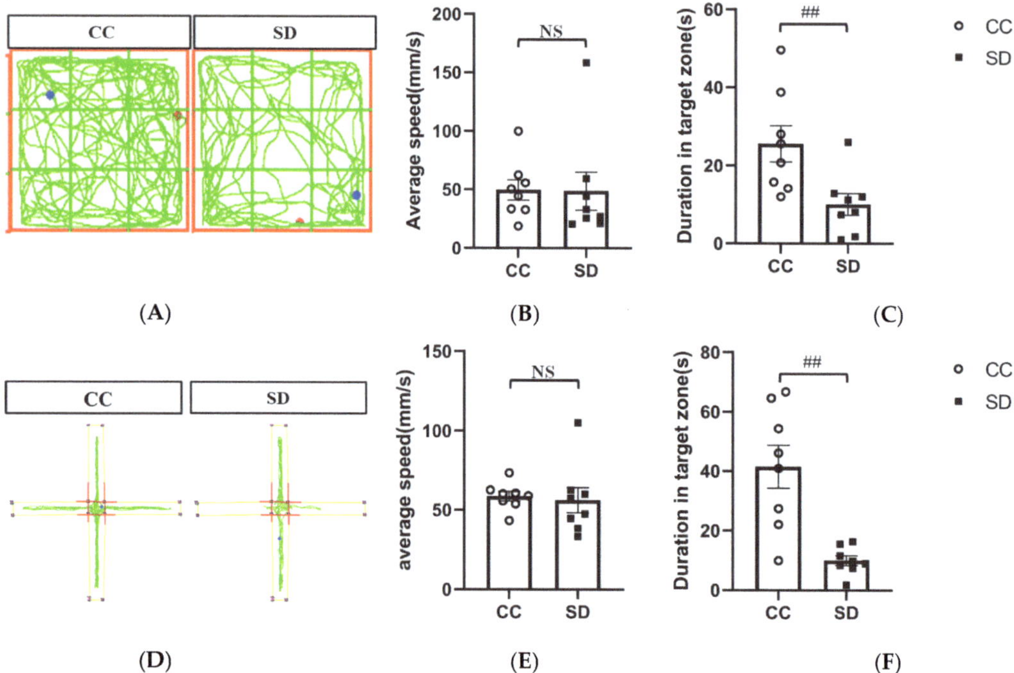

Figure 2. Effects of sleep deprivation on emotional behavior in mice. (**A**) Representative track plot of mice in the open-field test. (**B,C**) Results of the open-field test. CC, $n = 8$; SD, $n = 8$. (**D**) Representative track plot of mice in the elevated plus maze experiment. (**E,F**) Results of the elevated plus maze experiment. CC, $n = 8$; SD, $n = 8$. Data shown are mean ± SEM. Data were analyzed by one-way ANOVA (**B,C,E,F**).; ## $p < 0.01$ vs. CC; NS, no significance.

3.2. Effect of SD on the Transcriptome of the Prefrontal Cortex

To further investigate the molecular mechanisms of SD-induced anxiety-like emotional behavior in mice, we performed a transcriptome sequencing analysis of mouse prefrontal cortex tissue. SD led to the upregulation of 297 genes and downregulation of 578 genes (Figure 3A). The results of enrichment analysis showed that these differentially expressed genes (DEGs) were strongly associated with neurogenic inflammation, substance abuse, pathological neuralgia, and mood disorders (Figure 3B). The top 50 DEGs with respect to degree of interaction were extracted using cytoHubba to generate a protein–protein interaction (PPI) network (Figure 3C). KEGG-based functional enrichment analysis of the top 50 DEGs revealed that the DEGs were closely associated with amoebiasis, hepatitis C, and the MAPK pathway. In particular, the Tnf/Nras/Map3k6 genes were significantly enriched in the MAPK pathway. GO-based Biological Process (GOBP) analysis showed that the DEGs were closely associated with epithelial cell proliferation, positive regulation of acute inflammatory responses, and negative regulation of cytokine production. A GO-based Molecular Function (GOMF) analysis indicated that the DEGs were closely associated with motor activity, protein–hormone receptor activity, and protein tyrosine kinase activity. Transcriptome sequencing of the prefrontal cortex of the mouse brain indicated that, compared with CC mice, SD mice showed differential gene expression closely associated with the MAPK pathway (Figure 3).

3.3. Effect of SD on the Activation of the p38/MSK1/NFκB Pathway

To investigate whether the p38/MSK1/NFκB pathway is involved in the molecular mechanism of SD leading to anxiety-like mood changes in mice, we analyzed the prefrontal cortex tissues of mice with immunofluorescence staining for p-p38, p-NFκBp65, and IBA1 (a marker of microglia activation). The results showed that the prefrontal cortex tissues of SD mice had more numerous activated microglia than those of CC mice. In addition, the expression of p-p38 and p-NFκBp65 was higher in SD mice, and some of the fluorescently labeled cells were co-localized with activated microglia (Figure 4A,B). Western blotting experiments were performed to confirm whether SD activated the p38/MSK1/NFκB pathway and the results indicated no differences in total expression levels of the proteins (p38, MSK, and NFκBp65) in the prefrontal cortex tissues of SD vs. CC mice. However, the expression levels of their respective phosphorylated (activated) forms—p-p38, p-MSK, and p-NFκBp65—were significantly higher ($p < 0.01$) in SD mice, indicating that the SD model promotes the activation of the p38/MSK1/NFκB pathway (Figure 4C).

3.4. Effect of SD on the Activation of the p38/MSK1/NFκB Pathway

To investigate whether the p38/MSK1/NFκB pathway is involved in the molecular mechanism of SD leading to anxiety-like mood changes in mice, we analyzed the prefrontal cortex tissues of mice with immunofluorescence staining for p-p38, p-NFκBp65, and IBA1 (a marker of microglia activation). The results showed that the prefrontal cortex tissues of SD mice had more numerous activated microglia than those of CC mice. In addition, the expression of p-p38 and p-NFκBp65 was higher in SD mice, and some of the fluorescently labeled cells were co-localized with activated microglia (Figure 4A,B). Western blotting experiments were performed to confirm whether SD activated the p38/MSK1/NFκB pathway and the results indicated no differences in total expression levels of the proteins (p38, MSK, and NFκBp65) in the prefrontal cortex tissues of SD vs. CC mice. However, the expression levels of their respective phosphorylated (activated) forms—p-p38, p-MSK, and p-NFκBp65—were significantly higher ($p < 0.01$) in SD mice, indicating that the SD model promotes the activation of the p38/MSK1/NFκB pathway (Figure 4C).

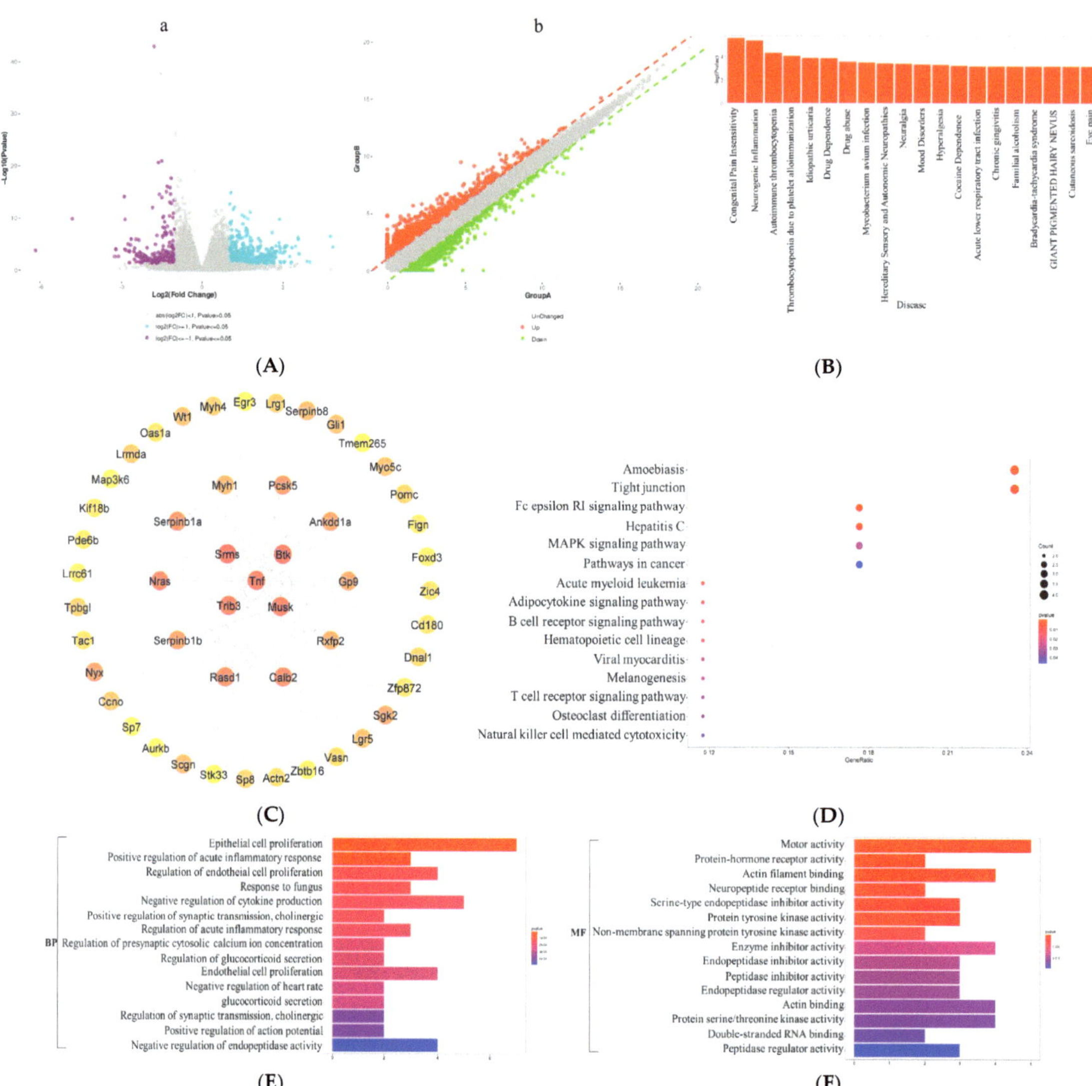

Figure 3. Effects of sleep deprivation on the transcriptome of the prefrontal cortex in mice. (**A**) Volcano map (**a**) and scatter map (**b**) showed differentially expressed genes (DEGs) in the prefrontal cortex of SD vs. CC mice ($n = 4$). (**B**) DEGs disease annotation analysis in the prefrontal cortex of SD vs. CC mice. (**C**) Top 50 DEGs in the prefrontal cortex of SD vs. CC mice. (**D**) KEGG-based enrichment analysis of the top 50 DEGs in the prefrontal cortex of SD vs. CC mice. (**E**) GO-based Biological Process analysis of the top 50 DEGs. (**F**) GO-based Molecular Function analysis of the top 50 DEGs.

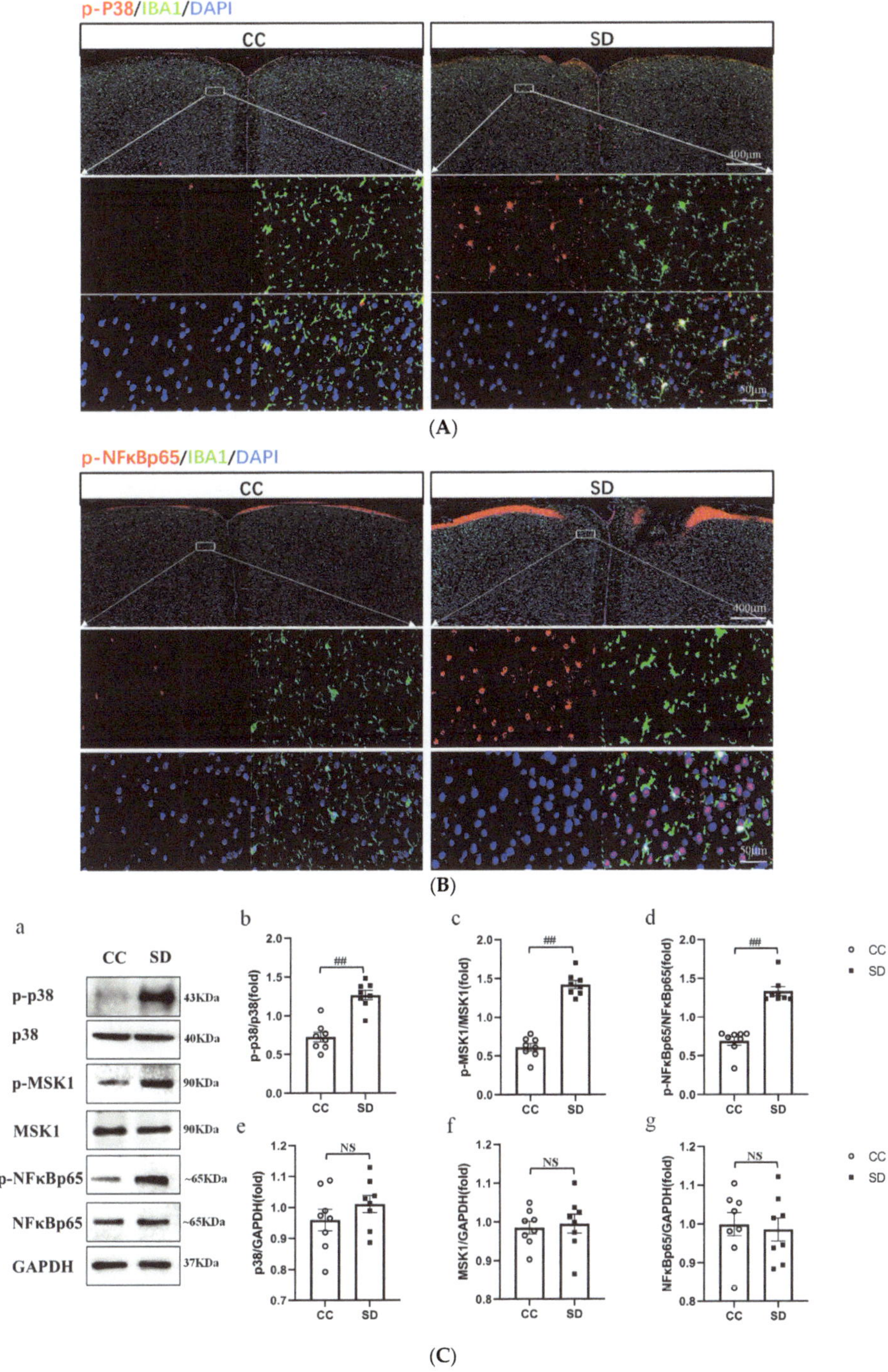

Figure 4. Effects of sleep deprivation on the activation of the p38/MSK1/ NF-κB pathway. (**A**) Immunofluorescence in cells co-labeled with p-p38 (red) and IBA1 (green). (**B**) Immunofluorescence

in cells co-labeled with p-NFκBp65 (red) and IBA1 (green). The white dashed arrow points to a higher magnification of the area in the white box. Scale bars: low magnification, 400 μm; high magnification, 50 μm. (**C**) (**a**) Western blot analysis of proteins in the p38/MSK1/NFκB pathway in the prefrontal cortex of mice; (**b–d**) Expression levels of phosphorylated proteins after standardization with respect to total protein expression; (**e,f**) Total protein expression after standardization with respect to GAPDH expression. CC, $n = 8$; SD, $n = 8$. Data shown are mean ± SEM. Data were analyzed by one-way ANOVA (**b–d,f,g**). ## $p < 0.01$ vs. CC; NS, no significance.

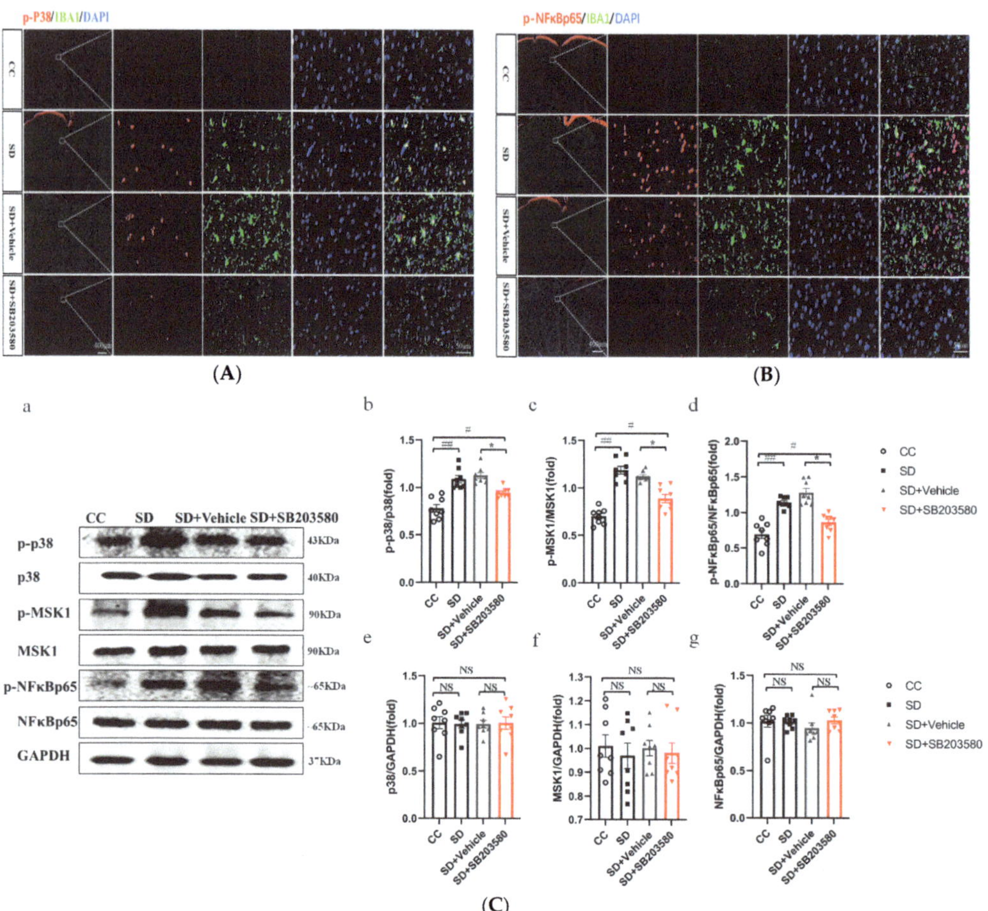

Figure 5. SB203580 inhibits the activation of the p38/MSK1/NF-κB pathway. (**A**) Immunofluorescence in cells co-labeled with p-p38 (red) and IBA1 (green). (**B**) Immunofluorescence in cells co-labeled with p-NFκBp65 (red) and IBA1 (green). The white dashed arrow points to a higher magnification of the area in the white box. Scale bars: low magnification, 400 μm; high magnification, 50 μm. (**C**) (**a**) Western blot analysis of proteins in the p38/MSK1/NFκB pathway in the prefrontal cortex of mice; (**b–d**) Expression levels of phosphorylated proteins after standardization with respect to total protein expression; (**e–g**) Total protein expression after standardization with respect to GAPDH expression. CC, $n = 8$; SD, $n = 8$. Data shown are mean ± SEM. Data were analyzed by one-way ANOVA (**b–d,f,g**). ## $p < 0.01$ vs. CC; # $p < 0.05$ vs. CC; * $p < 0.05$ vs. SD; NS, no significance.

3.5. SB203580 Ameliorates Oxidative Stress and Inflammatory Reactions in the Prefrontal Cortex of Sleep-Deprived Mice

ELISA was used to detect the secretion of inflammatory markers and the expression of oxidative stress-related factors in the cortical tissues of mice. These data were used to evaluate the effects of SB203580 on oxidative stress and inflammatory responses in the prefrontal cortex of SD mice. The levels of pro-inflammatory factors (IL-1β [F = 5.488], IL-6 [F = 4.462], and TNF-α [F = 5.255]) in the cortical tissues of mice were significantly higher in the SD group than in the CC group ($p < 0.01$) and lower in the SD + SB203580 group than in the SD + Vehicle group ($p < 0.05$). However, there were no statistically significant differences in levels of the anti-inflammatory factor, IL-10 [F = 1.243] ($p > 0.05$). The expression levels of oxidative stress-related factors (iNOS [F = 5.508] and COX-2 [F = 18.643]) were significantly higher in the SD group than in the CC group ($p < 0.01$) and lower in the SD + SB203580 group than in the SD + Vehicle group ($p < 0.05$). Total SOD activity [F = 11.883] in the cerebral cortex tissue was significantly lower in the SD group than in the CC group ($p < 0.01$) and higher in the SD + SB203580 group than in the SD + Vehicle group ($p < 0.05$). Taken together, this suggests that SB203580 alleviates the SD-induced increase in inflammatory responses and oxidative stress in mice (Figure 6).

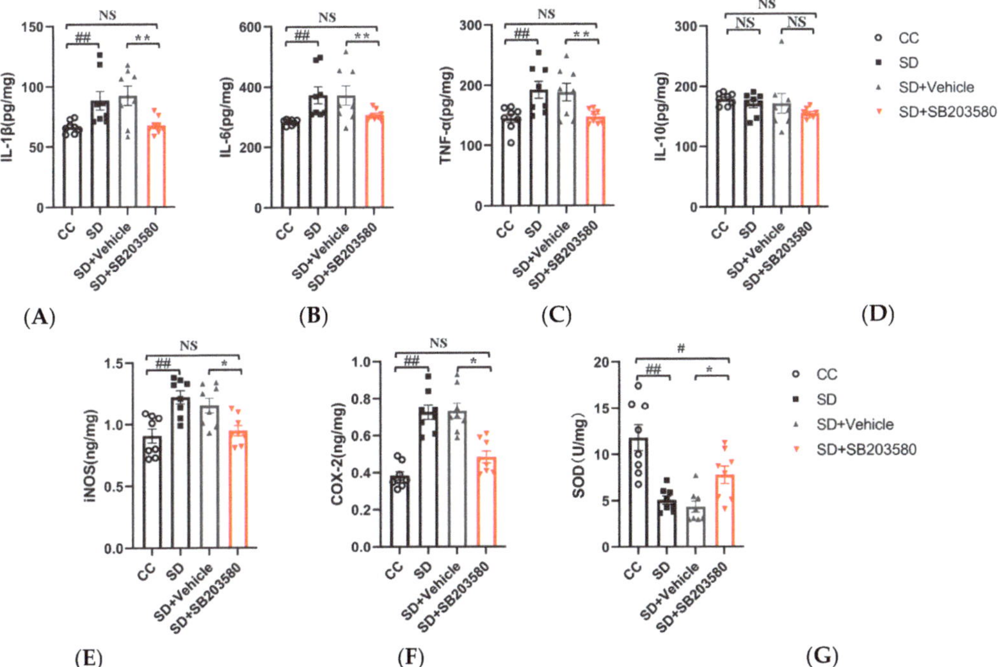

Figure 6. SB203580 alleviates oxidative stress and inflammatory responses in the prefrontal cortex of sleep-deprived mice. (**A–F**) ELISA to detect the expression of IL-1, IL-6, TNF-α, IL-10, iNOS, and COX-2 in the prefrontal cortex of mice ($n = 8$). (**G**) SOD activity ($n = 8$). Data shown are mean ± SEM. Data were analyzed by one-way ANOVA (**A–G**). ## $p < 0.01$ vs. CC; # $p < 0.05$ vs. CC; ** $p < 0.01$ vs. SD + Vehicle; * $p < 0.05$ vs. SD + Vehicle; NS, no significance.

3.6. Effect of Dex on Anxiety-like Behaviors in Sleep-Deprived Mice

Dex improves memory impairment [23] and mood disturbances [22] in sleep-deprived mice by attenuating inflammatory responses. Since Dex has known sedative effects, we determined whether it could improve anxiety-like emotional behaviors in the SD model. The effect of Dex on emotional behavior in sleep-deprived mice was evaluated using the open-field and elevated plus maze experiments. In the open-field experiment [F = 4.880], SD mice spent less time in the central square than CC mice ($p < 0.01$). In contrast, mice in the SD + Dex group spent more time in the central square than mice in the SD + saline group ($p < 0.05$). In the elevated plus maze experiment [F = 10.523], time spent in the open arms was significantly lower in the SD group than in the CC group ($p < 0.01$) but higher in the SD + Dex group than in the SD + saline group ($p < 0.05$). These results indicated that Dex effectively improves anxiety-like behaviors in SD mice. There was no significant change in movement speed between groups, indicating that Dex does not affect the mobility of mice (Figure 7). Results for non-target zones are presented in the Supplementary Material (Supplementary Figure S1).

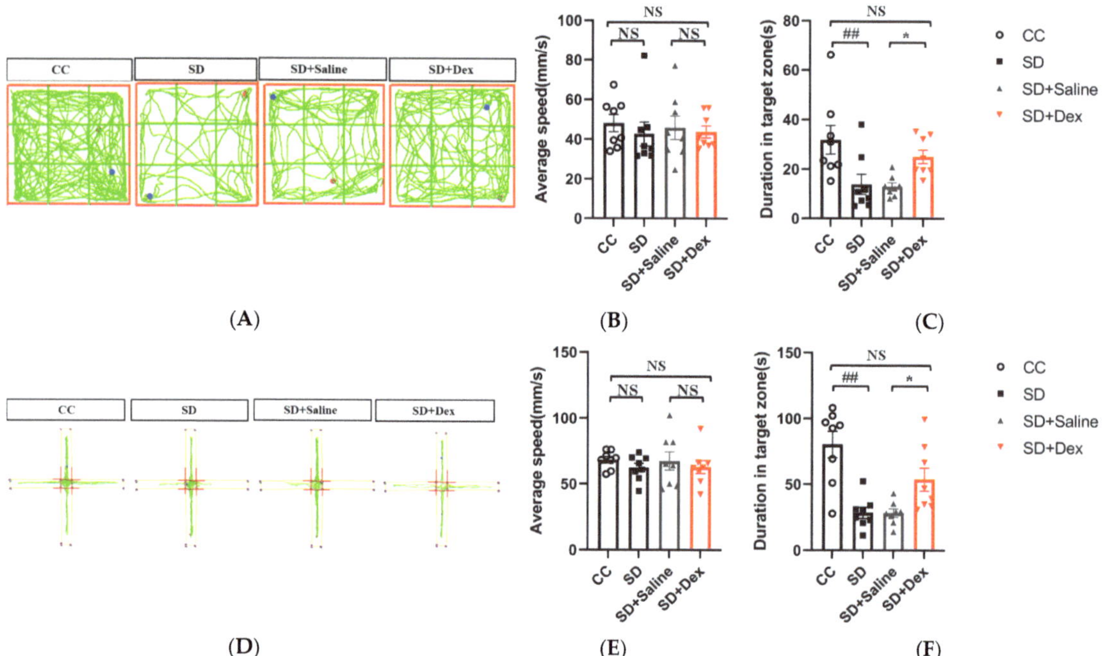

Figure 7. Effects of dexmedetomidine on anxiety-like behaviors in sleep-deprived mice. (**A**) Representative track plot of mice in the open-field test. (**B,C**) Results of the open-field test (*n* = 8). (**D**) Representative track plot of mice in the elevated plus maze test. (**E,F**) Results of the elevated plus maze test (*n* = 8). Data shown are mean ± SEM. Data were analyzed by one-way ANOVA (**B,C,E,F**). NS, no significance; ## $p < 0.01$ vs. CC; * $p < 0.05$ vs. SD + saline.

3.7. Dex Acts by Inhibiting the Activation of the p38/MSK1/NFκBp65 Pathway

To investigate whether Dex acts by inhibiting the activation of the p38/MSK1/NFκB pathway, we first performed immunofluorescence staining for p-p38, p-NFκBp65, and IBA1 in the prefrontal cortex tissues of mice in each group. The prefrontal cortex tissues of SD mice had more numerous activated microglia than those of mice in the CC group. The expression levels of p-p38 and p-NFκBp65 were higher in SD mice, and some of the fluorescently labeled cells co-localized with the activated microglia. The SD + Dex group had fewer activated microglia and fewer p-p38- and p-NFκBp65-positive cells than the SD + saline group (Figure 8A,B). Western blotting analysis showed no between-group differences in the total expression levels of the p38 [F = 0.737], MSK [F = 0.568], and NFκBp65 [F = 0.167] proteins in the prefrontal cortex tissues of mice ($p > 0.05$). However, the expression levels of the phosphorylated forms of these proteins (p-p38 [F = 52.053], p-MSK [F = 36.140], and p-NFκBp65 [F = 10.855]) were significantly higher in the SD group than in the control group ($p < 0.01$) and significantly lower in the SD + Dex group than in the SD + saline group ($p < 0.05$) (Figure 8C).

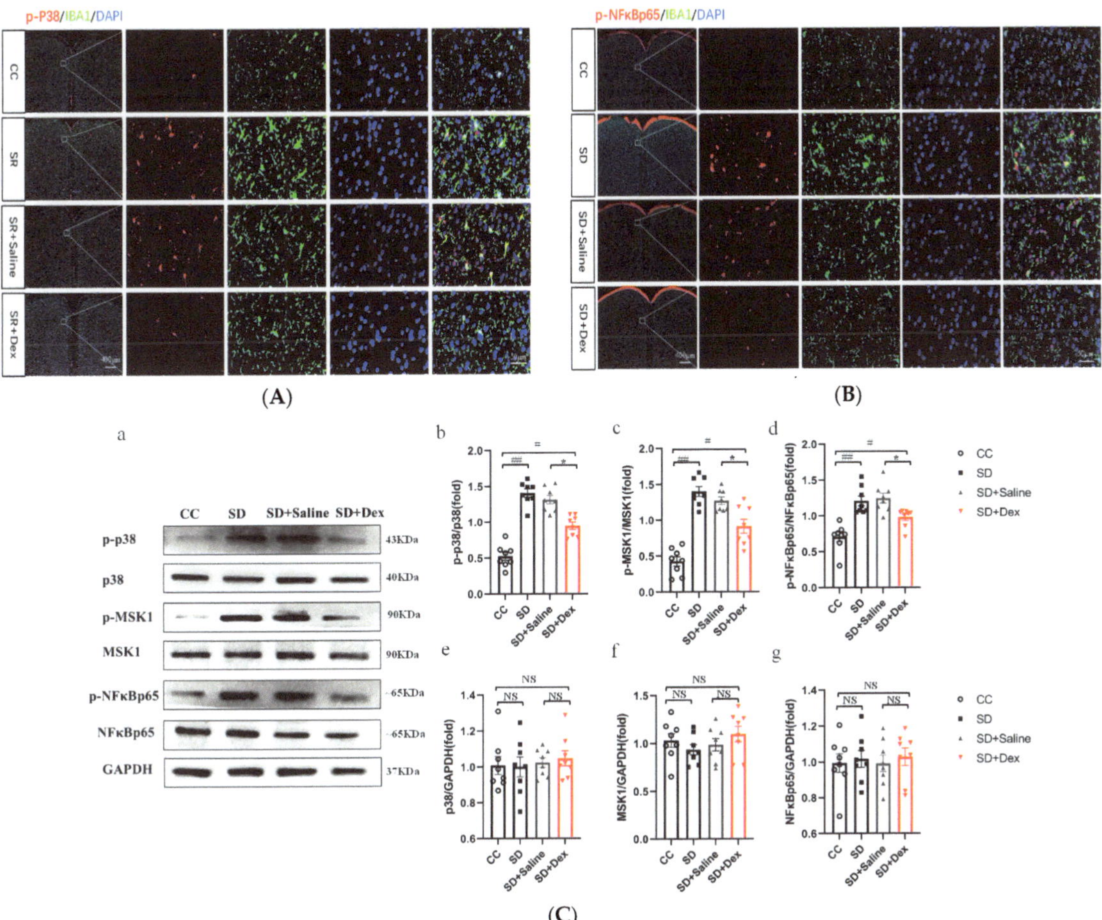

Figure 8. Dexmedetomidine inhibits the activation of the p38/MSK1/NF-κB pathway. (**A**) Immunofluorescence in cells co-labeled with p-p38 (red) and IBA1 (green). (**B**) Immunofluorescence in cells co-labeled with p-NFκBp65 (red) and IBA1 (green). The white dashed arrow points to a higher magnification of the area in the white box. Scale bars: low magnification, 400 μm; high magnification, 50 μm. (**C**) (**a**) Western blot analysis of proteins in the p38/MSK1/NFκB pathway in the prefrontal cortex; (**b–d**) Expression levels of phosphorylated proteins after standardization with respect to total protein expression; (**e,f**) Total protein expression after standardization with respect to GAPDH expression. CC, $n = 8$; SD, $n = 8$. Data shown are mean ± SEM. Data were analyzed by one-way ANOVA (**b–d,f,g**). ## $p < 0.01$ vs. CC; # $p < 0.05$ vs. CC; * $p < 0.05$ vs. SD + saline; NS, no significance.

3.8. Dex Ameliorates Oxidative Stress and Inflammatory Reactions in the Prefrontal Cortex of Sleep-Deprived Mice

ELISA was used to detect the secretion of inflammatory markers and expression of oxidative stress-related factors in the cortical tissues of mice. These data were used to evaluate the effects of Dex on oxidative stress and inflammatory responses in the prefrontal cortex of SD mice. The levels of pro-inflammatory factors (IL-1β [F = 9.289], IL-6 [F = 4.523], and TNF-α [F = 8.018]) in the cortical tissues of mice were significantly higher in the SD group than in the CC group ($p < 0.01$) and lower in the SD + Dex group than in the SD + saline group ($p < 0.05$). However, there were no statistically significant differences in levels of the anti-inflammatory factor, IL-10 [F = 1.121] ($p > 0.05$). The expression levels of oxidative stress-related factors (iNOS [F = 9.649] and COX-2 [F = 6.804]) were significantly higher in the SD group than in the CC group ($p < 0.01$) and lower in the SD + Dex group than in the SD + saline group ($p < 0.05$), with the difference in iNOS levels being highly significant ($p < 0.05$). Total SOD activity [F = 52.316] in the cerebral cortex tissue was significantly lower in the SD group than in the CC group ($p < 0.01$) and higher in the SD + Dex group than in the SD + Saline group ($p < 0.05$). Taken together, this suggests that Dex alleviates the SD-induced increase in inflammatory responses and oxidative stress in mice (Figure 9).

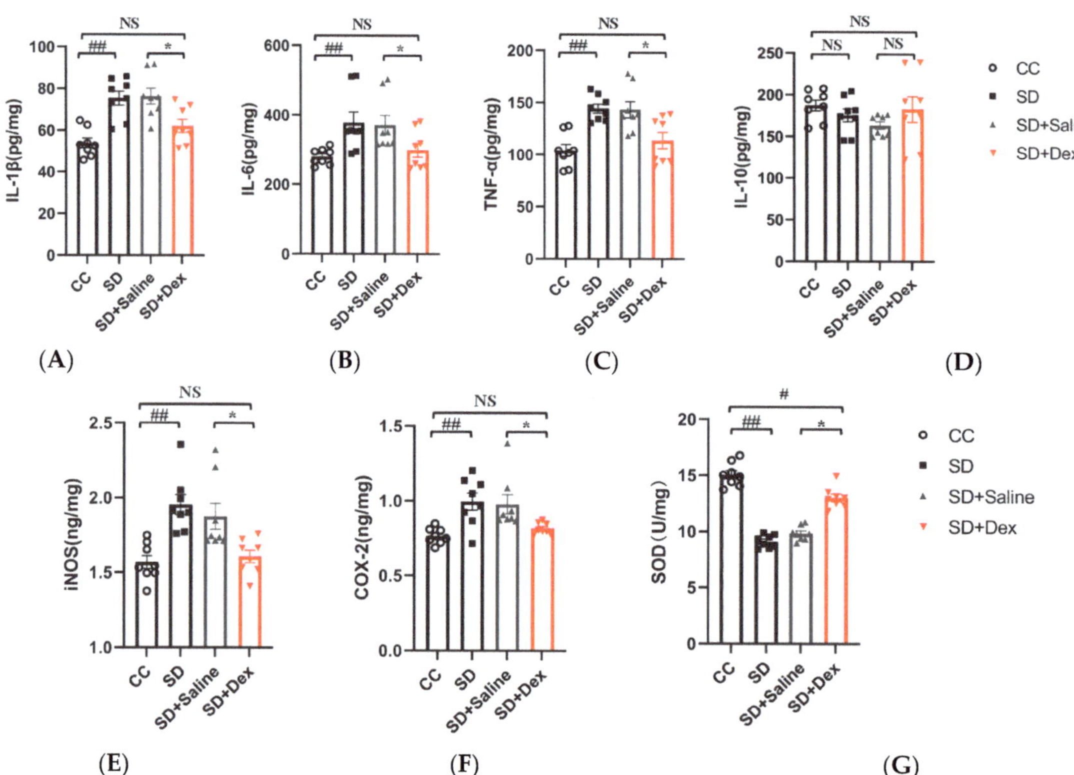

Figure 9. Dexmedetomidine alleviates oxidative stress and inflammatory responses in the prefrontal cortex of sleep-deprived mice. (**A–F**) ELISA to detect the expression of IL-1, IL-6, TNF-α, IL-10, iNOS, and COX-2 in the prefrontal cortex of mice ($n = 8$). (**G**) SOD activity ($n = 8$). Data shown are mean ± SEM. Data were analyzed by one-way ANOVA (**A–G**). # $p < 0.05$ vs. CC; ## $p < 0.01$ vs. CC; * $p < 0.05$ vs. SD + saline; NS, no significance.

4. Discussion

REM SD protocols (or paradoxical SD protocols) are the most frequently used methods of SD [28]. The modified multiple platform (MMP) method [11,29,30] is a common form of REM SD that overcomes the shortcomings of unstable experimental animal populations. However, the animals are still affected by stress from movement restriction, resulting in environmental confounding factors. Researchers have recently developed a novel SD protocol for mice, consisting of a cylinder and built-in bar. The mice are allowed to move freely in the cylinder with ad libitum access to food and water. To induce SD, the bar inside the cylinder is rotated continuously at a constant speed [31]. This type of SD neither isolates the mice nor restricts their mobility. The results showed that this sleep deprivation model induced anxiety-like behaviors in mice (Figure 2), which was consistent with previous findings [10,32].

Evidence from human studies suggests that the prefrontal cortex (PFC) plays an important role in sleep [33] and anxiety [34]. The impairment of PFC activity after sleep deprivation is closely related to the anxiety induced by sleep deprivation and can predict the degree of anxiety amplification [35]. Moreover, it has been demonstrated that the activation of microglia in the PFC underlies anxiety-like behaviors in sleep-deprived mice [10]. Therefore, to further investigate the molecular mechanisms underlying SD-induced anxiety-like emotional behaviors in mice, we performed transcriptome sequencing analysis using the prefrontal cortex tissue of mice. Several DEGs with strong interactions were significantly enriched in the MAPK pathway and were closely associated with positive regulation of acute inflammatory response and negative regulation of cytokine production (Figure 3). Further validation of the sequencing results revealed that SD leads to the activation of the p38/MSK1/NFκB pathway (Figure 4). The MAPK family includes ERK1/2, JNK, p38, and ERK5. Of these, JNK and p38 can be activated by intracellular and extracellular stress (including changes in environmental factors such as UV, heat, and hyperosmotic stress) and inflammatory cytokines [36]. As mentioned above, the p38 MAPK pathway appears to be a key link between pathological microglia activation and deleterious inflammation in CNS disease [37]. MSK1 and 2 are two related kinases that are activated downstream of p38 and Erk1/2 [38], and MSK1 can prolong the activation of NFκB [39]. The p38 and ERK-MAPK signaling pathways are involved in the SD-induced activation of the NLRP3/pyroptosis axis [25]. Insufficient sleep also induces other pro-inflammatory factors such as IL-1β and TNF-α, which can activate p38 MAPK and are involved in the inhibition of neural precursor cells [24]. Our results confirmed that SD leads to the activation and phosphorylation of p38 and MSK1, further leading to NFκB activation, oxidative stress, and inflammatory responses. We further confirmed the important role of the p38/MSK1/NFκB pathway in SD-induced neuroinflammatory response using the p38 inhibitor SB203580 (Figures 5 and 6).

Recent studies have shown that Dex can alleviate lung injury in septic mice by regulating the p38 MAPK signaling pathway [40]. Dex also alleviates lipopolysaccharide-induced apoptosis in hippocampal neurons by reducing the level of inflammatory factors (such as IL-1β, TNF-α, and IL-6) through phosphorylation of the p38 MAPK pathway [41]. These lines of evidence demonstrate that Dex plays a neuroprotective role by modulating the p38 MAPK pathway. Dex attenuates the SD-induced exacerbation of postoperative immunosuppression [29] and improves memory impairment [23] and depression in sleep-deprived mice by attenuating the inflammatory response [22]. Our results showed that Dex ameliorated anxiety in SD mice by inhibiting the activation of the p38/MSK1/NFκB pathway in microglia (Figures 7 and 8).

Long-term SD (e.g., due to short sleep duration or sleep disorders) can lead to chronic systemic low-grade inflammation. It is also associated with several diseases with inflammatory aspects such as diabetes, atherosclerosis, and neurodegeneration [7]. Astrocytic phagocytosis of synaptic elements (mainly the presynaptic components of large synapses) is increased after both acute and chronic SD (compared with sleep and wakefulness). Moreover, low levels of sustained microglia activation can lead to abnormal responses to secondary injury. Thus, chronic SD initiated by the microglia may render the brain vulnerable to further injury [9]. In the present study, we found that microglia activation was caused by SD (Figure 4) and inhibited by Dex (Figure 8). The cytokines IL-1β and TNF-α are both involved in the regulation of sleep homeostasis [42,43]. Animal studies have shown that most pro-inflammatory cytokines promote NREM sleep, whereas anti-inflammatory cytokines reduce NREM sleep; in addition, the inhibition of the activity of certain inflammatory cytokines improves sleep quantity and quality [7]. Recent studies have shown that

MMP-induced sleep deprivation for 72 h increases the levels of pro-inflammatory cytokines (IL-1β, IL-6, and TNF-α) and reduces the levels of anti-inflammatory cytokines (IL-4 and IL-10) in rat hippocampal tissue [44]. MMP-induced sleep deprivation for 20 h per day for 7 days resulted in increased levels of TNF-a, IL-1β, and IL-6 in mice. Similar results were obtained in the present study, where SD led to increased secretion of the pro-inflammatory factors IL-1β, IL-6, and TNF-α (but had no effect on IL-10 levels; Figure 9). Dex has been shown to ameliorate SD-induced decreases in short-term memory and spatial learning in rats by inhibiting the SD-induced production of inflammatory mediators (TNF-α and IL-6) [23]. Dex also inhibits the production of TNF-α, IL-1β, and IL-18, the phosphorylation of ERK1/2 and P38, and the activation of caspase-1, and reduces pyroptosis [45]. These results are consistent with the findings of this study, where we show that Dex administration reduced the SD-induced secretion of pro-inflammatory factors.

One of the functions of sleep is to promote antioxidant mechanisms. This may be an adaptive response to sleep deficiency/deprivation, which can induce oxidative stress [28]. Recent findings suggest that the duration of sleep fragmentation is a major factor in the development of anxiety-related behaviors and that these effects are mediated through oxidative stress in the brain [46]. Chronically sleep-deprived rats exhibited reduced SOD activity in the hippocampus and brainstem [47]. In contrast, acutely sleep-deprived rats (6 h) exhibited reduced glutathione (GSH) levels in the cortex, brainstem, and forebrain and enhanced glutathione peroxidase (GPx) activity in the hippocampus and cerebellum after mild treatment [48]. These findings are consistent with the results of the present study (Figure 9G). However, studies on other SD animal models have not reported any changes in oxidative stress markers or antioxidant capacity in the peripheral blood or brain regions after SD [49,50]. These inconsistencies may be associated with the different methods and durations of sleep deprivation and the mouse strains used in the SD experiments.

The present study has some limitations. On one hand, we found that the effect of SB203580 on anxiety behavior was not statistically significant (Supplementary Figure S2). The possible reason was that the duration of inhibitor use was not enough. We refer to previous literature [26] using SB203580 for up to 14 days. Another possible reason is that there are other important signaling pathways in the anxiety behavior caused by SD, which is also the direction of our future research. On the other hand, only male mice were used in this experiment and we did not test for any between-sex differences in the ameliorating effects of Dex on emotional behavior in sleep-deprived mice. However, many drugs are known to have sex-dependent effects [51], and the associations between stress, sleep deprivation, and inflammation appear to be stronger in females than in males. Although sex is typically considered a confounder, future studies should investigate differences between the sexes in a more systematic manner [52].

5. Conclusions

In summary (Figure 10), the present study demonstrates that a novel sleep deprivation instrument can trigger anxiety-like behaviors in mice. Transcriptome sequencing and other experiments confirmed that SD leads to the activation of the p38/MSK1/NFκB pathway. In addition, Dex inhibits the activation of the p38/MSK1/NFκB pathway, thus attenuating SD-induced inflammatory responses and oxidative stress in the cerebral cortex of mice. These insights provide a theoretical basis for using Dex in the treatment of patients with insomnia and insomnia-induced mood disorders. Moreover, we provide potential research targets for further investigation of the molecular mechanisms underlying the Dex-induced improvement of SD-induced anxiety-like emotional behaviors.

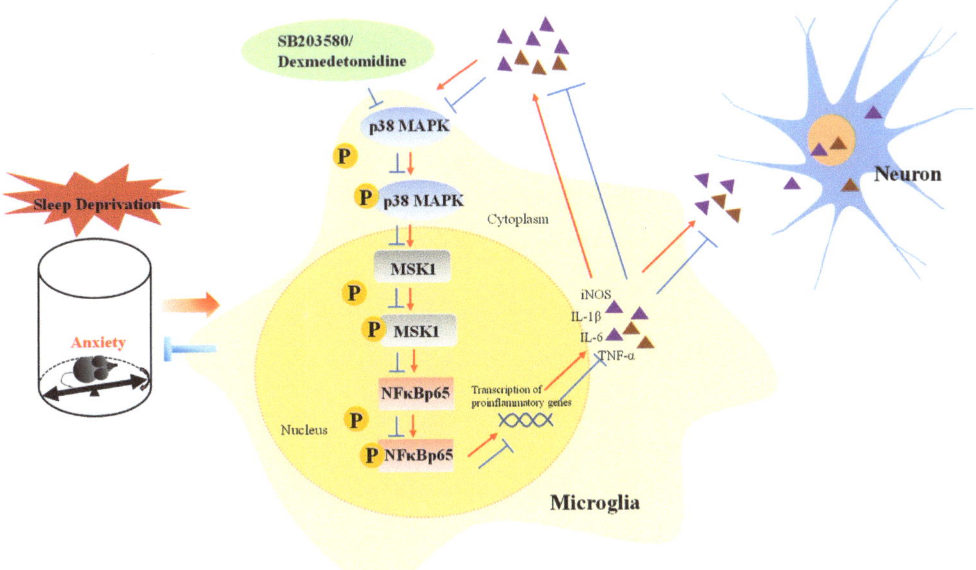

Figure 10. Mechanism of Dex in relieving anxiety caused by sleep deprivation. In this study, a mouse model of sleep deprivation was established using an interference rod device, which induced anxiety in mice. The mechanism may be related to the activation of the P38/MSK1/NFκB signaling pathway in the damaged cortex of SD mice. At the same time, microglia are significantly activated and central inflammation ensues (the schematic is shown with red arrows). Dexmedetomidine can inhibit the inflammatory response, inhibit the activation of the P38/MSK1/NFκB signaling pathway, and finally alleviate the anxiety-like behavior of SD mice (the schematic is shown with blue lines).

Supplementary Materials: The following supporting information can be downloaded at: https://www.mdpi.com/article/10.3390/brainsci13071058/s1, Figure S1: Information about the speeds and times spent in the non-target zones; Figure S2: (A,B) Results of the open-field test. (C,D) Results of the elevated plus maze test.

Author Contributions: J.L. and J.W. conceived the project. H.Z. and B.D. designed the study. X.S. and C.G. directed the study. J.L., H.Z. and X.W. were involved in data analysis and drafted the manuscript. S.X., Z.X. and L.S. revised the manuscript. P.L., Z.J. provided the assistance and support during the study. All authors have read and agreed to the published version of the manuscript. J.L., H.Z. and B.D. contributed equally. X.S., J.W. and C.G. are the co-corresponding authors of this article.

Funding: This work was supported by the Social Talents Fund of Tangdu Hospita (2021SHRC067), the discipline innovation and development plan of Tangdu hospital (2021LCYJ046), and the Key R&D Program of Shaanxi Province (2022SF-189).

Institutional Review Board Statement: Experimental protocols were approved by the Medical Experimental Animal Administrative Committee of Air Force Medical University (No. IACUC-20210963).

Informed Consent Statement: Not applicable.

Data Availability Statement: The analysis data used to support the findings of this study are available from the corresponding author upon request.

Conflicts of Interest: The authors declare no conflict of interest.

References

1. Pires, G.N.; Bezerra, A.G.; Tufik, S.; Andersen, M.L. Effects of acute sleep deprivation on state anxiety levels: A systematic review and meta-analysis. *Sleep Med.* **2016**, *24*, 109–118. [CrossRef] [PubMed]
2. Wang, Z.; Chen, W.H.; Li, S.X.; He, Z.M.; Zhu, W.L.; Ji, Y.B.; Wang, Z.; Zhu, X.M.; Yuan, K.; Bao, Y.P.; et al. Gut microbiota modulates the inflammatory response and cognitive impairment induced by sleep deprivation. *Mol. Psychiatry* **2021**, *26*, 6277–6292. [CrossRef]
3. Fullagar, H.H.; Skorski, S.; Duffield, R.; Hammes, D.; Coutts, A.J.; Meyer, T. Sleep and athletic performance: The effects of sleep loss on exercise performance, and physiological and cognitive responses to exercise. *Sports Med.* **2015**, *45*, 161–186. [CrossRef] [PubMed]
4. Graves, L.; Pack, A.; Abel, T. Sleep and memory: A molecular perspective. *Trends Neurosci.* **2001**, *24*, 237–243. [CrossRef] [PubMed]
5. Spiegel, K.; Leproult, R.; L'Hermite-Baleriaux, M.; Copinschi, G.; Penev, P.D.; Van Cauter, E. Leptin levels are dependent on sleep duration: Relationships with sympathovagal balance, carbohydrate regulation, cortisol, and thyrotropin. *J. Clin. Endocrinol. Metab.* **2004**, *89*, 5762–5771. [CrossRef]
6. Manchanda, S.; Singh, H.; Kaur, T.; Kaur, G. Low-grade neuroinflammation due to chronic sleep deprivation results in anxiety and learning and memory impairments. *Mol. Cell. Biochem.* **2018**, *449*, 63–72. [CrossRef]
7. Besedovsky, L.; Lange, T.; Haack, M. The Sleep-Immune Crosstalk in Health and Disease. *Physiol. Rev.* **2019**, *99*, 1325–1380. [CrossRef]
8. Chennaoui, M.; Gomez-Merino, D.; Drogou, C.; Geoffroy, H.; Dispersyn, G.; Langrume, C.; Ciret, S.; Gallopin, T.; Sauvet, F. Effects of exercise on brain and peripheral inflammatory biomarkers induced by total sleep deprivation in rats. *J. Inflamm.* **2015**, *12*, 56. [CrossRef]
9. Bellesi, M.; de Vivo, L.; Chini, M.; Gilli, F.; Tononi, G.; Cirelli, C. Sleep Loss Promotes Astrocytic Phagocytosis and Microglial Activation in Mouse Cerebral Cortex. *J. Neurosci.* **2017**, *37*, 5263–5273. [CrossRef]
10. Liu, H.; Huang, X.; Li, Y.; Xi, K.; Han, Y.; Mao, H.; Ren, K.; Wang, W.; Wu, Z. TNF signaling pathway-mediated microglial activation in the PFC underlies acute paradoxical sleep deprivation-induced anxiety-like behaviors in mice. *Brain Behav. Immun.* **2022**, *100*, 254–266. [CrossRef]
11. Xue, R.; Wan, Y.H.; Sun, X.Q.; Zhang, X.; Gao, W.; Wu, W. Nicotinic Mitigation of Neuroinflammation and Oxidative Stress After Chronic Sleep Deprivation. *Front. Immunol.* **2019**, *10*, 2546. [CrossRef]
12. Mantz, J.; Josserand, J.; Hamada, S. Dexmedetomidine: New insights. *Eur. J. Anaesthesiol.* **2011**, *28*, 3–6. [CrossRef]
13. Mei, B.; Li, J.; Zuo, Z. Dexmedetomidine attenuates sepsis-associated inflammation and encephalopathy via central α2A adrenoceptor. *Brain Behav. Immun.* **2021**, *91*, 296–314. [CrossRef]
14. Unchiti, K.; Leurcharusmee, P.; Samerchua, A.; Pipanmekaporn, T.; Chattipakorn, N.; Chattipakorn, S.C. The potential role of dexmedetomidine on neuroprotection and its possible mechanisms: Evidence from in vitro and in vivo studies. *Eur. J. Neurosci.* **2021**, *54*, 7006–7047. [CrossRef]
15. Liaquat, Z.; Xu, X.; Zilundu, P.L.M.; Fu, R.; Zhou, L. The Current Role of Dexmedetomidine as Neuroprotective Agent: An Updated Review. *Brain Sci.* **2021**, *11*, 846. [CrossRef]
16. Akeju, O.; Brown, E.N. Neural oscillations demonstrate that general anesthesia and sedative states are neurophysiologically distinct from sleep. *Curr. Opin. Neurobiol.* **2017**, *44*, 178–185. [CrossRef]
17. Zhang, Z.F.; Su, X.; Zhao, Y.; Zhong, C.L.; Mo, X.Q.; Zhang, R.; Wang, K.; Zhu, S.N.; Shen, Y.E.; Zhang, C.; et al. Effect of mini-dose dexmedetomidine supplemented intravenous analgesia on sleep structure in older patients after major noncardiac surgery: 1 A randomized trial. *Sleep Med.* **2023**, *102*, 9–18. [CrossRef]
18. Chen, Z.; Tang, R.; Zhang, R.; Jiang, Y.; Liu, Y. Effects of dexmedetomidine administered for postoperative analgesia on sleep quality in patients undergoing abdominal hysterectomy. *J. Clin. Anesth.* **2017**, *36*, 118–122. [CrossRef]
19. Zeng, W.; Chen, L.; Liu, X.; Deng, X.; Huang, K.; Zhong, M.; Zhou, S.; Zhan, L.; Jiang, Y.; Liang, W. Intranasal Dexmedetomidine for the Treatment of Pre-operative Anxiety and Insomnia: A Prospective, Randomized, Controlled, and Clinical Trial. *Front. Psychiatry* **2022**, *13*, 816893. [CrossRef]
20. Alexopoulou, C.; Kondili, E.; Diamantaki, E.; Psarologakis, C.; Kokkini, S.; Bolaki, M.; Georgopoulos, D. Effects of dexmedetomidine on sleep quality in critically ill patients: A pilot study. *Anesthesiology* **2014**, *121*, 801–807. [CrossRef]
21. Oto, J.; Yamamoto, K.; Koike, S.; Onodera, M.; Imanaka, H.; Nishimura, M. Sleep quality of mechanically ventilated patients sedated with dexmedetomidine. *Intensive Care Med.* **2012**, *38*, 1982–1989. [CrossRef] [PubMed]
22. Moon, E.J.; Ko, I.G.; Kim, S.E.; Jin, J.J.; Hwang, L.; Kim, C.J.; An, H.; Lee, B.J.; Yi, J.W. Dexmedetomidine Ameliorates Sleep Deprivation-Induced Depressive Behaviors in Mice. *Int. Neurourol. J.* **2018**, *22* (Suppl. 3), S139–S146. [CrossRef]
23. Hwang, L.; Ko, I.G.; Jin, J.J.; Kim, S.H.; Kim, C.J.; Chang, B.; Rho, J.H.; Moon, E.J.; Yi, J.W. Dexmedetomidine ameliorates memory impairment in sleep-deprived mice. *Anim. Cells Syst.* **2019**, *23*, 371–379. [CrossRef]
24. Cui, L.; Xue, R.; Zhang, X.; Chen, S.; Wan, Y.; Wu, W. Sleep deprivation inhibits proliferation of adult hippocampal neural progenitor cells by a mechanism involving IL-17 and p38 MAPK. *Brain Res.* **2019**, *1714*, 81–87. [CrossRef] [PubMed]
25. Fan, K.; Yang, J.; Gong, W.Y.; Pan, Y.C.; Zheng, P.; Yue, X.F. NLRP3 inflammasome activation mediates sleep deprivation-induced pyroptosis in mice. *PeerJ* **2021**, *9*, e11609. [CrossRef]

26. Wang, C.; Li, Y.; Yi, Y.; Liu, G.; Guo, R.; Wang, L.; Lan, T.; Wang, W.; Chen, X.; Chen, S.; et al. Hippocampal microRNA-26a-3p deficit contributes to neuroinflammation and behavioral disorders via p38 MAPK signaling pathway in rats. *J. Neuroinflamm.* **2022**, *19*, 283. [CrossRef]

27. Huang, H.; Liu, A.; Wu, H.; Ansari, A.R.; Wang, J.; Huang, X.; Zhao, X.; Peng, K.; Zhong, J.; Liu, H. Transcriptome analysis indicated that Salmonella lipopolysaccharide-induced thymocyte death and thymic atrophy were related to TLR4-FOS/JUN pathway in chicks. *BMC Genom.* **2016**, *17*, 322. [CrossRef]

28. Villafuerte, G.; Miguel-Puga, A.; Rodriguez, E.M.; Machado, S.; Manjarrez, E.; Arias-Carrion, O. Sleep deprivation and oxidative stress in animal models: A systematic review. *Oxidative Med. Cell. Longev.* **2015**, *2015*, 234952. [CrossRef]

29. Wang, G.; Wu, X.; Zhu, G.; Han, S.; Zhang, J. Dexmedetomidine alleviates sleep-restriction-mediated exaggeration of postoperative immunosuppression via splenic TFF2 in aged mice. *Aging* **2020**, *12*, 5318–5335. [CrossRef]

30. Wang, X.; Wang, Z.; Cao, J.; Dong, Y.; Chen, Y. Melatonin Alleviates Acute Sleep Deprivation-Induced Memory Loss in Mice by Suppressing Hippocampal Ferroptosis. *Front. Pharmacol.* **2021**, *12*, 708645. [CrossRef]

31. Xie, J.; Zhao, Z.Z.; Li, P.; Zhu, C.L.; Guo, Y.; Wang, J.; Deng, X.M.; Wang, J.F. Senkyunolide I Protects against Sepsis-Associated Encephalopathy by Attenuating Sleep Deprivation in a Murine Model of Cecal Ligation and Puncture. *Oxid. Med. Cell. Longev.* **2021**, *2021*, 6647258. [CrossRef]

32. Tai, F.; Wang, C.; Deng, X.; Li, R.; Guo, Z.; Quan, H.; Li, S. Treadmill exercise ameliorates chronic REM sleep deprivation-induced anxiety-like behavior and cognitive impairment in C57BL/6J mice. *Brain Res. Bull.* **2020**, *164*, 198–207. [CrossRef]

33. Mashour, G.A.; Pal, D.; Brown, E.N. Prefrontal cortex as a key node in arousal circuitry. *Trends Neurosci.* **2022**, *45*, 722–732. [CrossRef]

34. Kenwood, M.M.; Kalin, N.H.; Barbas, H. The prefrontal cortex, pathological anxiety, and anxiety disorders. *Neuropsychopharmacology* **2022**, *47*, 260–275. [CrossRef]

35. Ben Simon, E.; Rossi, A.; Harvey, A.G.; Walker, M.P. Overanxious and underslept. *Nat. Hum. Behav.* **2020**, *4*, 100–110. [CrossRef]

36. Hotamisligil, G.S.; Davis, R.J. Cell Signaling and Stress Responses. *Cold. Spring Harb. Perspect. Biol.* **2016**, *8*, a006072. [CrossRef]

37. Bachstetter, A.D.; Van Eldik, L.J. The p38 MAP Kinase Family as Regulators of Proinflammatory Cytokine Production in Degenerative Diseases of the CNS. *Aging Dis.* **2010**, *1*, 199–211.

38. Arthur, J.S. MSK activation and physiological roles. *Front. Biosci.* **2008**, *13*, 5866–5879. [CrossRef]

39. Gorska, M.M.; Liang, Q.; Stafford, S.J.; Goplen, N.; Dharajiya, N.; Guo, L.; Sur, S.; Gaestel, M.; Alam, R. MK2 controls the level of negative feedback in the NF-kappaB pathway and is essential for vascular permeability and airway inflammation. *J. Exp. Med.* **2007**, *204*, 1637–1652. [CrossRef]

40. Shao, Y.; Chen, X.; Liu, Y.; Chen, X.; Li, C.; Wang, L.; Zhao, W. Dexmedetomidine alleviates lung injury in sepsis mice through regulating P38 MAPK signaling pathway. *Panminerva Med.* **2021**, *63*, 563–564. [CrossRef]

41. Chen, Y.; Li, L.; Zhang, J.; Cui, H.; Wang, J.; Wang, C.; Shi, M.; Fan, H. Dexmedetomidine Alleviates Lipopolysaccharide-Induced Hippocampal Neuronal Apoptosis via Inhibiting the p38 MAPK/c-Myc/CLIC4 Signaling Pathway in Rats. *Mol. Neurobiol.* **2021**, *58*, 5533–5547. [CrossRef] [PubMed]

42. Krueger, J.M.; Majde, J.A. Cytokines and sleep. *Int. Arch. Allergy Immunol.* **1995**, *106*, 97–100. [CrossRef] [PubMed]

43. Opp, M.R. Cytokines and sleep. *Sleep Med. Rev.* **2005**, *9*, 355–364. [CrossRef]

44. Kang, X.; Jiang, L.; Lan, F.; Tang, Y.Y.; Zhang, P.; Zou, W.; Chen, Y.J.; Tang, X.Q. Hydrogen sulfide antagonizes sleep deprivation-induced depression- and anxiety-like behaviors by inhibiting neuroinflammation in a hippocampal Sirt1-dependent manner. *Brain Res. Bull.* **2021**, *177*, 194–202. [CrossRef] [PubMed]

45. Ji, X.; Guo, Y.; Zhou, G.; Wang, Y.; Zhang, J.; Wang, Z.; Wang, Q. Dexmedetomidine protects against high mobility group box 1-induced cellular injury by inhibiting pyroptosis. *Cell Biol. Int.* **2019**, *43*, 651–657. [CrossRef]

46. Grubac, Z.; Sutulovic, N.; Suvakov, S.; Jerotic, D.; Puskas, N.; Macut, D.; Rasic-Markovic, A.; Simic, T.; Stanojlovic, O.; Hrncic, D. Anxiogenic Potential of Experimental Sleep Fragmentation Is Duration-Dependent and Mediated via Oxidative Stress State. *Oxid. Med. Cell. Longev.* **2021**, *2021*, 2262913. [CrossRef]

47. Ramanathan, L.; Gulyani, S.; Nienhuis, R.; Siegel, J.M. Sleep deprivation decreases superoxide dismutase activity in rat hippocampus and brainstem. *Neuroreport* **2002**, *13*, 1387–1390. [CrossRef]

48. Ramanathan, L.; Hu, S.; Frautschy, S.A.; Siegel, J.M. Short-term total sleep deprivation in the rat increases antioxidant responses in multiple brain regions without impairing spontaneous alternation behavior. *Behav. Brain Res.* **2010**, *207*, 305–309. [CrossRef]

49. D'Almeida, V.; Hipolide, D.C.; Azzalis, L.A.; Lobo, L.L.; Junqueira, V.B.; Tufik, S. Absence of oxidative stress following paradoxical sleep deprivation in rats. *Neurosci. Lett.* **1997**, *235*, 25–28. [CrossRef]

50. Gopalakrishnan, A.; Ji, L.L.; Cirelli, C. Sleep deprivation and cellular responses to oxidative stress. *Sleep* **2004**, *27*, 27–35. [CrossRef]

51. Vutskits, L.; Clark, J.D.; Kharasch, E.D. Reporting Laboratory and Animal Research in ANESTHESIOLOGY: The Importance of Sex as a Biologic Variable. *Anesthesiology* **2019**, *131*, 949–952. [CrossRef]

52. Dolsen, E.A.; Crosswell, A.D.; Prather, A.A. Links Between Stress, Sleep, and Inflammation: Are there Sex Differences? *Curr. Psychiatry Rep.* **2019**, *21*, 8. [CrossRef]

Disclaimer/Publisher's Note: The statements, opinions and data contained in all publications are solely those of the individual author(s) and contributor(s) and not of MDPI and/or the editor(s). MDPI and/or the editor(s) disclaim responsibility for any injury to people or property resulting from any ideas, methods, instructions or products referred to in the content.

Article

The Crosstalk between the EGFR and IFN-γ Pathways and Synergistic Roles in Survival Prediction and Immune Escape in Gliomas

Xingang Zhou [1,†], Tingyu Liang [2,†], Yulu Ge [3,†], Yu Wang [2] and Wenbin Ma [2,*]

1   Department of Pathology, Beijing Ditan Hospital, Capital Medical University, Beijing 100015, China; zhouxg1980@126.com
2   Department of Neurosurgery, Center for Malignant Brain Tumors, National Glioma MDT Alliance, Peking Union Medical College Hospital, Chinese Academy of Medical Sciences and Peking Union Medical College, Beijing 100730, China; lzn13391610953@126.com (T.L.); ywang@pumch.cn (Y.W.)
3   Eight-Year Medical Doctor Program, Chinese Academy of Medical Sciences and Peking Union Medical College, Beijing 100730, China; ge-yl19@mails.tsinghua.edu.cn
*   Correspondence: mawb2001@hotmail.com; Tel.: +86-13701364566
†   These authors contributed equally to this work.

Citation: Zhou, X.; Liang, T.; Ge, Y.; Wang, Y.; Ma, W. The Crosstalk between the EGFR and IFN-γ Pathways and Synergistic Roles in Survival Prediction and Immune Escape in Gliomas. *Brain Sci.* **2023**, *13*, 1349. https://doi.org/10.3390/brainsci13091349

Academic Editors: Junhui Wang, Hongxing Wang, Jing Sun and Lucia Lisi

Received: 6 August 2023
Revised: 6 September 2023
Accepted: 15 September 2023
Published: 20 September 2023

Copyright: © 2023 by the authors. Licensee MDPI, Basel, Switzerland. This article is an open access article distributed under the terms and conditions of the Creative Commons Attribution (CC BY) license (https:// creativecommons.org/licenses/by/ 4.0/).

Abstract: Glioma is the most common primary malignant brain tumor. The poor prognosis of gliomas, especially glioblastoma (GBM), is associated with their unique molecular landscape and tumor microenvironment (TME) features. The epidermal growth factor receptor (EGFR) gene is one of the frequently altered loci in gliomas, leading to the activation of the EGFR signaling pathway and thus, promoting the genesis of gliomas. Whether there exist factors within the TME that can lead to EGFR activation in the context of gliomas is currently unexplored. In total, 702 samples from The Cancer Genome Atlas (TCGA) and 325 samples from The Chinese Glioma Genome Atlas (CGGA) were enrolled in this study. Gene signatures related to EGFR signaling and interferon-γ (IFN-γ) response were established via the LASSO-COX algorithm. Gene Set Enrichment Analysis (GSEA) and Gene Ontology (GO) analysis were applied for function exploration. Kaplan–Meier (KM) curves and single sample GSEA (ssGSEA) of immune cell subpopulations were performed to analyze the prognosis and TME characteristics of different subgroups. Moreover, Western blotting (WB) and flow cytometry (FCM) demonstrated the correlation between IFN-γ and EGFR signaling activation and the subsequent induction of programmed death ligand 1 (PD-L1) expression. An EGFR signaling-related risk score was established, and a higher score was correlated with poorer prognosis and a more malignant phenotype in gliomas. Biological function analysis revealed that a higher EGFR-related score was significantly associated with various cytokine response pathways, especially IFN-γ. Long-term (7 days) exposure to IFN-γ (400 ng/mL) induced the activation of EGFR signaling in the u87 cell line. Next, an IFN-γ response-related risk score was established; the combination of these two scores could be used to further reclassify gliomas into subtypes with different clinical features and TME features. Double high-risk samples tended to have a poorer prognosis and more immunosuppressive TME. Additionally, FCM discovered that the activation of EGFR signaling via EGF (100 ng/mL) could trigger PD-L1 protein expression. This research indicates that IFN-γ, an inflammatory cytokine, can activate the EGFR pathway. The combination of EGFR signaling and IFN-γ response pathway can establish a more precise classification of gliomas.

Keywords: glioma; EGFR; IFN-γ; tumor microenvironment; classification

1. Introduction

Glioma is the most common primary malignant brain tumor, accounting for approximately 80–85% of malignant brain tumors in adults [1]. It possesses high heterogeneity and consists of multiple subtypes of tumor cells. Each cell subtype is genetically and functionally different with a unique immunological landscape, such as differences in microglia or macrophage composition and T cell infiltration [2,3]. Its striking cellular heterogeneity, combined with its aggressive nature, contributes broadly to the failure of immunotherapy and molecular targeted therapies. According to the fifth edition of the World Health Organization classification of tumors of the central nervous system (WHO CNS5) [4], the median survival time of glioblastoma (GBM) did not exceed 15 months, with surgical resection, radiotherapy, and chemotherapy [5].

Epidermal growth factor receptor (EGFR) is a transmembrane tyrosine kinase located on the chromosome band 7p12. Amplification of EGFR has been observed in approximately 34–39% of GBMs [6–9], which always leads to the overexpression of EGFR and subsequent activation of downstream signaling pathways. Given the important role of EGFR amplification in glioma progression, EGFR amplification has been incorporated as one of the criteria for molecular classification of GBM in the WHO CNS5 classification [4]. Activation and autophosphorylation of EGFR result in the recruitment of downstream pathway proteins [10]. The downstream pathways of EGFR signaling not only contribute to DNA synthesis and cell proliferation [11] but also exert an influence on the tumor microenvironment (TME). For instance, EGFR signaling has been implicated in promoting macrophage infiltration within the tumor, via the chemokine ligand 2 (CCL2) [12]. Moreover, studies have shown that the activation of EGFR induces the secretion of programmed death ligand 1 (PD-L1) to inhibit the function of T cells [13]. In summary, studies reveal that EGFR signaling can reshape the TME [14], while the role of TMEs and cytokines in the activation of EGFR signaling remain unclear.

Interferon-γ (IFN-γ) is an important component within the TME that exhibits a dual role in glioma progression. First, this cytokine demonstrates the ability to directly inhibit the proliferation and invasion of glioma cells [15,16]. Second, IFN-γ in the TME is necessary for immune cells to maintain their tumor-killing activity. Studies have shown that the absence of IFN-γ in chimeric antigen receptor (CAR) T cells may hamper the in vivo antitumor activity and the activation of host immune cells [17]. Additionally, the presence of the IFN-γ receptor signaling pathway within GBMs is essential for CAR T therapy [18]. Furthermore, compared to adjuvant therapy alone, neoadjuvant programmed cell death protein 1 (PD-1) blockade is associated with the upregulation of IFN-γ-related gene expression to elevate the antitumor activity [19]. Thus, IFN-γ is an important cytokine induced by immunotherapy. However, various immunotherapies have failed to improve glioma clinical outcomes. It is noteworthy that IFN-γ also serves as a significant inducer of PD-L1 expression in the TME. The increased PD-L1 expression, in turn, leads to T cell dysfunction and apoptosis, thus contributing to the suppression of inflammatory responses and facilitating tumor immune evasion [13,20].

Studies have reported some communication between the IFN-γ and EGFR signaling pathways. In A431 cells, an epidermoid carcinoma cell line, IFN-γ induced a rapid and reversible tyrosine phosphorylation of EGFR [21]. In ovarian cancer cell lines, although IFN-γ reduced cell proliferation by 30–40%, it strikingly increased the EGFR expression, including cell surface receptors and total cellular receptors [22]. However, in a breast cancer cell line MDA468, IFN-γ inhibited cell proliferation while reducing the number of available EGFR binding sites, without any change in the EGFR affinity [23]. As for glioma, no studies have reported that IFN-γ stimulation could directly modulate EGFR expression or EGFR activity. Since IFN-γ can both inhibit the malignant phenotype of glioma cells and maintain the killing activity of immune cells, and EGFR serves as an important factor in glioma progression, it is necessary to explore the effect of IFN-γ on EGFR in gliomas. The relationship between these two factors is instructive for glioma treatment. For example, if IFN-γ can increase EGFR activity, IFN γ may upregulate PD-L1 expression by elevating the activity of the EGFR pathway. Therefore, inhibition of the EGFR pathway may somewhat reduce PD-L1 expression caused by IFN-γ stimulation, thus enhancing the tumor-killing activity of infiltrated immune cells.

This research aimed to explore the relationship between IFN-gamma and EGFR in the context of gliomas, followed by an investigation of their effects on the TME. The analysis of

RNA sequencing data sourced from the Cancer Genome Atlas (TCGA) and The Chinese Glioma Genome Atlas (CGGA) databases revealed a noteworthy correlation between the activation of the EGFR pathway and IFN-γ pathway, and the activation of these two pathways significantly affected the immune infiltration and the immune checkpoint gene expression within the TME. To validate this finding, experiments were conducted using the U87 cell line, followed by validating the results that IFN-γ can activate the EGFR pathway and activation of the EGFR pathway can upregulate PD-L1 expression. The results further support the notion that inflammatory molecules within TME could potentially influence the specific molecular mechanisms underlying tumor progression. Through our current research, we provide the promising possibility of a combination of immunotherapy and EGFR-targeted medicine in gliomas.

2. Methods

2.1. Samples and Datasets

RNA sequencing data and corresponding clinical information of 702 glioma samples from TCGA-GBM and TCGA-LGG (https://portal.gdc.cancer.gov/, accessed on 1 May 2023) and 325 glioma samples from CGGA-325 (https://www.cgga.org.cn, accessed on 1 May 2023) were retrospectively enrolled in this study. TCGA mainly covers European and American races. Therefore, we selected CGGA, the largest glioma database in China, to validate the results obtained from TCGA. All samples with RNA sequencing data and clinical data were enrolled without selection. Clinical information included age, gender, histology, WHO grade, overall survival, chromosome 1p19q codeletion status, and isocitrate dehydrogenase (IDH) mutation status. Overall survival was estimated from the date of diagnosis to the date of last follow-up or the date of death. These two cohorts were independent with no patients overlapping, and were widely used in glioma research due to comprehensive data. In total, 186 EGFR signaling pathway-related genes were obtained from the Genecards Database (https://www.genecards.org/, accessed on 1 May 2023) [24].

This study was conducted based on data from publicly available database sources and based on experiments using cell lines. Ethical approval was therefore not required.

2.2. Construction of EGFR-Related and IFN-γ-Related Prognostic Gene Signatures

In total, 186 EGFR signaling pathway-related genes were obtained from the Genecards Database. In all the following processes, genes were selected based on statistical significance, namely p-value less than 0.05. To analyze whether expression, a continuous variable, had an effect on survival, univariate Cox regression analysis was performed in the TCGA and CGGA cohorts separately. Genes with significantly different expression patterns based on EGFR amplification status in IDH1-wildtype GBMs were screened out. In total, 23 genes that were retained in the results of all the above processes were selected.

A risk model was established using the TCGA dataset, after which the parameters obtained from this risk model were substituted into the CGGA dataset for validation. To prevent overfitting in the generation of a risk model with too many variables, the least absolute shrinkage and selection operator (LASSO) machine learning algorithm [25] was employed to dimensionally downscale the 23 genes. To find independent prognostic factors in the presence of multiple variables, multivariate Cox regression analysis was used for the remaining 9 genes. Finally, 4 genes that independently affected prognosis were retained and used to construct a risk model. The risk score for all patients was determined by summing the regression coefficients of the selected genes multiplied by the corresponding expression values.

For the IFN-γ-related risk score, the process was similar to that just used for EGFR. In total, 200 IFN-γ response genes were involved, followed by selection via multivariate Cox regression analysis, LASSO, and univariate Cox regression analysis to construct IFN-γ-related prognostic gene signatures. In summary, 4 genes and 8 genes were used to establish EGFR-related and IFN-γ-related prognostic gene signatures, respectively.

2.3. Biological Function and Signaling Pathway Analysis

Patients from both datasets were divided into high-risk and low-risk groups based on the median risk score as a threshold. Pearson correlation analysis was conducted to determine genes that were positively correlated with the risk score ($R > 0.4$, $p < 0.05$). The two variables were considered to be relatively strongly correlated when the R-value was greater than 0.4, and genes selected for the ensuing pathway enrichment could not be too few, with 0.4 used as the cut-off criterion. The TCGA and CGGA cohorts were analyzed separately. The correlation results were used individually for Gene Set Enrichment Analysis (GSEA). The intersection of positively correlated gene sets from the two cohorts was taken for gene ontology (GO) analysis. This was due to the different data entry requirements of GSEA and GO, where GSEA requires a gene set with correlation coefficients, while GO only requires a gene set containing the gene name. The database of all pathway enrichments came from the MSigDB database version 7.2. The pathway enrichment was conducted using the ClusterProfiler R package [26]. This package is conducive to enrichment analysis based on a given gene set. PROGENy was used to calculate the pathways activity score (N = 11) [27]. The PROGENy algorithm can infer the activation of 11 tumorigenesis-related signaling pathways based on gene expression.

2.4. Comprehensive Analysis of Immune and Molecular Characteristics

The ssGSEA method was used for immune cell infiltration and inflammation activity [28]. Also, the ESTIMATE R package was used to evaluate the purity of gliomas and the proportion of infiltrating immune cells and stromal cells [29].

2.5. Cell Culture for Western Blot (WB) and Flow Cytometry (FCM)

The GBM cell line u87 was collected from the Institute of Biochemistry and Cell Biology, Chinese Academy of Science, and was cultured in Dulbecco's modified Eagle's medium (DMEM, Corning, NY, USA) supplemented with 10% fetal bovine serum (FBS, Gibco, Grand Island, NY, USA) and 1% penicillin–streptomycin (Gibco). To analyze the function of IFN-γ on EGFR signaling activation, the u87 cell was exposed to IFN-γ (400 ng/mL, PeproTech, Cranbury, NJ, USA) for 7 days. Next, total protein was extracted for WB assay, and whole-cell lysates were prepared on ice in RIPA buffer. A microplate spectrophotometer (Infinite M200 PRO, Tecan, Männedorf, Switzerland) was used to determine the protein concentration via Coomassie Brilliant Blue. An amount of 40 mg of total protein from cell lysates was loaded on a 10% SDS-PAGE gel and then transferred to the PVDF membrane (Merck Millipore, Burlington, MA, USA). After 5% skimmed milk closure, the primary antibody was diluted with 1X TBST (Tris-buffered saline with Tween-20). Primary antibodies included EGFR (Abcam ab52894, 1:1000), p-EGFR (Abcam ab32430, 1:1000), and glyceraldehyde 3-phosphate dehydrogenase (GAPDH) (Proteintech 60004-1-Ig, 1:10,000). The membrane was incubated with primary antibodies at 4 °C overnight and then with horseradish peroxidase-conjugated secondary antibodies (ZSGB-BIO, Beijing, China) for 1 h at room temperature. The ECL Western Blotting Detection System (Bio-Rad, Hercules, CA, USA) was used to visualize protein signals. GAPDH was used as the loading control to quantify relative protein levels. Additionally, for flow cytometry (FCM) analysis, after exposure to Epidermal Growth Factor (EGF) (100 ng/mL for 48 h), 1×10^5 u87 cells in EGF treatment and negative control (NC) groups were harvested and washed twice before PD-L1 antibody (BioLegend, PE anti-human PD-L1, No. 393608) staining for 30 min. Unbound antibodies were then washed out by PBS, and PD-L1 protein expression on the tumor cell surface was tested by FCM.

2.6. Statistical Analysis

All statistical analyses were performed using R software (version 4.1.1; https://www.r-project.org/, accessed on 1 May 2023), a free software environment for statistical calculation and graphics. The log-rank test and Kaplan–Meier method were used to evaluate survival time and calculate survival differences, which were conducted via R package survival [30].

The correlation between risk score and gene expression was tested using Pearson correlation analysis. The continuous variables between the two groups were compared using Student's *t*-test. For categorical variables, Fisher's exact and Chi-square tests were used for group comparison. A *p*-value of less than 0.05 was considered statistically significant.

3. Results

3.1. Establishment of an EGFR Pathway-Related Prognostic Gene Signature

A total of 186 genes involved in the EGFR signaling pathway were included based on information obtained from the Genecards Database. First, 74 prognosis-related genes were identified using the univariate Cox regression. Among these genes, 23 genes that showed significant differences in expression based on the EGFR gene amplification status were sorted out (Figure 1A). In total, 4 final candidate genes, MAP3K14, RHOC, STAT1, and VAV3 (Figure 1B–D) were determined using the LASSO algorithm and subsequent multivariate Cox regression analysis. The risk score was constructed using these genes and their corresponding Cox regression coefficients via the formula: $0.23 \times \mathrm{MAP3K14}^{\mathrm{expr}} + 0.21 \times \mathrm{RHOC}^{\mathrm{expr}} + 0.20 \times \mathrm{STAT1}^{\mathrm{expr}} + 0.23 \times \mathrm{VAV3}^{\mathrm{expr}}$. Moreover, there was a significant difference in the expression of four candidate genes and the risk score between the EGFR amplified group and the non-amplified group ($p < 0.05$, Figure 1E). The overall process of constructing EGFR-related signatures is demonstrated in Supplementary Figure S1.

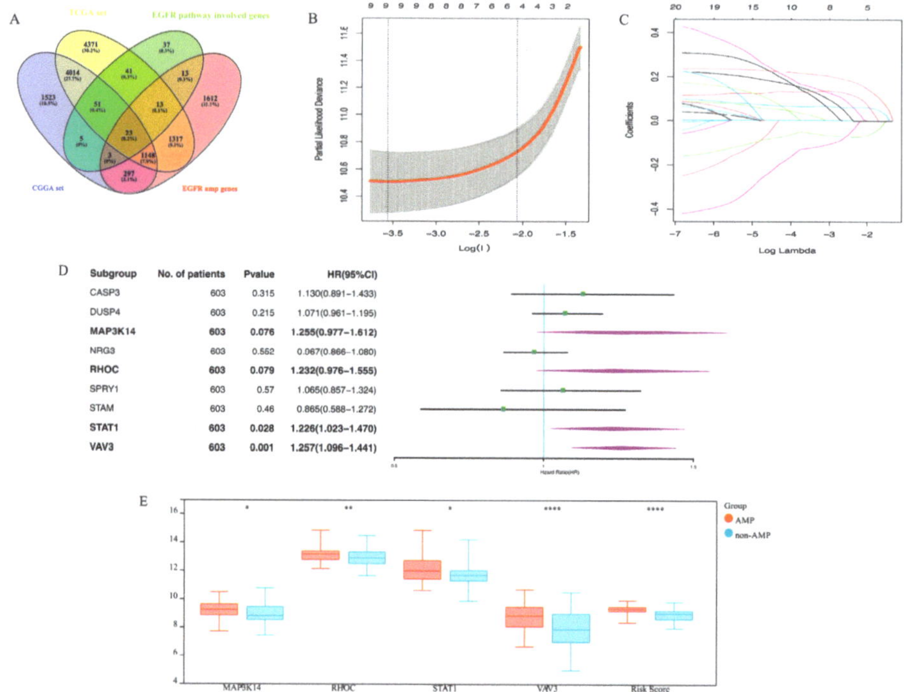

Figure 1. The establishment of EGFR pathway-related prognostic gene signature. (**A**). The screening process of the 23 EGFR pathway-related prognostic genes; (**B,C**). LASSO coefficient profiles of the remaining 23 genes; (**D**). Multivariate Cox regression analysis of the remaining 9 genes in the TCGA cohort; (**E**). The 4 left genes and the risk score distribution in the EGFR amplification and non-amplification groups (* means $p < 0.05$; ** means $p < 0.01$; **** means $p < 0.0001$).

3.2. Clinic Pathological Features Related to the EGFR Pathway-Related Prognostic Gene Signature in Gliomas

To investigate the clinical and pathological relevance of the gene signature, the correlation between risk score and various clinicopathological factors was evaluated in this study. Figure 2A,B illustrates the ordering of patients based on the risk score in both datasets. The results revealed that risk score was positively associated with age at diagnosis, and lower risk score was significantly associated with low-grade gliomas, IDH mutation, and chro 1p/19q codeletion. This suggests that the EGFR pathway-based signature predicts the malignant phenotype. For further evaluation of the correlation between this gene signature and patient survival time, all glioma patients were classified into high-risk or low-risk groups according to the median cut-off point. Compared to the high-risk group in the TCGA set, the low-risk group had a significantly better prognosis (log-rank $p < 0.0001$, Figure 2C). To enhance the practicality of clinical application for individual glioma patients, a nomogram model was developed, which incorporated prognostic factors including risk score to predict overall survival at 1, 3, and 5 years (Figure 2E). The model demonstrated that the nomogram had superior predictive ability and could facilitate clinical decision-making. In addition, the above conclusions were validated in the CGGA independent cohort (Figure 2D,F).

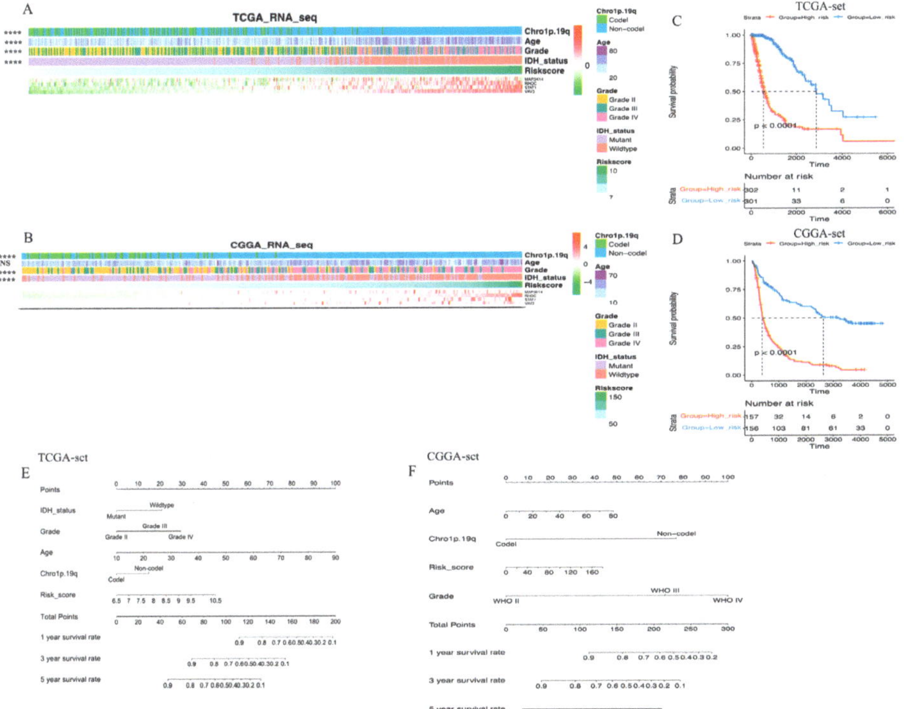

Figure 2. The risk score distribution and prognosis prediction of EGFR pathway-related risk score. (**A,B**). The expression pattern of risk score and 4 genes in glioma samples in the TCGA set (**A**) and the CGGA set (**B**); (**C,D**). Kaplan–Meier curves present risk score is negatively related with prognosis in the TCGA set (**C**) and the CGGA set (**D**); (**E,F**). The model, including risk score, demonstrates that the nomogram possesses superior predictive ability and can facilitate clinical decision-making in the TCGA set (**E**) and the CGGA set (**F**) (******** means $p < 0.0001$, NS means no significance).

3.3. Biological Processes and Signaling Pathway Analysis

Principal component analysis (PCA) of the transcriptome demonstrated that high-risk and low-risk groups had different transcriptomic expression profiles in the TCGA cohort (Figure 3A). The CGGA cohort had a similar PCA result (Figure not shown). Given that the transcriptome tumor underlies biologic characteristics, for example, higher oncogene transcription leads to a more malignant phenotype, the PCA result suggests that high-risk and low-risk groups may have different biological profiles.

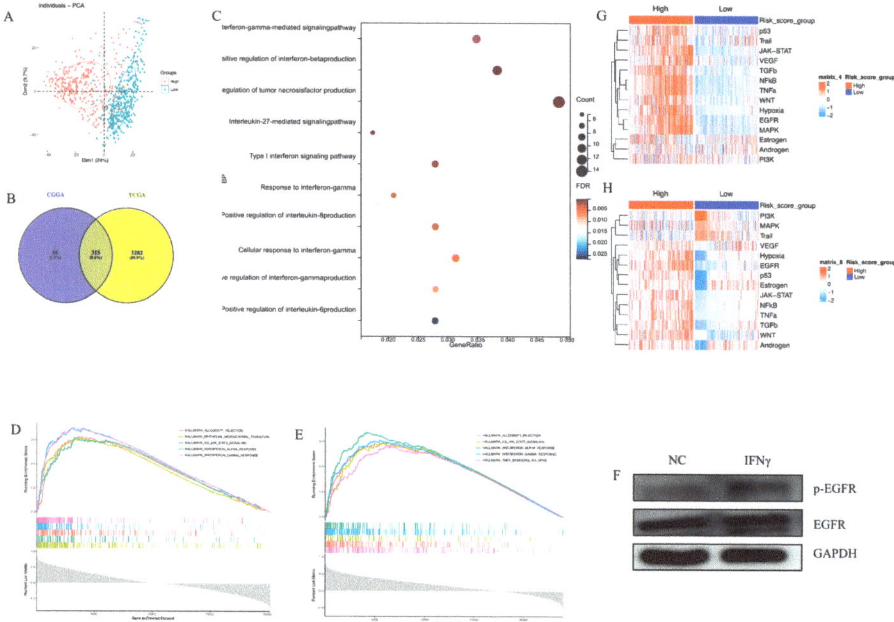

Figure 3. The biological function of EGFR pathway-related prognostic gene signature. (**A**). There is separation between the high-risk and low-risk groups in the TCGA cohort; (**B**). The 315 overlapped genes positively associated with risk score ($R > 0.4$, $p < 0.0001$) between the TCGA and CGGA sets; (**C**). BP analysis indicates that risk score is positively related with multiple cytokines pathways; (**D,E**). GSEA reveals the relationship between risk score and IFN-γ response pathway in the TCGA set (**D**) and the CGGA set (**E**); (**F**). WB assay demonstrates that long-term exposure to IFN-γ (400 ng/mL, 7 days) can trigger the activation of EGFR pathway in u87 GBM cell line; (**G,H**). Heatmap of signaling pathway activity scores by PROGENy in the TCGA set (**G**) and the CGGA set (**H**).

To further explore the distinct biology, Pearson correlation analysis between risk score and other genes in whole genome gene profiling was conducted. In total, 315 positively correlated genes ($R > 0.4$, $p < 0.05$), intersecting with the TCGA and CGGA populations, were selected for GO analysis (Figure 3B). Biological process (BP) analysis pointed out that multiple cytokine-related pathways were positively correlated with the EGFR pathway-related risk score, especially IFN-γ-related signaling (Figure 3C). GSEA analysis in the TCGA and CGGA cohorts showed that the gene signature was enriched in IFN-γ response signaling in both cohorts (Figure 3D,E). The long-term exposure to IFN-γ (400 ng/mL, 7 days) triggered the activation of the EGFR pathway in the u87 GBM cell line (Figure 3F), which provided further evidence for the findings obtained by pathway enrichment.

Finally, NF-κB, JAK-STAT, and TNFα signaling pathways were significantly activated in the high-risk group (Figure 3G,H). All the results revealed that the EGFR pathway-related prognostic gene signature was linked with various malignant pathways and tumor malignant behaviors.

3.4. Establishment of an IFN-γ-Related Prognostic Gene Signature

Based on the results of the pathway analysis, the combination between the EGFR-related signature and the IFN-γ response pathway was investigated. The process of setting the EGFR-related risk score was repeated, but based on 200 IFN-γ response genes (Figure 4A–C). In total, eight genes were left to construct an IFN-γ-related prognostic gene signature. The overall process of constructing IFN-γ-related signatures is demonstrated in Supplementary Figure S2. Next, the combined function of the above two signatures was explored. Prognostic analysis subsequently revealed distinct survival outcomes among these patient groups. The patients with low scores exhibited the most favorable prognosis, those with high and low scores displayed intermediate prognosis, and those with high scores demonstrated the poorest prognosis in both the TCGA and CGGA RNA-seq sets (Figure 4D,E, $p < 0.0001$). This indicated that the combination of these two signatures had a strong prognostic predictive power. Notably, a strong positive correlation was observed between the EGFR-related signature and the IFN-γ-related signature (Figure 4F,G; TCGA: $R = 0.88$, $p < 0.0001$; CGGA: $R = 0.30$, $p < 0.0001$). To further explore this relationship, glioma patients were stratified based on the median values of both scores. The majority of the patients exhibited either high or low scores in both signatures, and only a small proportion of patients displayed a combination of high and low scores (Figure 4H).

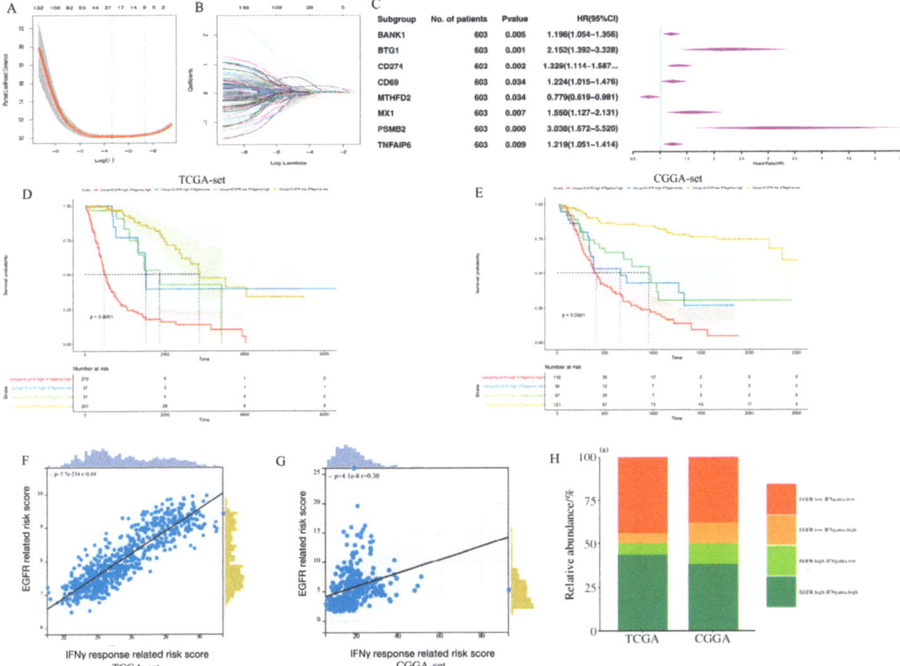

Figure 4. The establishment of IFN-γ response pathway signature and the relationship between EGFR pathway-related and IFN-γ response pathway prognostic gene signature. (**A,B**). LASSO coefficient profiles of IFN-γ response genes; (**C**). Multivariate Cox regression analysis of the selected genes in the TCGA set; (**D,E**). The combination of EGFR pathway-related and IFN-γ response pathway signatures can divide glioma samples into 4 groups with distinct prognosis in the TCGA set (**D**) and the CGGA set (**E**); (**F,G**). The relationship between the above two signatures in the TCGA set (**F**) and the CGGA set (**G**); (**H**). The majority of patients exhibit either high or low scores in both signatures, with only a small proportion of patients displaying a combination of one high and one low score in the TCGA and CGGA sets.

3.5. Immune Cell Infiltration and Inflammatory Profiles Related to the Gene Signature

The TME characteristics of gliomas were assessed using the ESTIMATE algorithm [29]. The results showed that in the TCGA dataset, a higher score was observed in the patients with high scores (Figure 5A), while immune and stromal scores were also higher in the patients with high scores (Figure 5B,C). The patients with low scores had higher tumor purity (Figure 5D), which predicted a better prognosis [31]. Similar results were found in the CGGA dataset (Figure 5E–H). For the distribution of immune cell subsets, various immune cells were enriched in the high-risk group, such as memory $CD4^+$ and $CD8^+$ T cells (Figure 6A). However, exhausted T cells and regulatory T cells were also enriched in the high-risk group. Additionally, high-risk patients expressed higher levels of immune checkpoint molecules (Figure 6B), such as PD-1, PD-L1, and IDO1, which further suggests stronger immunosuppression in the high-risk group. To further validate the results, flow cytometry (FCM) was used to analyze the protein expression of PD-L1 on the surface of the u87 cell line. The result revealed that EGF (100 ng/mL, 48 h), an agonist of EGFR signaling, could remarkably upregulate PD-L1 protein expression (Figure 6C).

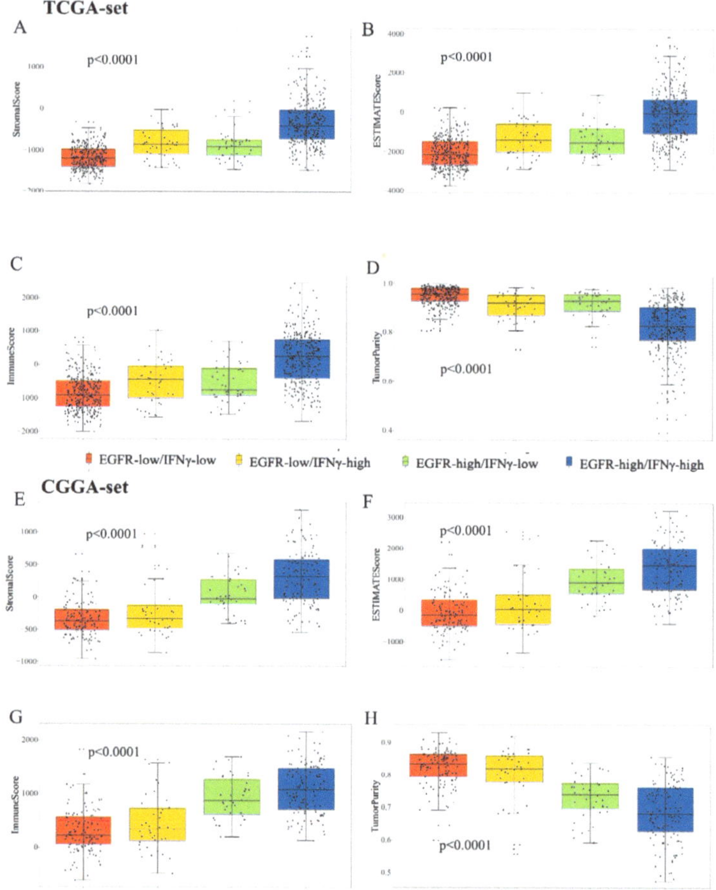

Figure 5. Differences among immune phenotypes of combination of the two signatures in terms of four glioma immune microenvironment signatures. (**A–D**). The comparison of the ESTIMATE, immune and stromal scores and tumor purity among distinct groups in the TCGA cohort; (**E–H**). The validation of above results in the CGGA cohort.

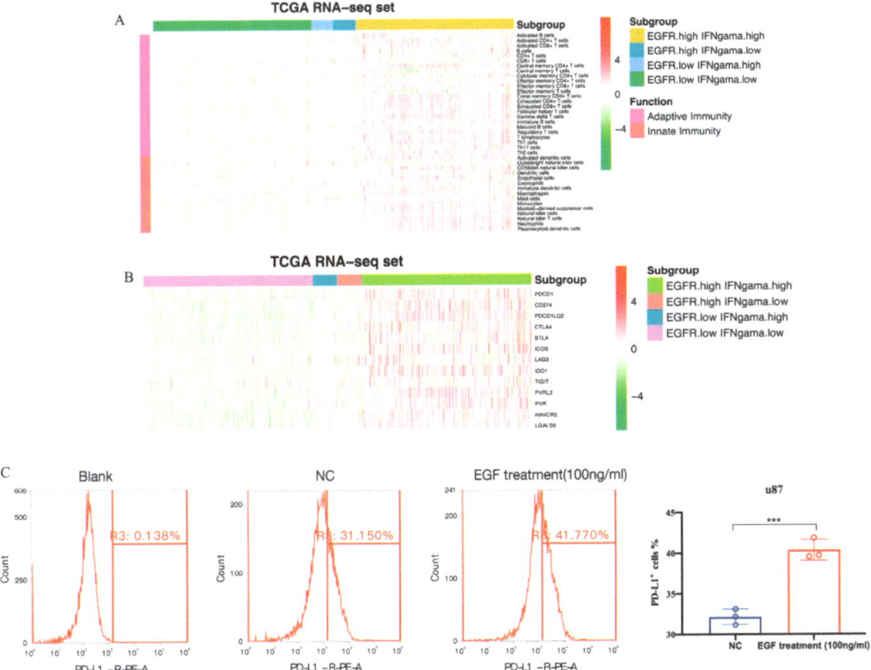

Figure 6. Immune cell infiltration and inflammatory profiles related to the gene signatures. (**A**). The distribution of various immune cell subpopulations in distinct groups; (**B**). The distribution of immune checkpoint molecules in different groups; (**C**). The flow cytometry verifies the activation of EGFR pathway via EGF (100 ng/mL 48 h) can upregulate PD-L1 expression on the surface of u87 cell line (*** means $p < 0.001$).

4. Discussion

Recently, the discovery of the lymphatic system in the central nervous system (CNS) has challenged the notion that CNS is an immune-privileged site [32,33], and an increasing number of studies have focused on the crosstalk between glioma progression and the TME. The transmitters, chemokines, and cytokines within the TME could not only enhance tumor progression and invasion, as well as immune evasion but also enhance resistance to therapy [14]. For instance, CCL2 overexpression could reduce TMZ-induced apoptosis by activating AKT signaling, thus leading to TMZ resistance [34]. The activation of the oncogenic pathway in glioma cells could alter cell secretion and thus TME composition. The EGFR pathway plays a key role in the secretion of multiple cytokines and the infiltration of immune cells. EGFR variant III (EGFRvIII) could potentiate IL-1β and IL-6 secretion in GBM cells [35]. For immune cell composition, activation of EGFR signaling could induce CCL2 expression and then elevate the infiltration of tumor-associated macrophages (TAMs) in the TME [12]. It has been found that EGFR-induced changes in the TME can in turn increase the malignant phenotype or drug resistance of glioma cells, which further illustrates the potential therapeutic benefit of targeting EGFR. As a result, it is necessary to have a deeper understanding of the modulation of the EGFR pathway, and the upstream regulation mechanism of EGFR signaling in gliomas remains unclear.

In this study, a comprehensive analysis of RNA-seq data from TCGA and CGGA was conducted to construct an EGFR pathway-related prognostic gene signature, including four genes (MAP3K14, RHOC, STAT1, and VAV3), with a good prediction for clinical outcomes. All the above four genes were reported to be oncogenic in gliomas. Guo et al. discovered that VAV3 can regulate GBM cell proliferation, invasion, and cancer stem-like cell self-renewal [36]. In addition, RHOC was reported to be involved in the downstream of EGFR signaling, and knock-down RHOC can inhibit EGF-induced VEGF expression [37]. Moreover, STAT1 plays a key role in the glioma malignant phenotype, and STAT1 downregulation can inhibit the aggressiveness of GBM cells by regulating the epithelial–mesenchymal transition (EMT) [38]. Furthermore, MAP3K14 can promote cell invasion by regulating mitochondrial dynamics and trafficking [39]. Based on PROGENy (Figure 3G,H), this signature related to four genes can reflect the activation of the EGFR signaling pathway.

There was a positive correlation between the activation of the EGFR-related pathway and the activation of the IFN-γ response pathway. This correlation may be explained by the fact that the downstream of the EGFR and IFN-γ receptors have some overlapping with each other. In hepatocellular carcinoma cells, EGF and IFN-γ both activated PD-L1 expression via the MAPK signaling pathway, which can be blocked by the MEK inhibitor selumetinib [40]. In melanoma, IFN-γ activated the JAK-STAT-IRF1 pathway, resulting in IRF1 binding to PD-L1 promoter [41]. As for EGFR signaling, it triggered the activation of STAT transcription factors [11]. Research has reported that EGF could induce STAT1 expression to exacerbate the IFN-γ-mediated PD-L1 axis in EGFR-positive cancer cell lines, excluding glioma, and blockade of EGFR by afatinib inhibited EGF- and IFN-γ-mediated PD-L1 expression [42]. This downstream overlapping provides a plausible explanation for the positive correlation between the IFN-γ-related and EGFR-related risk scores.

Notably, this study revealed that IFN-γ can indeed upregulate the phosphorylated form of EGFR, indicating that this pro-inflammatory cytokine can serve as an upstream inducer of the EGFR pathway in gliomas. Moreover, the results verified that activation of EGFR can upregulate PD-L1 expression, and thus, the upregulation of EGFR activity may be one of the pathways by which IFN-γ stimulates PD-L1 expression. Although IFN-γ has been reported to lead to increased EGFR activity in other cancer types [21,22], it is notable that this relationship persists in glioma, a CNS tumor with an immune microenvironment completely different from that of peripheral solid tumors. This direct relationship is expected to be one of the therapeutic targets. For example, IFN-γ is one of the indicators that rise in immune cell therapy. Given the dual nature of IFN-γ effects, if its ability to promote PD-L1 expression is inhibited, the efficacy of immune cells can be raised in theory. Based on the findings, inhibition of EGFR may help achieve this goal, which provides a preliminary theoretical basis for the combination of immune cell therapy and EGFR-targeted drugs. A study reported that the blockade of EGFR by afatinib resulted in decreased STAT1 and IRF-1 levels, and disabled the IFN-γ-STAT1-mediated PD-L1 axis in vitro and in vivo for oral cancer and lung cancer [42]. Whether this result can be reproduced in gliomas deserves to be explored in future studies.

In total, 1027 glioma samples were divided into four groups based on EGFR and IFN-γ related signatures. This shows that the combination of two risk scores could predict immune infiltrating cells within the glioma TME. Patients in the high-risk group exhibited low-purity tumors and increased infiltration of immune cells, which indicates that high-risk gliomas induce more immune responses and attract more immune cells to infiltrate due to their greater proliferation and invasion properties. However, it is also important to note that the highest level of inhibitory immune cell infiltration and the highest level of immune checkpoint molecular expression were also observed in the high-risk group. This indicates a strong immunosuppressive TME, which can be reflected by significantly more exhausted T cells in the high-risk group. High activation of these two pathways in the high-risk group may contribute to the high expression of immune checkpoints such as PD-L1. High activation of EGFR may also lead to a more suppressed TME [12,43]. This paradoxical observation suggests that the host has a potentially strong immune response to high-risk gliomas, but it is inhibited by stronger immunosuppressive mechanisms, which highlights the need to reverse the inhibitory TME to unleash the full potential of the intrinsic antitumor immune response. Therefore, from the perspective of immune cell infiltration, it can be speculated that high-risk patients may exhibit a more effective antitumor immune response when receiving immunotherapy, as long as the inhibitory TME is adequately modulated.

The combination of EGFR and IFN-γ related scores is helpful for clinical decision-making. If a patient has a high-risk score, it suggests that this patient may have a poorer prognosis, and a more aggressive treatment should be taken. In addition, as discussed earlier, patients with high-risk scores may be good candidates for immunotherapy when the TME is adjusted. Based on the findings, EGFR-targeted drugs may be a good choice for patients with high-risk scores to have their TME modulated. Thus, patients with high-risk scores may be well suited for a combination of immunotherapy and EGFR-targeted agents, which is currently rarely investigated in gliomas and deserves further exploration.

There are several limitations in this research. First, preliminary in vitro experiments were only conducted to validate the activation of EGFR by IFN-γ. The specific mechanism by which IFN-gamma leads to the elevation of phosphorylated EGFR was not explored, which is important for elucidating the clear interaction between the two and deserves further investigation in vitro and in vivo in future studies. Second, the expression of immune checkpoint molecules and immune cell content in the TME was calculated by RNA expression data. Future studies could validate the conclusions at the protein level using patient specimens and immunostaining techniques.

In conclusion, this study provides novel insights into the relationship between IFN-γ-related and EGFR-related pathways in glioma patients. Moreover, the results demonstrate that IFN-γ could be an upstream inducer of EGFR signaling activation. Based on this relationship, EGFR-related and IFN-γ-related signatures were jointly used to divide the glioma patients. High-risk patients tended to have poorer prognosis and more inhibitory TME. Therefore, these patients may be more suitable for immune checkpoint blockade (ICB) therapy or other immunotherapeutic approaches, which warrants further validation for a clinical cohort in a follow-up study.

Supplementary Materials: The following supporting information can be downloaded at: https://www.mdpi.com/article/10.3390/brainsci13091349/s1, Supplementary Figure S1. The process of constructing EGFR-related signature. Supplementary Figure S2. The process of constructing IFN-γ response-related signature.

Author Contributions: T.L. and X.Z. draft the work. Y.G. analyzes the data. T.L. draw a sketch. Y.W. and W.M. supervise the analysis process. All authors have read and agreed to the published version of the manuscript.

Funding: This work was funded by the Beijing Municipal Natural Science Foundation (7202150) and the National High Level Hospital Clinical Research Funding (2022-PUMCH-A-019) for Y.W., and by the National High Level Hospital Clinical Research Funding (2022-PUMCH-B-113), the Tsinghua University-Peking Union Medical College Hospital Initiative Scientific Research Program (2019ZLH101) and the Beijing Municipal Natural Science Foundation (19JCZDJC64200[Z]) for W.M.

Institutional Review Board Statement: Not applicable.

Informed Consent Statement: Not applicable.

Data Availability Statement: All data and materials used are available from the corresponding author upon reasonable request.

Conflicts of Interest: The authors declare that they have no competing interest.

References

1. Ostrom, Q.T.; Cioffi, G.; Waite, K.; Kruchko, C.; Barnholtz-Sloan, J.S. CBTRUS Statistical Report: Primary Brain and Other Central Nervous System Tumors Diagnosed in the United States in 2014–2018. *Neuro Oncol.* **2021**, *23*, iii1–iii105. [CrossRef]
2. Yu, M.W.; Quail, D.F. Immunotherapy for Glioblastoma: Current Progress and Challenges. *Front. Immunol.* **2021**, *12*, 676301. [CrossRef]
3. Patel, A.P.; Tirosh, I.; Trombetta, J.J.; Shalek, A.K.; Gillespie, S.M.; Wakimoto, H.; Cahill, D.P.; Nahed, B.V.; Curry, W.T.; Martuza, R.L.; et al. Single-cell RNA-seq highlights intratumoral heterogeneity in primary glioblastoma. *Science* **2014**, *344*, 1396–1401. [CrossRef] [PubMed]

4.   Louis, D.N.; Perry, A.; Wesseling, P.; Brat, D.J.; Cree, I.A.; Figarella-Branger, D.; Hawkins, C.; Ng, H.K.; Pfister, S.M.; Reifenberger, G.; et al. The 2021 WHO Classification of Tumors of the Central Nervous System: A summary. *Neuro Oncol.* **2021**, *23*, 1231–1251. [CrossRef] [PubMed]

5.   Jiang, T.; Nam, D.H.; Ram, Z.; Poon, W.S.; Wang, J.; Boldbaatar, D.; Mao, Y.; Ma, W.; Mao, Q.; You, Y.; et al. Clinical practice guidelines for the management of adult diffuse gliomas. *Cancer Lett.* **2021**, *499*, 60–72. [CrossRef]

6.   Järvelä, S.; Helin, H.; Haapasalo, J.; Järvelä, T.; Junttila, T.T.; Elenius, K.; Tanner, M.; Haapasalo, H.; Isola, J. Amplification of the epidermal growth factor receptor in astrocytic tumours by chromogenic in situ hybridization: Association with clinicopathological features and patient survival. *Neuropathol. Appl. Neurobiol.* **2006**, *32*, 441–450. [CrossRef] [PubMed]

7.   Houillier, C.; Lejeune, J.; Benouaich-Amiel, A.; Laigle-Donadey, F.; Criniere, E.; Mokhtari, K.; Thillet, J.; Delattre, J.Y.; Hoang-Xuan, K.; Sanson, M. Prognostic impact of molecular markers in a series of 220 primary glioblastomas. *Cancer* **2006**, *106*, 2218–2223. [CrossRef]

8.   Saadeh, F.S.; Mahfouz, R.; Assi, H.I. EGFR as a clinical marker in glioblastomas and other gliomas. *Int. J. Biol. Markers* **2018**, *33*, 22–32. [CrossRef]

9.   Stichel, D.; Ebrahimi, A.; Reuss, D.; Schrimpf, D.; Ono, T.; Shirahata, M.; Reifenberger, G.; Weller, M.; Hanggi, D.; Wick, W.; et al. Distribution of EGFR amplification, combined chromosome 7 gain and chromosome 10 loss, and TERT promoter mutation in brain tumors and their potential for the reclassification of IDHwt astrocytoma to glioblastoma. *Acta Neuropathol.* **2018**, *136*, 793–803. [CrossRef]

10. Oprita, A.; Baloi, S.C.; Staicu, G.A.; Alexandru, O.; Tache, D.E.; Danoiu, S.; Micu, E.S.; Sevastre, A.S. Updated Insights on EGFR Signaling Pathways in Glioma. *Int. J. Mol. Sci.* **2021**, *22*, 587. [CrossRef]

11. Oda, K.; Matsuoka, Y.; Funahashi, A.; Kitano, H. A comprehensive pathway map of epidermal growth factor receptor signaling. *Mol. Syst. Biol.* **2005**, *1*, 2005.0010. [CrossRef] [PubMed]

12. An, Z.; Knobbe-Thomsen, C.B.; Wan, X.; Fan, Q.W.; Reifenberger, G.; Weiss, W.A. EGFR Cooperates with EGFRvIII to Recruit Macrophages in Glioblastoma. *Cancer Res.* **2018**, *78*, 6785–6794. [CrossRef] [PubMed]

13. Litak, J.; Mazurek, M.; Grochowski, C.; Kamieniak, P.; Roliński, J. PD-L1/PD-1 Axis in Glioblastoma Multiforme. *Int. J. Mol. Sci.* **2019**, *20*, 5347. [CrossRef]

14. Barthel, L.; Hadamitzky, M.; Dammann, P.; Schedlowski, M.; Sure, U.; Thakur, B.K.; Hetze, S. Glioma: Molecular signature and crossroads with tumor microenvironment. *Cancer Metastasis Rev.* **2022**, *41*, 53–75. [CrossRef] [PubMed]

15. Cheng, S.M.; Xing, B.; Li, J.C.; Cheung, B.K.; Lau, A.S. Interferon-gamma regulation of TNFalpha-induced matrix metalloproteinase 3 expression and migration of human glioma T98G cells. *Int. J. Cancer* **2007**, *121*, 1190–1196. [CrossRef] [PubMed]

16. George, J.; Banik, N.L.; Ray, S.K. Knockdown of hTERT and concurrent treatment with interferon-gamma inhibited proliferation and invasion of human glioblastoma cell lines. *Int. J. Biochem. Cell Biol.* **2010**, *42*, 1164–1173. [CrossRef]

17. Alizadeh, D.; Wong, R.A.; Gholamin, S.; Maker, M.; Aftabizadeh, M.; Yang, X.; Pecoraro, J.R.; Jeppson, J.D.; Wang, D.; Aguilar, B.; et al. IFNγ Is Critical for CAR T Cell-Mediated Myeloid Activation and Induction of Endogenous Immunity. *Cancer Discov.* **2021**, *11*, 2248–2265. [CrossRef]

18. Larson, R.C.; Kann, M.C.; Bailey, S.R.; Haradhvala, N.J.; Llopis, P.M.; Bouffard, A.A.; Scarfó, I.; Leick, M.B.; Grauwet, K.; Berger, T.R.; et al. CAR T cell killing requires the IFNγR pathway in solid but not liquid tumours. *Nature* **2022**, *604*, 563–570. [CrossRef]

19. Cloughesy, T.F.; Mochizuki, A.Y.; Orpilla, J.R.; Hugo, W.; Lee, A.H.; Davidson, T.B.; Wang, A.C.; Ellingson, B.M.; Rytlewski, J.A.; Sanders, C.M.; et al. Neoadjuvant anti-PD-1 immunotherapy promotes a survival benefit with intratumoral and systemic immune responses in recurrent glioblastoma. *Nat. Med.* **2019**, *25*, 477–486. [CrossRef]

20. Qian, J.; Wang, C.; Wang, B.; Yang, J.; Wang, Y.; Luo, F.; Xu, J.; Zhao, C.; Liu, R.; Chu, Y. The IFN-γ/PD-L1 axis between T cells and tumor microenvironment: Hints for glioma anti-PD-1/PD-L1 therapy. *J. Neuroinflamm.* **2018**, *15*, 290. [CrossRef]

21. Burova, E.; Vassilenko, K.; Dorosh, V.; Gonchar, I.; Nikolsky, N. Interferon gamma-dependent transactivation of epidermal growth factor receptor. *FEBS Lett.* **2007**, *581*, 1475–1480. [CrossRef] [PubMed]

22. Boente, M.P.; Berchuck, A.; Rodriguez, G.C.; Davidoff, A.; Whitaker, R.; Xu, F.J.; Marks, J.; Clarke-Pearson, D.L.; Bast, R.C., Jr. The effect of interferon gamma on epidermal growth factor receptor expression in normal and malignant ovarian epithelial cells. *Am. J. Obstet. Gynecol.* **1992**, *167*, 1877–1882. [CrossRef]

23. Chakravarthy, A.; Chen, L.C.; Mehta, D.; Hamburger, A.W. Modulation of epidermal growth factor receptors by gamma interferon in a breast cancer cell line. *Anticancer Res.* **1991**, *11*, 347–351.

24. Safran, M.; Dalah, I.; Alexander, J.; Rosen, N.; Iny Stein, T.; Shmoish, M.; Nativ, N.; Bahir, I.; Doniger, T.; Krug, H.; et al. GeneCards Version 3: The human gene integrator. *Database* **2010**, *2010*, baq020. [CrossRef] [PubMed]

25. Tibshirani, R. The lasso method for variable selection in the Cox model. *Stat. Med.* **1997**, *16*, 385–395. [CrossRef]

26. Yu, G.; Wang, L.G.; Han, Y.; He, Q.Y. clusterProfiler: An R package for comparing biological themes among gene clusters. *Omics* **2012**, *16*, 284–287. [CrossRef]

27. Schubert, M.; Klinger, B.; Klünemann, M.; Sieber, A.; Uhlitz, F.; Sauer, S.; Garnett, M.J.; Blüthgen, N.; Saez-Rodriguez, J. Perturbation-response genes reveal signaling footprints in cancer gene expression. *Nat. Commun.* **2018**, *9*, 20. [CrossRef] [PubMed]

28. Hänzelmann, S.; Castelo, R.; Guinney, J. GSVA: Gene set variation analysis for microarray and RNA-seq data. *BMC Bioinform.* **2013**, *14*, 7. [CrossRef] [PubMed]

29. Yoshihara, K.; Shahmoradgoli, M.; Martínez, E.; Vegesna, R.; Kim, H.; Torres-Garcia, W.; Treviño, V.; Shen, H.; Laird, P.W.; Levine, D.A.; et al. Inferring tumour purity and stromal and immune cell admixture from expression data. *Nat. Commun.* **2013**, *4*, 2612. [CrossRef] [PubMed]
30. Therneau, T.M. A Package for Survival Analysis in R. 2020. Available online: https://CRAN.R-project.org/package=survival (accessed on 1 May 2023).
31. Zhang, C.; Cheng, W.; Ren, X.; Wang, Z.; Liu, X.; Li, G.; Han, S.; Jiang, T.; Wu, A. Tumor Purity as an Underlying Key Factor in Glioma. *Clin. Cancer Res.* **2017**, *23*, 6279–6291. [CrossRef]
32. Feng, E.; Liang, T.; Wang, X.; Du, J.; Tang, K.; Wang, X.; Wang, F.; You, G. Correlation of alteration of HLA-F expression and clinical characterization in 593 brain glioma samples. *J. Neuroinflamm.* **2019**, *16*, 33. [CrossRef]
33. Zhao, B.; Wang, Y.; Wang, Y.; Dai, C.; Wang, Y.; Ma, W. Investigation of Genetic Determinants of Glioma Immune Phenotype by Integrative Immunogenomic Scale Analysis. *Front. Immunol.* **2021**, *12*, 557994. [CrossRef]
34. Qian, Y.; Ding, P.; Xu, J.; Nie, X.; Lu, B. CCL2 activates AKT signaling to promote glycolysis and chemoresistance in glioma cells. *Cell Biol. Int.* **2022**, *46*, 819–828. [CrossRef]
35. Gurgis, F.M.; Yeung, Y.T.; Tang, M.X.; Heng, B.; Buckland, M.; Ammit, A.J.; Haapasalo, J.; Haapasalo, H.; Guillemin, G.J.; Grewal, T.; et al. The p38-MK2-HuR pathway potentiates EGFRvIII-IL-1beta-driven IL-6 secretion in glioblastoma cells. *Oncogene* **2015**, *34*, 2934–2942. [CrossRef]
36. Miao, R.; Huang, D.; Zhao, K.; Li, Y.; Zhang, X.; Cheng, Y.; Guo, N. VAV3 regulates glioblastoma cell proliferation, migration, invasion and cancer stem-like cell self-renewal. *Mol. Med. Rep.* **2023**, *27*, 94. [CrossRef]
37. Nicolas, S.; Abdellatef, S.; Haddad, M.A.; Fakhoury, I.; El-Sibai, M. Hypoxia and EGF Stimulation Regulate VEGF Expression in Human Glioblastoma Multiforme (GBM) Cells by Differential Regulation of the PI3K/Rho-GTPase and MAPK Pathways. *Cells* **2019**, *8*, 1397. [CrossRef]
38. Zhao, L.; Li, X.; Su, J.; Wang Gong, F.; Lu, J.; Wei, Y. STAT1 determines aggressiveness of glioblastoma both in vivo and in vitro through wnt/beta-catenin signalling pathway. *Cell Biochem. Funct.* **2020**, *38*, 630–641. [CrossRef]
39. Jung, J.U.; Ravi, S.; Lee, D.W.; McFadden, K.; Kamradt, M.L.; Toussaint, L.G.; Sitcheran, R. NIK/MAP3K14 Regulates Mitochondrial Dynamics and Trafficking to Promote Cell Invasion. *Curr. Biol.* **2016**, *26*, 3288–3302. [CrossRef] [PubMed]
40. Xing, S.; Chen, S.; Yang, X.; Huang, W. Role of MAPK activity in PD-L1 expression in hepatocellular carcinoma cells. *J. Buon* **2020**, *25*, 1875–1882.
41. Garcia-Diaz, A.; Shin, D.S.; Moreno, B.H.; Saco, J.; Escuin-Ordinas, H.; Rodriguez, G.A.; Zaretsky, J.M.; Sun, L.; Hugo, W.; Wang, X.; et al. Interferon Receptor Signaling Pathways Regulating PD-L1 and PD-L2 Expression. *Cell Rep.* **2017**, *19*, 1189–1201. [CrossRef]
42. Cheng, C.C.; Lin, H.C.; Tsai, K.J.; Chiang, Y.W.; Lim, K.H.; Chen, C.G.; Su, Y.W.; Peng, C.L.; Ho, A.S.; Huang, L.; et al. Epidermal growth factor induces STAT1 expression to exacerbate the IFNr-mediated PD-L1 axis in epidermal growth factor receptor-positive cancers. *Mol. Carcinog.* **2018**, *57*, 1588–1598. [CrossRef] [PubMed]
43. Sidorov, M.; Dighe, P.; Woo, R.W.L.; Rodriguez-Brotons, A.; Chen, M.; Ice, R.J.; Vaquero, E.; Jian, D.; Desprez, P.Y.; Nosrati, M.; et al. Dual Targeting of EGFR and MTOR Pathways Inhibits Glioblastoma Growth by Modulating the Tumor Microenvironment. *Cells* **2023**, *12*, 547. [CrossRef] [PubMed]

Disclaimer/Publisher's Note: The statements, opinions and data contained in all publications are solely those of the individual author(s) and contributor(s) and not of MDPI and/or the editor(s). MDPI and/or the editor(s) disclaim responsibility for any injury to people or property resulting from any ideas, methods, instructions or products referred to in the content.

Article

EEG Biofeedback Decreases Theta and Beta Power While Increasing Alpha Power in Insomniacs: An Open-Label Study

Huicong Wang [1,2,3,4,5,†], Yue Hou [1,2,3,4,5,6,7,†], Shuqin Zhan [1,2,3,4,5], Ning Li [1,2,3,4,5], Jianghong Liu [1,2,3,4,5], Penghui Song [1,2,3,4,5], Yuping Wang [1,2,3,4,5,6,7,*] and Hongxing Wang [1,2,3,4,5,*]

[1] Department of Neurology, Xuanwu Hospital, Capital Medical University, Beijing 100053, China; 15652523402@163.com (H.W.); houyuechina@sina.cn (Y.H.); shqzhan@hotmail.com (S.Z.); lining945110@163.com (N.L.); liujh@xwhosp.org (J.L.); songpenghui0104@163.com (P.S.)
[2] Beijing Key Laboratory of Neuromodulation, Beijing 100053, China
[3] Center of Epilepsy, Beijing Institute for Brain Disorders, Capital Medical University, Beijing 100053, China
[4] Center for Sleep and Consciousness Disorders, Beijing Institute for Brain Disorders, Beijing 100053, China
[5] Collaborative Innovation Center for Brain Disorders, Capital Medical University, Beijing 100053, China
[6] Hebei Hospital of Xuanwu Hospital, Capital Medical University, Shijiazhuang 050030, China
[7] Neuromedical Technology Innovation Center of Hebei Province, Shijiazhuang 050030, China
* Correspondence: doctorwangyuping@163.com (Y.W.); wanghongxing@xwh.ccmu.edu.cn (H.W.); Tel.: +86-10-83198273 (Y.W.); +86-10-83198650 (H.W.); Fax: +86-10-83157841 (Y.W.); +86-10-83167306 (H.W.)
[†] These authors contributed equally to this paper.

Abstract: Insomnia, often associated with anxiety and depression, is a prevalent sleep disorder. Biofeedback (BFB) treatment can help patients gain voluntary control over physiological events such as by utilizing electroencephalography (EEG) and electromyography (EMG) power. Previous studies have rarely predicted biofeedback efficacy by measuring the changes in relative EEG power; therefore, we investigated the clinical efficacy of biofeedback for insomnia and its potential neural mechanisms. We administered biofeedback to 82 patients with insomnia, of whom 68 completed 10 sessions and 14 completed 20 sessions. The average age of the participants was 49.38 ± 12.78 years, with 26 men and 56 women. Each biofeedback session consisted of 5 min of EMG and 30 min of EEG feedback, with 2 min of data recorded before and after the session. Sessions were conducted every other day, and four scale measures were taken before the first, fifth, and tenth sessions and after the twentieth session. After 20 sessions of biofeedback treatment, scores on the Pittsburgh Sleep Quality Index (PSQI) were significantly reduced compared with those before treatment (-5.5 ± 1.43, t $= -3.85$, $p = 0.006$), and scores on the Beck Depression Inventory (BDI-II) (-7.15 ± 2.43, t $= -2.94$, $p = 0.012$) and the State-Trait Anxiety Inventory (STAI) (STAI-S: -12.36 ± 3.40, t $= -3.63$, $p = 0.003$; and STAI-T: -9.86 ± 2.38, t $= -4.41$, $p = 0.001$) were significantly lower after treatment than before treatment. Beta and theta power were significantly reduced after treatment, compared with before treatment (F $= 6.25$, $p = 0.014$; and F $= 11.91$, $p = 0.001$). Alpha power was increased after treatment, compared with before treatment, but the difference was not prominently significant ($p > 0.05$). EMG activity was significantly decreased after treatment, compared with before treatment (F $= 2.11$, $p = 0.015$). Our findings suggest that BFB treatment based on alpha power and prefrontal EMG relieves insomnia as well as anxiety and depression and may be associated with increased alpha power, decreased beta and theta power, and decreased EMG power.

Keywords: biofeedback; insomnia; EMG; EEG; alpha power

Citation: Wang, H.; Hou, Y.; Zhan, S.; Li, N.; Liu, J.; Song, P.; Wang, Y.; Wang, H. EEG Biofeedback Decreases Theta and Beta Power While Increasing Alpha Power in Insomniacs: An Open-Label Study. *Brain Sci.* **2023**, *13*, 1542. https://doi.org/10.3390/brainsci13111542

Academic Editors: Chantal Delon-Martin and Alan Pegna

Received: 28 September 2023
Revised: 27 October 2023
Accepted: 30 October 2023
Published: 2 November 2023

Copyright: © 2023 by the authors. Licensee MDPI, Basel, Switzerland. This article is an open access article distributed under the terms and conditions of the Creative Commons Attribution (CC BY) license (https://creativecommons.org/licenses/by/4.0/).

1. Introduction

Insomnia has significant long-term health consequences [1], with prevalence ranging from 4–36% among teens to 9–50% among adults [2], and comorbid insomnia-related conditions such as depression and anxiety are common in patients [3]. The American Academy of Sleep Medicine clinical practice guidelines and the European guidelines for treating insomnia recommend cognitive behavioral therapy (CBT-I) as the first-line treatment for chronic insomnia in adults [4]. If CBT-I is unavailable or ineffective for adult insomnia, benzodiazepines (BZDs), non-BZD receptor agonists, melatonin receptor agonists, antidepressants, and antipsychotics should be considered. However, these drugs will produce side effects [5], emphasizing the need for more non-pharmaceutical interventions for insomnia.

The disruption of hyperarousal, cortical activation, cognition, and somatic dysfunction are common clinical features of insomnia [6]. Hyperarousal can be detected by elevated cortisol levels, heightened muscle tension, high heart rate (HR) variability (HRV), and self-reporting. Increased high-frequency electroencephalography (EEG) activity (beta and gamma), decreased delta activity, and increased rapid eye movement EEG in states of excessive arousal are EEG indicators [7]. Therefore, changing the above indicators through different interventions is not only expected to have therapeutic effects on insomnia but can also be used to monitor the effectiveness of the interventions.

Biofeedback (BFB) is a non-invasive behavioral therapy that helps patients gain voluntary control over physiological events. Previous research has shown that BFB treatment can benefit disorders such as epilepsy [8], migraine [9], stroke [10], chronic insomnia [11,12], anxiety [13], attention-deficit/hyperactivity disorder [14,15], autism spectrum disorder [16], and major depression [17]. As a BFB method, neurofeedback training allows individuals to independently adjust specific brain activities, thereby changing their cognitive functions [18]. BFB is based on the idea that autonomic responses may be conditioned by instrumentation and includes biological monitors and sensors such as electromyography (EMG), EEG, electrodermal activity, skin temperature, HR, HRV, and end-tidal carbon dioxide [19]. Relative power may be a more stable and sensitive method for detecting non-rapid eye movement EEG signals in patients with insomnia [20]. Alpha signals are observed when a person is awake, calm, prepared, meditating, or relaxed [21]. Increasing alpha power can reduce symptoms of anxiety and depression [22] and improve working and episodic memory [23]. Increased forehead muscle tone is considered a sign of psychoemotional tension or stress [24]. EMG feedback reduces muscle tension and arousal associated with some types of insomnia and promotes sleep onset [11]. Initial pressure in insomniacs correlated positively with improved sleep, based on EMG findings [25]. In fact, some patients with insomnia can be treated using BFB to reduce muscle tension in the center of the forehead [26]. Therefore, the combination of the increased alpha frequency band of EEG and decreased frontal EMG power caused by BFB treatment can further increase the effectiveness of BFB in treating insomnia.

Previous studies have not analyzed the changes in EEG relative power after BFB treatment to assess its effectiveness. Therefore, we hypothesized that by increasing EEG alpha BFB treatment in insomnia patients, other frequency bands of brain power could be altered to achieve relief. We evaluated the clinical effectiveness of BFB treatment using the Pittsburgh Sleep Quality Index (PSQI) to measure insomnia and the Beck Depression Inventory (BDI-II) and State-Trait Anxiety Inventory (STAI) to measure emotional status. We analyzed the related results of EEG and frontal EMG power by monitoring the relative power before and after the BFB treatment. This study also explored possible BFB treatments to improve clinical understanding of this therapy and provides a reference for clinical practice.

2. Methods

2.1. Participants

We recruited 135 right-handed patients with primary insomnia from the Department of Neurology, Xuanwu Hospital, Capital Medical University, between 2014 and 2023, and excluded 14 patients who did not meet the inclusion criteria. A total of 121 patients with insomnia were treated using BFB. During this treatment, 82 patients completed more than 10 BFB sessions and 14 completed more than 20 sessions (Figure 1). Patients with incomplete data and those receiving fewer than 10 sessions were excluded. Participants were instructed to maintain a sleep diary upon returning home.

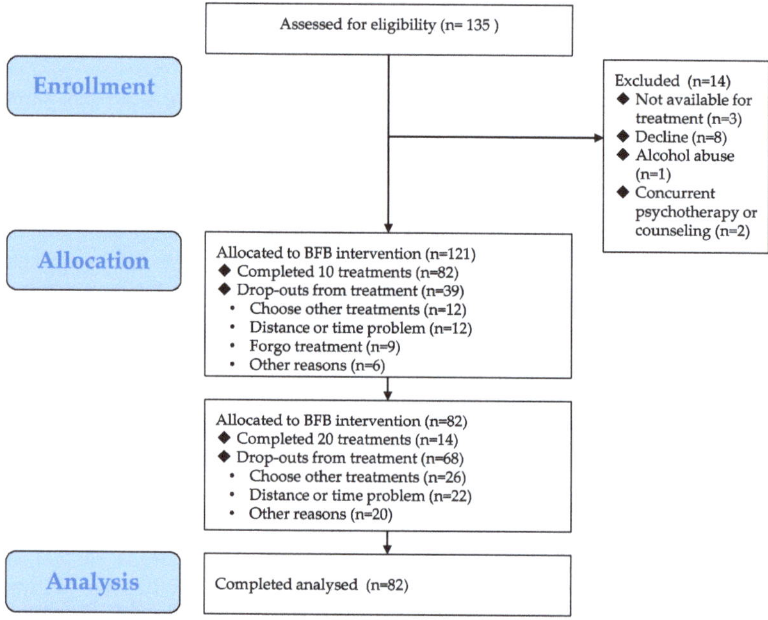

Figure 1. Flow diagram of the study. BFB, biofeedback.

The inclusion criteria were as follows: (1) a diagnosis of chronic insomnia (for more than 3 months) according to the Diagnostic and Statistical Manual of Mental Disorders, fourth edition; (2) age $\geq$ 18 years and non-perimenopausal women; (3) PSQI > 6 points [27]; (4) patients currently taking insomnia medication (they did not need to discontinue their medication, but their dose remained stable for 1 month before the experiment); (5) a neurological examination revealing no positive findings; (6) accessible audio-visual equipment to complete the questionnaires and examinations required for the study; and (7) an informed consent form signed by the patient or their family member. Exclusion criteria were as follows: (1) patients with a history of other mental illnesses, alcohol or drug abuse or dependence, or low intelligence; (2) clinical evidence of neurological or other physical diseases, including respiratory, cardiac, renal, hepatic, and endocrine disorders; (3) women who were pregnant or breastfeeding; (4) patients undergoing psychotherapy or counseling concurrently; (5) insomnia caused by other organic diseases; and (6) medical conditions that, in the investigator's opinion, precluded participation in the study.

2.2. Data Collection

Sociodemographic and clinical data were collected from patients, including age, sex, education level (coded as 1, illiterate; 2, primary school; 3, junior high school; 4, senior high school; 5, junior college; 6, bachelor's degree; and 7, master's degree), insomnia medication use and duration, and disease duration (number of years since diagnosis). We excluded illiterate patients to improve cooperation and included only those with primary or higher education. Moreover, the participants completed the study records and provided informed consent.

2.3. BFB Treatment

We conducted the BFB treatment at Xuanwu Hospital using BioNeuro Infiniti Bio 3000C V6.0.3 (T.T. Thought Technology Co., Montreal, QC, Canada). The eight-channel

ProComp Infiniti encoder was used to collect data. The EEG signal was sampled at 256 Hz, with a bandpass of 0.5–70 Hz and a time constant of 0.3. The reference and ground electrodes were placed in the binaural mastoid with a pair of ear clips, and the impedance values were adjusted to below 5 kΩ. The signal electrodes of the EEG sensor were placed at the Cz point according to the International 10/20 System [28]. The EMG sensor was attached to the forehead with a headband.

Before and after the BFB treatment, EEG/EMG power was measured for 2 min while the participants were at rest. A 5 min period of decreasing frontal EMG BFB was followed by a 30 min period of increasing alpha power (8–12 Hz) neurofeedback training. Ten BFB sessions lasted 1 month and were conducted on Mondays, Wednesdays, and Fridays during the first and third weeks, and on Tuesdays and Thursdays during the second and fourth weeks. To avoid movement artifacts, participants were instructed to remain as still as possible during the experiment. All participants received the same treatment (Figure 2).

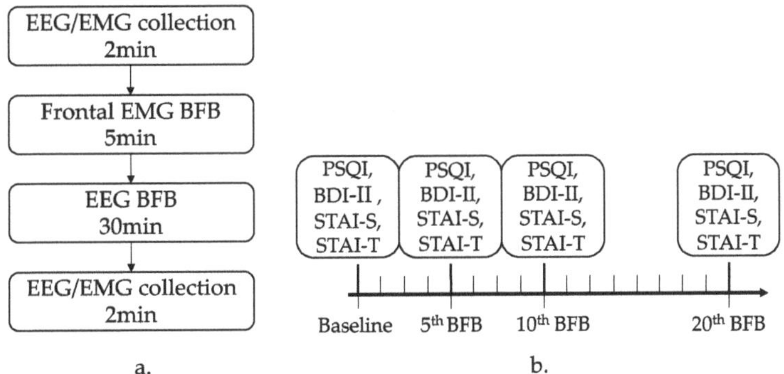

Figure 2. Experimental program schematic. (**a**) Biofeedback treatment flow chart for one session. (**b**) All process sketches including biofeedback treatment sessions and scale assessments. BDI-II, Beck Depression Inventory; BFB, biofeedback; EEG, electroencephalography; EMG, electromyography; PSQI, Pittsburgh Sleep Quality Index; STAI-S/T, State-Trait Anxiety Inventory.

We adjusted the feedback thresholds based on the baseline measures at each session using the following formulae [26]: reinforcer: alpha mean amplitude (standard deviation/4); and inhibitor: EMG mean amplitude (standard deviation/2). All treatment sessions were conducted by the same experienced supervisor. During treatment, the therapist could adjust the threshold artificially and ensure that rewards were administered at 50–80% of the baseline [29]. A reward percentage was determined by the therapist and adjusted according to the participant's motivation. To ensure participants were retained in the study, more feedback was provided if they were not motivated to complete all the neurofeedback sessions. For example, when training to increase alpha, if their alpha power was above the threshold, i.e., when the patient was relaxed, they would continuously hear beautiful music and watch beautiful scenery videos. The music and videos would stop if the alpha power fell below the threshold. At that point, participants needed only to adjust their own state of being and wait until the body and brain were completely relaxed. As the alpha power increased, the participant would resume hearing the music and viewing the video. If the therapist set the threshold too high and the patient struggled to reach it, their motivation was greatly reduced, especially if the patient was very anxious and could not concentrate long enough to complete the therapy. If the threshold was set too low, the patient could easily maintain motivation and the point of the therapy was negated. Video and audio playback were also important; a noisy video was not conducive to maintaining a relaxed and calm mood, and a lagging video led to patient irritation.

2.4. Evaluation

Primary outcome: EEG alpha power, beta power, theta power, and prefrontal EMG power were recorded before and after each BFB treatment session. Secondary outcome: sleep conditions (PSQI), depression severity (BDI-II), and anxiety severity (STAI) were assessed at baseline and after the fifth, tenth, and twentieth treatments [22]. In addition, the Treatment Emergent Symptom Scale and the Adverse Event Scale were completed.

PSQI is a self-rated questionnaire that assesses sleep quality and disturbances over a 1-month time interval. Nineteen individual items generate seven "component" scores: subjective sleep quality, sleep latency, sleep duration, habitual sleep efficiency, sleep disturbances, use of sleeping medication, and daytime dysfunction. Each item is scored on a scale from 0 to 3 points, and the total score ranges from 0 to 21 points. PSQI > 6 points reflects poor sleep quality [27]. The Cronbach's α coefficient for internal consistency of the total PSQI score was 0.84, which showed high reliability [30].

The revised BDI-II [31] is a widely used measure for assessing the severity of depression in psychiatric patients and for screening for possible depression in normal populations, according to the DSM-IV criteria for the diagnosis of depressive disorders. The BDI-II is scored by summing the highest ratings for each of the 21 items. Each item is rated on a 4-point scale ranging from 0 to 3, and the total scores can range from 0 to 63. BDI-II total scores of 0 to 13 indicate "minimal" depression; totals of 14 to 19 indicate "mild" depression; totals of 20 to 28 indicate "moderate" depression; and totals of 29 to 63 indicate "severe" depression. The BDI-II proved to be internally consistent (Cronbach's α = 0.840) [32].

The STAI has 40 items, with 20 items each for the State-Trait Anxiety Inventory–State (STAI-S) and State-Trait Anxiety Inventory–Trait (STAI-T). The STAI-S assesses the current state of anxiety by asking respondents how they feel "right now" using items that measure subjective feelings of apprehension, tension, nervousness, worry, and autonomic nervous system activation/arousal. The STAI-T assesses relatively stable aspects of "anxiety proneness," including general states of calmness, confidence, and security. Internal consistency alpha coefficients are quite high, ranging from 0.86 for high school students to 0.95 for military personnel [33]. The Cronbach's α for the STAI-S was 0.950 and for the STAI-T was 0.926 in previous studies [34].

2.5. Data Preprocessing

Real-time online processing and display of results for the EEG and EMG signals were acquired using the amplifier. Unit time segmentation of the acquired signals was performed. Signal processing and analysis were performed in each segment. First, a 50 Hz trap was applied to the signal to filter out the industrial frequency noise, and then high- and low-pass filtering (0.5–45 Hz) was applied to remove unwanted signals using an infinite impulse response filter. Wavelet transform was performed on the pre-processed signal to decompose the alpha signal from 8 to 13 Hz and calculate the mean value. EMG signals were similarly captured, with high- and low-pass filtering per unit of time, and then averaged to calculate the display of real-time results [35–37].

2.6. Analysis

Data regarding age, disease course, educational background, and baseline scores (PSQI, BDI-II, and STAI-S/T) were analyzed using Wilcoxon rank-sum tests. Sex and insomnia medication use was analyzed using chi-square analysis. Linear mixed-effects models (LMMs) were used to analyze the PSQI, BDI-II, and STAI-S/T scores at baseline and after the fifth, tenth, and twentieth sessions. To assess the effect of the intervention over time, LMMs were used for repeated measures data. When repeated measures data were missing at random, LMMs provided accurate parameter estimates. Four time points were considered in the fitting of the models (baseline and after the fifth, tenth, and twentieth sessions).

As the primary outcome of our analysis of EEG alpha, beta, theta, and frontal EMG power, we defined the time coefficient as a measure of the effect of BFB over time and analyzed the results using analysis of repeated variance measures. The "post- vs. pre-treatment" coefficient represents the difference between before and after treatment, or the immediate effect, and the "time×treatment" coefficient describes the long-term trend. The change in EMGs was defined as the difference between the first EMG and the last EMG after treatment. The changes in scale (PSQI/BDI-II/ STAI-S/T) scores were defined as the first measured scale score minus the last measured scale score. Pearson correlation analyses were conducted for these two variables. Statistical analysis was conducted with SPSS Statistics for Windows, version 26.0. Differences were regarded as statistically significant at $p \leq 0.05$.

3. Results

3.1. Clinical Characteristics

In total, 82 individuals completed more than 10 BFB sessions, including 14 individuals who completed more than 20 sessions. The average age of the participants was 49.38 ± 12.78 years, with 26 men and 56 women. There was no statistically significant difference between the 14 individuals who completed 20 sessions and the remaining 68 individuals in terms of demographic characteristics such as sex, age, education, and pre-treatment baseline scale scores (Table 1).

Table 1. Demographic and clinical characteristics of the participants.

Variables [a]	10–20 BFB Sessions ($n = 68$)	>20 BFB Sessions ($n = 14$)	X^2/z	p-Value [b]
Sex (female/male)	46/22	10/4	0.094	0.158
Medication use (yes/no)	22/46	3/11	0.209	0.259
Age (years)	49.45 ± 13.24	48.95 ± 10.02	−0.813	0.416
Education (years)	13.63 ± 4.76	12.36 ± 6.08	−0.554	0.580
Duration (years)	5.29 ± 3.285	4.36 ± 1.99	−0.876	0.387
PSQI	14.77 ± 3.63	11.00 ± 4.41	−0.165	0.869
BDI-II	16.19 ± 8.12	10.85 ± 10.17	−0.751	0.453
STAI-S	42.22 ± 49.21	32.22 ± 14.88	−1.688	0.091
STAI-T	41.23 ± 11.21	35.86 ± 13.09	−0.341	0.733

[a] Data are presented as mean $\pm$ standard deviation. [b] The p-value was obtained using a two-sample, two-tailed t-test. BDI-II, Beck Depression Inventory; BFB, biofeedback; PSQI, Pittsburgh Sleep Quality Index; STAI-S/T, State-Trait Anxiety Inventory.

3.2. Sleep Scale (PSQI)

PSQI scores after the 20th, 10th, and 5th BFB sessions were significantly lower than those before treatment (-5.50 ± 1.43, t = -3.85, $p = 0.006$; -3.64 ± 0.57, t = -6.4, $p < 0.001$; and -2.00 ± 0.51, t = -3.99, $p < 0.001$, respectively). The PSQI score was significantly lower after the 10th session than after the 5th session (-2.42 ± 0.60, t = -4.03, $p < 0.001$). There was no statistical difference in the PSQI scores between the 20th and 5th sessions ($p = 0.107$) and between the 20th and 10th sessions ($p = 0.479$) (Figure 3a).

3.3. Emotional Scales

3.3.1. BDI-II

BDI-II scores were significantly lower after the 20th, 10th, and 5th sessions than before treatment (-7.15 ± 2.43, t = -2.94, $p = 0.012$; -6.1 ± 0.71, t = -8.55, $p < 0.001$; and -4.96 ± 0.64, t = -7.71, $p < 0.00$, respectively). BDI-II scores were significantly lower after the 10th session than after the 5th session (-1.79 ± 0.53, t = -3.35, $p = 0.001$). Other comparisons were not statistically significant (Figure 3b).

3.3.2. STAI (STAI-S and STAI-T)

STAI-S scores were significantly lower after the 20th and 10th sessions than before treatment (-12.36 ± 3.40, t = -3.63, $p = 0.003$; and -10.47 ± 4.78, t = -2.19, $p = 0.031$, respectively). Other comparisons did not show any significant differences (Figure 3c).

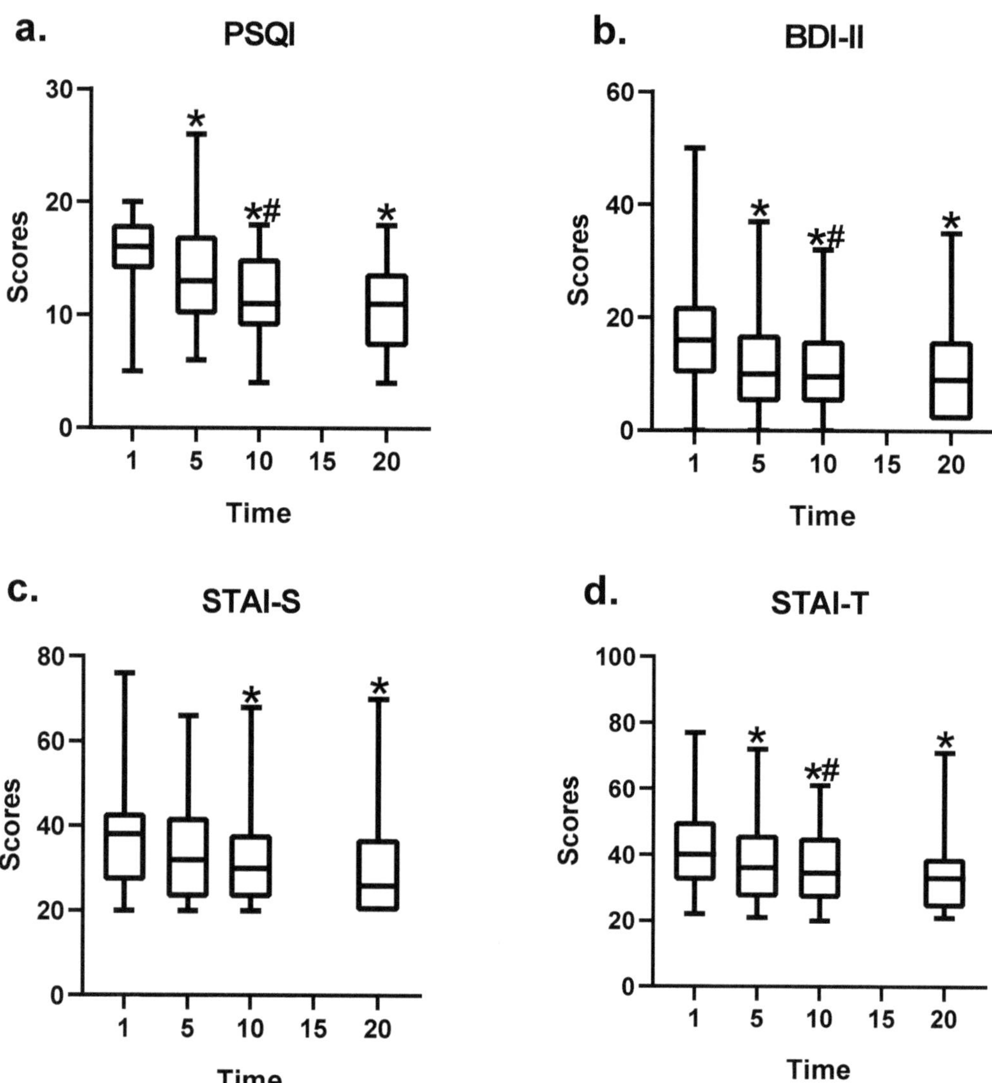

Figure 3. Changes in clinical scale scores before and after biofeedback sessions. (**a**) The PSQI scores before treatment and after the 5th, 10th, and 20th biofeedback sessions. (**b**) The BDI-II scores before treatment and after the 5th, 10th, and 20th biofeedback sessions. (**c**) The STAI-S scores before treatment and after the 5th, 10th, and 20th biofeedback sessions. (**d**) The STAI-T scores before treatment and after the 5th, 10th, and 20th biofeedback sessions. * Denotes a statistically significant difference from baseline ($p < 0.05$). # Denotes that the difference was statistically significant with comparison to the 5th BFB treatment sessions (median and Q1/Q3, $p < 0.05$). BDI-II, Beck Depression Inventory; PSQI, Pittsburgh Sleep Quality Index; STAI-S/T, State-Trait Anxiety Inventory.

STAI-T scores were significantly lower after the 20th, 10th, and 5th sessions than before treatment (-9.86 ± 2.38, $t = -4.41$, $p = 0.001$; -5.57 ± 1.01, $t = -5.53$, $p < 0.001$; and -4.88 ± 0.93, $t = -5.27$, $p < 0.001$, respectively). STAI-T scores were significantly lower after the 10th session than after the 5th session (-2.09 ± 0.68, $t = -3.07$, $p = 0.003$). Compared with the 5th session, there was no significant difference in STAI-T scores between the 20th and 10th sessions (Figure 3d).

3.4. EEG

In the analysis of the two-factor repeated variance measures, there were two variables: the number of BFB treatments, with 20 levels (20 BFB sessions), and the treatment effect, with 2 levels (before and after treatment). According to the immediate post-treatment effect, post-treatment beta power significantly decreased compared with pre-treatment (F = 6.25, p = 0.014), and post-treatment theta power statistically decreased compared with pre-treatment (F = 11.91, p = 0.001). Over time after treatment, beta power decreased significantly (F = 2.02, p = 0.035), and theta power decreased significantly (F = 2.15, p = 0.024). There was no significant interaction between the two variables. Alpha power increased more after treatment than before treatment but the difference was not prominently significant (p > 0.05) (Table 2).

Table 2. Improvement in the final EEG and EMG power compared with baseline after BFB treatment.

Item	Baseline [a] (Mean ± SD)	After 20 Sessions [a] (Mean ± SD)	Effect	df	F	p-Value [b]
Alpha	15.42 ± 9.07	16.24 ± 7.09	Time	9	0.95	0.483
			post- vs. pre-treatment	1	0.11	0.736
			Time × treatment	9	0.99	0.450
Beta	10.06 ± 5.07	9.06 ± 5.27	Time	9	2.02	0.035
			post- vs. pre-treatment	1	6.25	0.014
			Time × treatment	9	0.64	0.764
Theta	14.31 ± 10.53	6.76 ± 1.07	Time	9	2.15	0.024
			post- vs. pre-treatment	1	11.91	0.001
			Time × treatment	9	0.27	0.981
EMG	20.08 ± 58.73	2.44 ± 1.15	Time	9	2.91	0.002
			post- vs. pre-treatment	1	2.11	0.015
			Time × treatment	9	0.36	0.952

[a] The mean power of EEG and EMG are presented as mean ± standard deviation (SD) (µV). [b] The p-value was obtained by repeated measures analysis of variance. EEG, electroencephalography; EMG, electromyography.

Patients with insomnia who received BFB treatment experienced increased activation intensity of the alpha frequency band and decreased activation intensity of the beta and theta frequency bands. As treatment time increased, this effect did not decrease or increase (Figure 4a–c). Alpha power increased significantly after each of the first 10 biofeedback sessions, with a trough after the 11th and 12th BFB sessions and a peak after the 19th and 20th sessions (that is, the maximum value of alpha over the whole course). Beta power decreased steadily with only slight fluctuations after the treatments, with a peak after the 13th BFB treatment session and a trough after the 17th and 18th BFB treatment sessions. Theta, on the other hand, decreased significantly after each BFB treatment session, with a peak after the 2nd treatment and a trough after the 20th treatment (that is, the minimum value of theta over the whole course).

3.5. EMG

We used a repeated variance measures linear model with two within-test variables: the number of treatments with 20 levels and the treatment effect with 2 levels (before and after treatment). The main effect of treatment was prominent (F = 2.11, p = 0.015). EMG tension level was significantly lower after treatment than before treatment. The duration of treatment had a significant effect (F = 2.91, p = 0.002). There was no significant interaction between the two variables (p > 0.05) (Table 2).

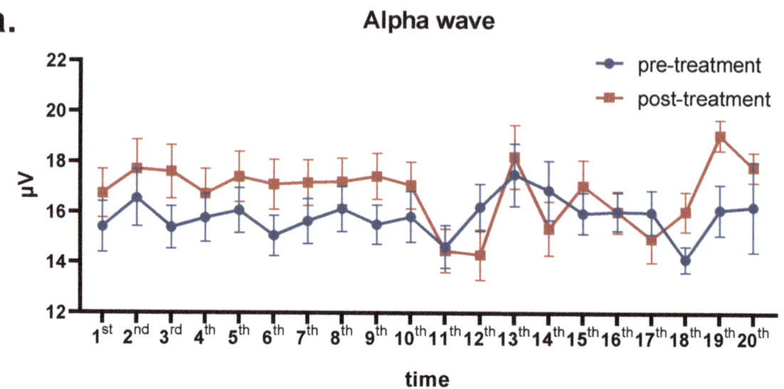

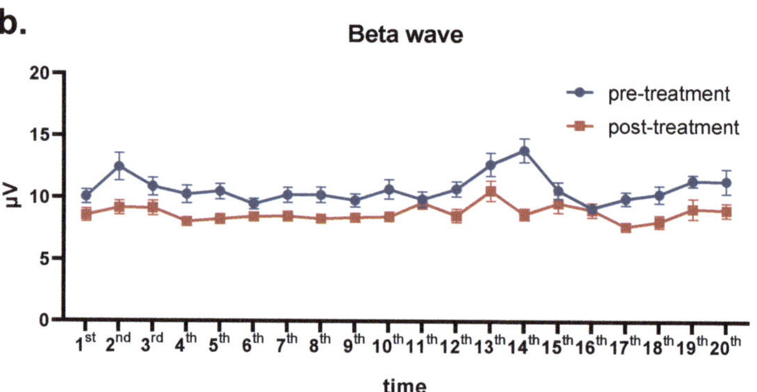

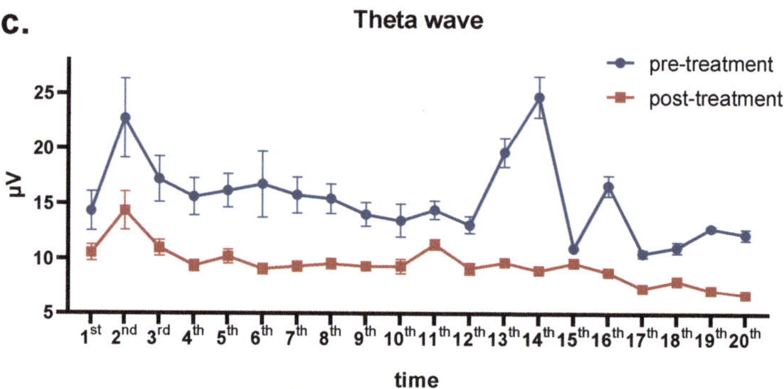

Figure 4. Changes in EEG waves pre- and post-treatment. (**a**) Alpha power before and after each of the 20 biofeedback sessions. (**b**) Beta power before and after each of the 20 biofeedback sessions. (**c**) Theta power before and after each of the 20 biofeedback sessions. Data presented as mean ± standard error. EEG, electroencephalography.

According to the study results, BFB treatment effectively decreased EMG activation in patients with sleep disorders. A longer treatment duration of more than 20 sessions may enhance this benefit (Figure 5).

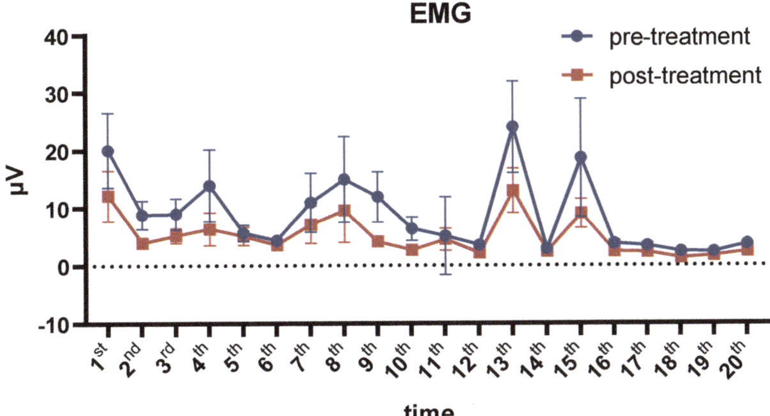

Figure 5. EMG power before and after each biofeedback session. EMG, electromyography. Data presented as mean ± standard error.

3.6. Regression Analysis

3.6.1. Reduced EMG Predicted Relief from Depression

Using the change in EMG as an independent variable and the change in BDI-II as the dependent variable, the regression analysis indicated that a decrease in EMG power could significantly predict a reduction in the BDI-II scores as an indicator of depression ($r^2 = 0.0478$, $p = 0.0482$) (Figure 6).

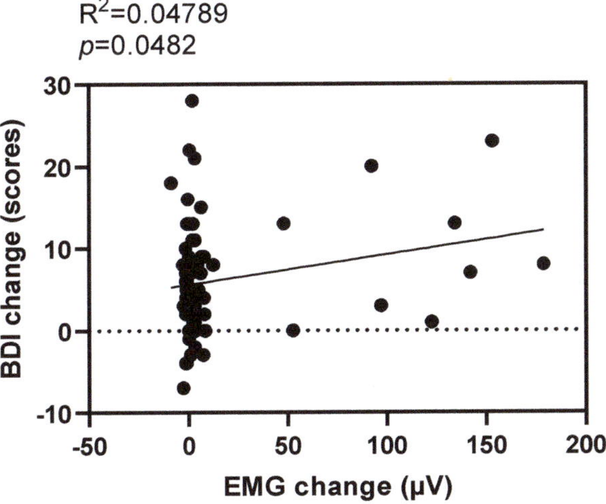

Figure 6. Regression results for EMG and BDI-II. The slash represents the slope. BDI-II, Beck Depression Inventory; EMG, electromyography.

3.6.2. Symptom Duration Was Negatively Correlated with Trait Anxiety Response

Each patient's illness lasted for a different length of time. The Pearson correlation analysis revealed that the longer the illness lasted, the smaller the decrease in state anxiety. A significant negative correlation was observed between symptom duration and state anxiety reduction after the 5th session ($r = -0.325$, $p = 0.014$) and after the 10th session ($r = -0.395$, $p = 0.003$) (Figure 7a,b).

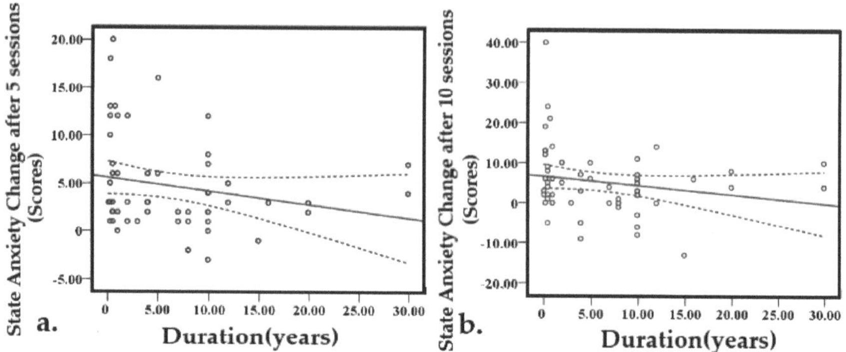

Figure 7. Correlation of insomnia duration and STAI. (**a**) State anxiety change after 5 sessions. (**b**) State anxiety change after 10 sessions. The dashed line represents the standard error and the slash represents the slope. STAI-S/T, State-Trait Anxiety Inventory.

4. Discussion

Our research demonstrated that by increasing alpha power and decreasing prefrontal EMG power, BFB treatment significantly reduced the symptoms of insomnia, anxiety, and depression in patients with insomnia, with noticeable improvements after 5 sessions of BFB treatment and further improvements after 10 or more sessions. To achieve a consistent treatment effect, a minimum of 10 sessions is required. A significant reduction in insomnia was observed, as was an easier transition to natural sleep, a shorter time to fall asleep, less waking while sleeping, and an improvement in the quality of sleep. After each BFB session, frontal EMG activation was almost always decreased. BFB sessions reduced EMG activity without it being affected by the number of treatments, suggesting that one session could reduce EMG activity and relieve muscle tension, regardless of whether the forehead muscles were tense before training. A decrease in frontal EMG predicted depression remission, and the longer the duration of insomnia, the fewer the anxiety symptoms alleviated by BFB treatment. According to previous research, progressive relaxation and EMG BFB significantly reduced sleep onset latency and depressive symptomatology, thus altering participants' attentional processes and thereby reducing cognitive arousal [38]. The standard theory states that muscle relaxation reduces anxiety by creating a physical state that counteracts the fight-or-flight response. Muscle relaxation lowers and alters HR, blood pressure, and stress hormone levels [39]. We assume that the high pre-treatment frontal EMG activity decreases after BFB treatment, which is associated with relief from insomnia and anxiety–depression.

In addition, we found that the pre-treatment brain power frequencies band tended to be lower during alpha power and higher during beta and theta power and that increasing the alpha power with BFB treatment could reduce beta and theta power. During thought, concentration, attention, nervousness, alertness, and excitement, beta power (20–30 Hz) is generated [40]. A previous meta-analysis of EEG power during periods of wakefulness demonstrated that absolute beta power increases significantly and powerfully, and absolute theta power significantly increases [21]. Cortical hyperexcitability is observed as an increased high-frequency EEG amplitude in patients with insomnia [41]. Neurofeedback may alleviate insomnia symptoms by reducing cortical hyperarousal in patients [11]. Theta power (4–8 Hz) is generally observed during wakefulness or the first stage of sleep [25], and its enhancement is associated with sleepiness [42]. It has also been observed during depression, anxiety, and distraction [22,43]. Patients with insomnia experience elevated waking theta power, sleepiness, and impaired cognitive performance, as well as drowsiness and fatigue related to overnight sleep disruptions or hypnotics [44,45]. Alpha and theta amplitudes increase, while beta amplitude decreases in insomniac patients with depression [46]. For depression, addiction, and anxiety, alpha-increasing and theta-decreasing neurofeedback treatments are the most popular stress reduction strategies, increasing creativity and relaxation, improving musical performance, and healing trauma responses [47]. Researchers have documented that increased sensorimotor rhythm feedback (12–15 Hz), suppressed theta power, and increased beta power under Cz may alleviate insomnia [48]. Therefore, a combination of increased alpha and decreased beta and theta power may alleviate insomnia and anxiety/depression symptoms.

Moreover, our study demonstrated that the therapist played an essential role in BFB treatment. Specifically, frontal EMG activity, which is high in individuals with insomnia before treatment, is difficult to change through self-regulation. Under the guidance of a therapist, these patients were able to relax and maintain their relaxation within one or two sessions after receiving EMG feedback training. Cortoos' remote BFB treatment of patients with insomnia without the presence of a trainer revealed little difference in frontal EMG power before and after treatment [26]. EEG feedback, in which therapists continuously adjust thresholds in real time according to a patient's condition, is also an important factor in treatment effectiveness and patient cooperation.

In total, 121 patients with insomnia were treated with BFB, and 82 of them attended more than 10 sessions without experiencing any adverse effects. Most patients fell asleep during the BFB treatment and felt well afterward. Patients with insomnia also expressed a high level of acceptance of BFB treatment; once they have mastered the technique and succeeded in relaxing, they are willing to continue participating and developing positive sleep habits. In addition to building confidence in patients with insomnia who are reluctant to take prescribed medications, this study's findings suggest that BFB may also improve sleep quality. The proportion of patients who completed all 20 sessions and had never taken insomnia medication was much higher than that of patients who completed only 10 sessions of BFB.

Our study has several limitations. First, patients were not followed, so it is unknown how long treatment efficacy can be maintained. Second, EEG BFB was restricted to a Cz single channel. Whole-brain EEG analysis, such as a topographic color map [11], could be used in the future to compare the changes in different brain power bands. Third, a sham BFB treatment as a control was not included in our study. Randomized controlled double-blind studies on BFB treatment are needed to define the effect of BFB treatment on insomnia. Lastly, the video graphics used in BFB treatment could be more varied and attractive to patients during follow-up sessions. Using brain–computer interface systems, adding new games [21], and using virtual reality devices may further reduce the attrition rate.

5. Conclusions

BFB treatment based on increasing alpha power and decreasing prefrontal EMG improves insomnia as well as anxiety and depression, with the symptom improvement becoming more pronounced and stable after more than 10 sessions. Biofeedback treatment that increases alpha power can effectively reduce beta and theta power, reducing alertness and promoting deep sleep. In addition, biofeedback treatment was effective in decreasing prefrontal EMG activation in patients with sleep disorders, and a reduction in prefrontal EMG can also predict an improved mood state. We will continue to control for other variables and conduct more extensive whole-brain EEG power analyses to explore the changes and effects produced by BFB treatment or combinations of EEG analyses at different locations and frequencies.

Author Contributions: H.W. (Huicong Wang) and H.W. (Hongxing Wang) contributed to the conception of the study; H.W. (Huicong Wang) and H.W. (Hongxing Wang) performed the experiment, drafted the manuscript, and provided statistical analysis; Y.H., Y.W. and H.W. (Hongxing Wang) performed the supervision and acquired funding; Y.H., S.Z., N.L., J.L. and P.S. enrolled patients with generalized anxiety disorder; and H.W. (Hongxing Wang) performed manuscript review and editing. All authors have read and agreed to the published version of the manuscript.

Funding: The work was supported in part by the National Key R&D Program of China (No. 2022YFC2503901, 2022YFC2503900, 2021YFC2501400, 2021YFC2501404), Beijing Health System Leading Talent Grant (No. 2022-02-10), National Natural Science Foundation of China (No. 82001388, 81771398), Beijing Hospitals Authority Youth Plan (No. QML20230803).

Institutional Review Board Statement: This study was approved by the local ethics review board of Xuanwu Hospital and was performed per the ethical standards laid down in the 1964 Declaration of Helsinki and its later amendments.

Informed Consent Statement: All patients gave their written informed consent before joining the study.

Data Availability Statement: The data that support the findings of this study are available on request from the corresponding author. The data are not publicly available due to privacy or ethical restrictions.

Acknowledgments: The authors thank all participants and their caregivers for supporting this study. We confirm that we have read the journal's guidelines on ethical publication issues and affirm that this report is consistent with those guidelines.

Conflicts of Interest: The authors declare that they have no conflict of interest.

References

1. Sutton, E.L. Insomnia. *Ann. Intern. Med.* **2021**, *174*, ITC33–ITC48. [CrossRef] [PubMed]
2. Chan, N.Y.; Chan, J.W.Y.; Li, S.X.; Wing, Y.K. Non-pharmacological Approaches for Management of Insomnia. *Neurother. J. Am. Soc. Exp. NeuroTher.* **2021**, *18*, 32–43. [CrossRef] [PubMed]
3. Cunnington, D.; Junge, M.F.; Fernando, A.T. Insomnia: Prevalence, consequences and effective treatment. *Med. J. Aust.* **2013**, *199*, S36–S40. [CrossRef] [PubMed]
4. Van Straten, A.; van der Zweerde, T.; Kleiboer, A.; Cuijpers, P.; Morin, C.M.; Lancee, J. Cognitive and behavioral therapies in the treatment of insomnia: A meta-analysis. *Sleep Med. Rev.* **2017**, *38*, 3–16. [CrossRef]
5. Dujardin, S.; Pijpers, A.; Pevernagie, D. Prescription Drugs Used in Insomnia. *Sleep Med. Clin.* **2022**, *17*, 315–328. [CrossRef]
6. Riemann, D.; Spiegelhalder, K.; Feige, B.; Voderholzer, U.; Berger, M.; Perlis, M.; Nissen, C. The hyperarousal model of insomnia: A review of the concept and its evidence. *Sleep Med. Rev.* **2010**, *14*, 19–31. [CrossRef]
7. Levenson, J.C.; Kay, D.B.; Buysse, D.J. The Pathophysiology of Insomnia. *Chest* **2015**, *147*, 1179–1192. [CrossRef]
8. Nagai, Y. Biofeedback and epilepsy. *Curr. Neurol. Neurosci. Rep.* **2011**, *11*, 443–450. [CrossRef]
9. Nestoriuc, Y.; Martin, A. Efficacy of biofeedback for migraine: A meta-analysis. *Pain* **2007**, *128*, 111–127. [CrossRef]
10. Rayegani, S.M.; Raeissadat, S.A.; Sedighipour, L.; Rezazadeh, I.M.; Bahrami, M.H.; Eliaspour, D.; Khosrawi, S. Effect of Neurofeedback and Electromyographic-Biofeedback Therapy on Improving Hand Function in Stroke Patients. *Top. Stroke Rehabil.* **2014**, *21*, 137–151. [CrossRef] [PubMed]
11. Kwan, Y.; Yoon, S.; Suh, S.; Choi, S. A Randomized Controlled Trial Comparing Neurofeedback and Cognitive-Behavioral Therapy for Insomnia Patients: Pilot Study. *Appl. Psychophysiol. Biofeedback* **2022**, *47*, 95–106. [CrossRef] [PubMed]
12. Lovato, N.; Miller, C.B.; Gordon, C.J.; Grunstein, R.R.; Lack, L. The efficacy of biofeedback for the treatment of insomnia: A critical review. *Sleep Med.* **2019**, *56*, 192–200. [CrossRef] [PubMed]
13. Alneyadi, M.; Drissi, N.; Almeqbaali, M.; Ouhbi, S. Biofeedback-Based Connected Mental Health Interventions for Anxiety: Systematic Literature Review. *JMIR mHealth uHealth* **2021**, *9*, e26038. [CrossRef] [PubMed]
14. Micoulaud-Franchi, J.-A.; McGonigal, A.; Lopez, R.; Daudet, C.; Kotwas, I.; Bartolomei, F. Electroencephalographic neurofeedback: Level of evidence in mental and brain disorders and suggestions for good clinical practice. *Neurophysiol. Clin.* **2015**, *45*, 423–433. [CrossRef]
15. Patil, A.U.; Madathil, D.; Fan, Y.-T.; Tzeng, O.J.L.; Huang, C.-M.; Huang, H.-W. Neurofeedback for the Education of Children with ADHD and Specific Learning Disorders: A Review. *Brain Sci.* **2022**, *12*, 1238. [CrossRef]
16. Holtmann, M.; Steiner, S.; Hohmann, S.; Poustka, L.; Banaschewski, T.; Bölte, S. Neurofeedback in autism spectrum disorders. *Dev. Med. Child Neurol.* **2011**, *53*, 986–993. [CrossRef]
17. Fernández-Alvarez, J.; Grassi, M.; Colombo, D.; Botella, C.; Cipresso, P.; Perna, G.; Riva, G. Efficacy of bio- and neurofeedback for depression: A meta-analysis. *Psychol. Med.* **2021**, *52*, 201–216. [CrossRef]
18. Liu, S.; Hao, X.; Liu, X.; He, Y.; Zhang, L.; An, X.; Song, X.; Ming, D. Sensorimotor rhythm neurofeedback training relieves anxiety in healthy people. *Cogn. Neurodyn.* **2021**, *16*, 531–544. [CrossRef]
19. Tolin, D.F.; Davies, C.D.; Moskow, D.M.; Hofmann, S.G. Biofeedback and Neurofeedback for Anxiety Disorders: A Quantitative and Qualitative Systematic Review. *Adv. Exp. Med. Biol.* **2020**, *1191*, 265–289.

20. Zhao, W.; Van Someren, E.J.W.; Li, C.; Chen, X.; Gui, W.; Tian, Y.; Lei, X. EEG spectral analysis in insomnia disorder: A systematic review and meta-analysis. *Sleep Med. Rev.* **2021**, *59*, 101457. [CrossRef]
21. Marzbani, H.; Marateb, H.R.; Mansourian, M. Methodological Note: Neurofeedback: A Comprehensive Review on System Design, Methodology and Clinical Applications. *Basic Clin. Neurosci. J.* **2016**, *7*, 143–158. [CrossRef] [PubMed]
22. Hou, Y.; Zhang, S.; Li, N.; Huang, Z.; Wang, L.; Wang, Y. Neurofeedback training improves anxiety trait and depressive symptom in GAD. *Brain Behav.* **2021**, *11*, e02024. [CrossRef] [PubMed]
23. Hsueh, J.-J.; Chen, T.-S.; Chen, J.-J.; Shaw, F.-Z. Neurofeedback training of EEG alpha rhythm enhances episodic and working memory. *Hum. Brain Mapp.* **2016**, *37*, 2662–2675. [CrossRef]
24. Bazanova, O.M.; Auer, T.; Sapina, E.A. On the Efficiency of Individualized Theta/Beta Ratio Neurofeedback Combined with Forehead EMG Training in ADHD Children. *Front. Hum. Neurosci.* **2018**, *12*, 3. [CrossRef] [PubMed]
25. Hauri, P. Treating Psychophysiologic Insomnia with Biofeedback. *Arch. Gen. Psychiatry* **1981**, *38*, 752–758. [CrossRef]
26. Cortoos, A.; De Valck, E.; Arns, M.; Breteler, M.H.M.; Cluydts, R. An Exploratory Study on the Effects of Tele-neurofeedback and Tele-biofeedback on Objective and Subjective Sleep in Patients with Primary Insomnia. *Appl. Psychophysiol. Biofeedback* **2009**, *35*, 125–134. [CrossRef]
27. Backhaus, J.; Junghanns, K.; Broocks, A.; Riemann, D.; Hohagen, F. Test-retest reliability and validity of the Pitts-burgh Sleep Quality Index in primary insomnia. *J. Psychosom. Res.* **2002**, *53*, 737–740. [CrossRef]
28. Jasper, H.H. The ten-twenty electrode system of the International Federation. *Electroencephalogr. Clin. Neurophysiol.* **1958**, *10*, 371–375.
29. Kouijzer, M.E.J.; van Schie, H.T.; Gerrits, B.J.L.; Buitelaar, J.K.; de Moor, J.M.H. Is EEG-biofeedback an Effective Treatment in Autism Spectrum Disorders? A Randomized Controlled Trial. *Appl. Psychophysiol. Biofeedback* **2012**, *38*, 17–28. [CrossRef]
30. Sohn, S.I.; Kim, D.H.; Lee, M.Y.; Cho, Y.W. The reliability and validity of the Korean version of the Pittsburgh Sleep Quality Index. *Sleep Breath.* **2011**, *16*, 803–812. [CrossRef]
31. Beck, A.T.; Steer, R.A.; Ball, R.; Ranieri, W.F. Comparison of Beck Depression Inventories-IA and-II in Psychiatric Outpatients. *J. Pers. Assess.* **1996**, *67*, 588–597. [CrossRef] [PubMed]
32. Maggi, G.; D'Iorio, A.; Aiello, E.N.; Poletti, B.; Ticozzi, N.; Silani, V.; Amboni, M.; Vitale, C.; Santangelo, G. Psychometrics and diagnostics of the Italian version of the Beck Depression Inventory-II (BDI-II) in Parkinson's disease. *Neurol. Sci.* **2023**, *44*, 1607–1612. [CrossRef] [PubMed]
33. Julian, L.J. Measures of anxiety: State-Trait Anxiety Inventory (STAI), Beck Anxiety Inventory (BAI), and Hospital Anxiety and Depression Scale-Anxiety (HADS-A). *Arthritis Care Res.* **2011**, *63* (Suppl. 11), S467–S472. [CrossRef] [PubMed]
34. Du, Q.; Liu, H.; Yang, C.; Chen, X.; Zhang, X. The Development of a Short Chinese Version of the State-Trait Anxiety Inventory. *Front. Psychiatry* **2022**, *13*, 854547. [CrossRef] [PubMed]
35. Guo, Y.; Zhao, X.; Zhang, X.; Li, M.; Liu, X.; Lu, L.; Liu, J.; Li, Y.; Zhang, S.; Yue, L.; et al. Effects on resting-state EEG phase-amplitude coupling in insomnia disorder patients following 1 Hz left dorsolateral prefrontal cortex rTMS. *Hum. Brain Mapp.* **2023**, *44*, 3084–3093. [CrossRef] [PubMed]
36. Naik, G.R.; Kumar, D.K. Subtle electromyographic pattern recognition for finger movements: A Pilot study using BSS techniques. *J. Mech. Med. Biol.* **2012**, *12*, 1250078. [CrossRef]
37. Greco, A.; Costantino, D.; Morabito, F.C.; Versaci, M. A Morlet Powerlet Classification Technique for ICA filtered sEMG experimental Data. In Proceedings of the International Joint Conference on Neural Networks, Portland, OR, USA, 20–24 July 2003; pp. 166–171.
38. Nicassio, P.M.; Boylan, M.B.; McCabe, T.G. Progressive relaxation, EMG biofeedback and biofeedback placebo in the treatment of sleep-onset insomnia. *Br. J. Med. Psychol.* **1982**, *55 Pt 2*, 159–166. [CrossRef]
39. Joseph, C.N.; Porta, C.; Casucci, G.; Casiraghi, N.; Maffeis, M.; Rossi, M.; Bernardi, L. Slow breathing improves arterial baroreflex sensitivity and decreases blood pressure in essential hypertension. *Hypertension* **2005**, *46*, 714–718. [CrossRef]
40. Egner, T.; Gruzelier, J. EEG Biofeedback of low beta band components: Frequency-specific effects on variables of attention and event-related brain potentials. *Clin. Neurophysiol.* **2004**, *115*, 131–139. [CrossRef]
41. Kwan, Y.; Baek, C.; Chung, S.; Kim, T.H.; Choi, S. Resting-state quantitative EEG characteristics of insomniac patients with depression. *Int. J. Psychophysiol.* **2018**, *124*, 26–32. [CrossRef]
42. Ogilvie, R.D. The process of falling asleep. *Sleep Med. Rev.* **2001**, *5*, 247–270. [CrossRef] [PubMed]
43. Beatty, J.; Greenberg, A.; Deibler, W.P.; O'Hanlon, J.F. Operant control of occipital theta rhythm affects performance in a radar monitoring task. *Science* **1974**, *183*, 871–873.
44. Åkerstedt, T.; Gillberg, M. Subjective and Objective Sleepiness in the Active Individual. *Int. J. Neurosci.* **1990**, *52*, 29–37. [CrossRef]
45. Horváth, M.; Frantík, E.; Kopriva, K.; Meissner, J. EEG theta activity increase coinciding with performance decrement in a monotonous task. *Act. Nerv. Super.* **1976**, *18*, 207–210.
46. Kaida, K.; Takahashi, M.; Akerstedt, T.; Nakata, A.; Otsuka, Y.; Haratani, T.; Fukasawa, K. Validation of the Ka-rolinska sleepiness scale against performance and EEG variables. *Clin. Neurophysiol.* **2006**, *117*, 1574–1581. [PubMed]

47. Raymond, J.; Varney, C.; Parkinson, L.A.; Gruzelier, J.H. The effects of alpha/theta neurofeedback on personality and mood. *Cogn. Brain Res.* **2005**, *23*, 287–292. [CrossRef]
48. Hammer, B.U.; Colbert, A.P.; Brown, K.A.; Ilioi, E.C. Neurofeedback for Insomnia: A Pilot Study of Z-Score SMR and Individualized Protocols. *Appl. Psychophysiol. Biofeedback* **2011**, *36*, 251–264. [CrossRef]

Disclaimer/Publisher's Note: The statements, opinions and data contained in all publications are solely those of the individual author(s) and contributor(s) and not of MDPI and/or the editor(s). MDPI and/or the editor(s) disclaim responsibility for any injury to people or property resulting from any ideas, methods, instructions or products referred to in the content.

Review

Neuroinflammation and Neurodegenerative Diseases: How Much Do We Still Not Know?

Carmela Rita Balistreri [1,*,†] and Roberto Monastero [2,†]

1 Cellular and Molecular Laboratory, Department of Biomedicine, Neuroscience and Advanced Diagnostics (Bi.N.D.), University of Palermo, 90134 Palermo, Italy
2 Unit of Neurology & Neuro-Physiopathology, Department of Biomedicine, Neuroscience, and Advanced Diagnostics (Bi.N.D), University of Palermo, Via La Loggia 1, 90129 Palermo, Italy; roberto.monastero@unipa.it
* Correspondence: carmelarita.balistreri@unipa.it
† These authors contributed equally.

Abstract: The term "neuroinflammation" defines the typical inflammatory response of the brain closely related to the onset of many neurodegenerative diseases (NDs). Neuroinflammation is well known, but its mechanisms and pathways are not entirely comprehended. Some progresses have been achieved through many efforts and research. Consequently, new cellular and molecular mechanisms, diverse and conventional, are emerging. In listing some of those that will be the subject of our description and discussion, essential are the important roles of peripheral and infiltrated monocytes and clonotypic cells, alterations in the gut–brain axis, dysregulation of the apelinergic system, alterations in the endothelial glycocalyx of the endothelial component of neuronal vascular units, variations in expression of some genes and levels of the encoding molecules by the action of microRNAs (miRNAs), or other epigenetic factors and distinctive transcriptional factors, as well as the role of autophagy, ferroptosis, sex differences, and modifications in the circadian cycle. Such mechanisms can add significantly to understanding the complex etiological puzzle of neuroinflammation and ND. In addition, they could represent biomarkers and targets of ND, which is increasing in the elderly.

Keywords: neuroinflammation; neurodegenerative diseases; emerging mechanisms

Citation: Balistreri, C.R.; Monastero, R. Neuroinflammation and Neurodegenerative Diseases: How Much Do We Still Not Know? *Brain Sci.* **2024**, *14*, 19. https://doi.org/10.3390/brainsci14010019

Academic Editors: Junhui Wang, Hongxing Wang and Jing Sun

Received: 27 November 2023
Revised: 12 December 2023
Accepted: 21 December 2023
Published: 23 December 2023

Copyright: © 2023 by the authors. Licensee MDPI, Basel, Switzerland. This article is an open access article distributed under the terms and conditions of the Creative Commons Attribution (CC BY) license (https://creativecommons.org/licenses/by/4.0/).

1. Introduction

The term "neuroinflammation" indicates representative pathological conditions induced in the brain by several (local and systemic) triggering factors (e.g., infections, trauma, ischemia, toxins, alterations in the microbiota–brain axis, etc.) and driving factors (e.g., genetic, vascular, and brain factors: for example, alterations in the expression of neurotrophins and components of the endothelial glycocalyx and/or endothelium) [1,2]. Neuroinflammation is evoked by typical immune cells residing in the brain and well known to have a key role in central nervous system (CNS) homeostasis and the development of neurodegenerative diseases (NDs), constituting their typical hallmark [3–5]. However, recent evidence suggests that both peripheral and infiltrating monocytes, as well as cells of clonotypic immunity, constitute other crucial actors of neuroinflammation [5,6]. Nevertheless, such evidence needs to be supported by larger studies, even if the growing data obtained in recent years appear promising and demonstrate the crucial contribution of this immune component in brain health and disease [3–6].

Other unconventional mechanisms related to the onset of neuroinflammation have recently emerged in the literature. Among these includes the altered relationship of the gut microbiota with the CNS, known as gut–brain microbiota (MGB) axis [7,8]. Of note also is autophagy, although the related mechanisms remain indefinable, and further investigation is necessary [9]. Interesting also is the contribution of ferroptosis, a novel cell death form caused by iron-dependent lipid peroxidation [10–12] and associated with the pathogenesis of many diseases, such as ND. Such has led some to suppose that variations in the iron metabolism's homeostasis, the consequent induction of oxidative stress, and the related inflammation in the CNS is involved in the onset of ferroptosis and neuronal health [10–12]. In such a process, the apelinergic system, mediated by ELA/APJ signaling, has also been recently documented to participate in this regulation [13]. Another nonclassical mechanism related to onset of neuroinflammation appears to be the modified expression of neurotrophins, such as BDNF [14].

Furthermore, it is emerging that neuroinflammation is also the result of the modulation in expression of genes encoding immune and injury's molecules. miRNAs [15] and epigenetic factors [16], including A-to-I RNA editing, M6A RNA methylation, and alternative splicing [15–17], have recently been revealed to have a fundamental role. Finally, circadian rhythm disorders have recently also been discovered to impact both the onset and development of neuroinflammation through the activation of glial cells and peripheral immune responses [18].

Insights have been achieved in identifying the mechanisms related to the complex neuroinflammation, although the complex puzzle is not complete and further studies are needed. Here, we describe and discuss the above mechanisms and others by reporting current clinical and experimental evidence.

2. Recent Evidence on Peripheral and Infiltrating Monocytes and Clonotypic Immune Cells in Neuroinflammation and Their Sex- and Gender-Mediated Modulation

Monocytes and clonotypic immune cells represent key actors of neuroinflammation, as recently underlined (see Figure 1) [18–21]. Such cells physiologically protect by pathogens, and particularly confer resistance against neurotropic viruses [18–21]. In addition, they contribute to CNS physiological functions and structure, by controlling the development, and improving cognitive function. In ND conditions, immune cells, i.e., monocytes and clonotypic cells, are deregulated. The deregulation commonly impacts both the levels and functions of clonotypic cells and monocytes, with the consequent evocation of abnormal immune responses [4]. Accordingly, monocytes are frequently altered, both peripherally and centrally, in quantity and quality with altered profiles and phenotypes. Data from recent human and animal studies report that monocytes, different populations of lymphocytes, and their mediators can evoke both self-protective and injurious mechanisms in ND, perturbing both their progression and risk of neuronal death. This displays a close interplay of peripheral immune cells with those residents in the CNS, which significantly influences the evolution of ND and consequent survival in ND cases. Accordingly, changes in number or functional quality of peripheral macrophages can modulate inflammation at the periphery along nerves and in the CNS [18–21]. Extracellular vehicles (EVs) from misfolded proteins and mediators of inflammation released by the cells appear to have a fundamental function in the inflammatory amplification [6], as recently reported by the literature.

2.1. Monocytes in Neuroinflammation

The participation of monocytes in the evocation of neuroinflammation has recently emerged from both clinical studies and experimental observations [21–25]. Accordingly, the presence of infiltrated peripheral immune cells in mouse models with degenerative conditions, including Parkinson's disease (PD), has been detected [21–26]. Infiltration of peripheral blood monocytes into the brain has also been observed in cerebrovascular diseases [27], as well as in patients with multiple sclerosis, where monocytes have been found to secrete some anti-neurodegenerative mediators [28]. In addition, in vitro studies have reported that high levels of chemokine CXCL12 from monocytes induce endothelial cell (EC) activation, thereby facilitating lymphocyte transmigration and validating the critical action of monocytes in the infiltration of immune cells in the brain [29]. Another study has established that brain immune infiltration originates from systemic inflammation [30]. Furthermore, circulating Ly-6C+ myeloid precursors have been observed to migrate into the CNS and have a pathogenic role in individuals affected by autoimmune demyelinating disease [31]. Moreover, monocytes have been demonstrated to phagocytose surplus brain proteins [32], such as amyloid-β peptide [33].

An updated view of neuroinflammation

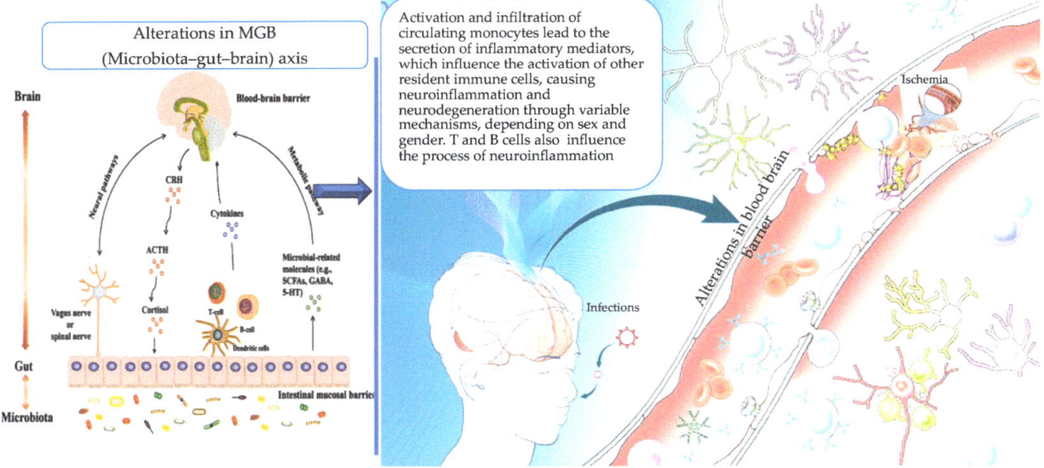

Figure 1. (by Biorender software version 04): Important drivers, i.e., alterations in the MGB axis, infections, and ischemia contribute to the activation and infiltration of circulating monocytes in the brain and their specific secretion of inflammation mediators. These latter influence the activation of other resident immune cells, thus leading to neuroinflammation and neurodegeneration. Similarly, clonotype cells influence neuroinflammation.

Other animal models, such as the murine stroke model, have evidenced a neuroprotective function of monocytes [34]. Precisely, they provide bioactive substances to brain cells [35,36]. In contrast, other investigations have reported the toxic action towards neuronal cells of monocytes by releasing saturated fatty acids, which can cause diverse pathologies, such as autoimmune disorders [21–25]. Thus, monocytes, through different mechanisms, actively contribute to the development of neuroinflammation, although some require further investigation.

2.2. Clonotypic Immune Cells in Neuroinflammation and ND

The role of clonotypic cells in neuroinflammation has only been theorized in previous research. However, recent evidence has largely established the key role of both T and B cell subsets and demonstrated their infiltration into the CNS by determining different effects depending on different subsets [37,38]. Therefore, regulatory T cells (Tregs) and Th2 cells have a neuroprotective effect. In contrast, Th1, Th17, cytotoxic T cells, natural killer (NK) cells mediate an accelerate progression of neuroinflammation, which can result in an exacerbated/accelerated neurodegeneration and an increased mortality risk [39–42]. Furthermore, they show different systemic levels. Precisely, circulating Th17 cells have shown higher levels in subjects with mild cognitive impairment (MCI) in cognitively normal subjects or those with non-Alzheimer's MCI [42]. In contrast, circulating Th1 cells have been demonstrated to have higher levels in subjects with AD [43]. Th2 cells and Th2-associated molecules have been observed to have lower levels in AD subjects [43]. Other studies have reported higher levels of circulating Th17 populations in individuals with AD [44,45]. Treg cells have also been discovered to have lower levels in patients with AD [46]. They are anti-inflammatory, with opposite effects to Th17 cells. Fu et al. have described low levels of Treg cells in AD subjects [47]. Contrarily, another investigation has observed no difference in Treg levels between MCI, AD, and healthy subjects, although Treg levels are associated positively with total tau and pTau181 in AD subjects [48].

Different T subsets with diverse functions have been detected in individuals affected by amyotrophic lateral sclerosis (ALS), characterized by the gradual degeneration of upper and lower motor neurons. T cells from superoxide dismutase (SOD)1-mutant mice have been observed to enhance and evoke survival of motoneuron cells (MNs) through a defensive neuroinflammatory response, likely mediated by interleukin 4 (IL-4). In contrast, motor impairment is accompanied by a decline in the functions of Treg cells, which inhibit microglia activation in SOD1-mutant mice [49]. Consequently, functions of neuroprotection mediated by the immune cells may happen in the early stage of the disease, although other studies are needed to confirm this. Moreover, disease progression is linked to numerous changes in the immune system, including the acquisition of an inflammatory phenotype of microglia cells [50], thymic involution [51], augmented levels of proinflammatory cytokines [52], and CNS leukocyte infiltration [53].

There is inadequate evidence and inconsistent data in the literature on the function of B lymphocytes, plasma cells, and antibodies in neuroinflammation and ND [54,55], and their contribution in AD pathogenesis needs further investigation. The diverse and obscure points of neuroinflammation and ND [56] could also be clarified. Furthermore, clinical trials on AD and other ND have failed to provide hopeful results [57] for diverse causes, including the relevant role of sex/gender dimorphism (which also justifies the differences observed in the onset, progression, and hallmarks of neuroinflammation in the various NDs), which will be described in the following section [58].

2.3. Considerations on Immune Cells Infiltrating the CNS and New Evidence on the Migration of Immune Cells outside the CNS

The mechanisms involved in the infiltration of peripheral immune cells into the CNS during neuroinflammation and ND have been gaining great interest in recent years, leading to the development of numerous therapies able to modulate immune cells at the BBB, the choroid plexus (ChP) epithelium, and glial barrier. For instance, natalizumab therapy, a drug inhibiting the adhesion and trafficking of monocytes and clonotypic cells across the BBB, has been used for almost two decades to treat MS [59,60].

Furthermore, fundamental CNS immune cell populations, i.e., dendritic cells (DCs), T cells, B cells, and other myeloid cell populations, have been found to migrate out of the CNS and mediate signals from the CNS to peripheral lymphatics [61]. This has been supported by recent evidence reporting the involvement of the meningeal lymphatic system not only in fluid homeostatic CNS functions but also in allowing immune cell migration and facilitating DC migration from the CNS to the meningeal borders and draining cervical lymph nodes [61].

However, work needs to explicate the function of each CNS-associated lymphatic region in overall CNS immunity. The results obtained would accelerate the development of new therapies to modulate the interplay between lymphocytes and leukocytes and consequently treat cases with CNS diseases.

2.4. Sex/Gender Dimorphism: An Important Modifier of Immune System and Physiology of Brain, and a Crucial Differential Driver in Diseases

The term "sex" indicates the diverse biological and physiological features of male and female individuals, the term "gender" refers to the social and cultural differences between men and women [58]. Biological, socioeconomic, and cultural differences impact the health and diseases of individuals. Dissimilarities in the anatomy and physiology of the body systems characterize women and men, while gender influences the norms that socially impose and specify roles and relationships among the individuals of a precise society and time. Hence, biological distinctions in the morphology and functions of the nervous system exist between the sexes, as proven by studies in human and animal models [58]. Specifically, amygdalae are larger in males than in females. While the dimorphism controls emotional memories in the female amygdala and implicates the involvement of the left region (visually predominant, positive, and negative emotions), in males, they activate the right region of the brain (negative emotional responses). Furthermore, prefrontal cortical regions have higher levels of estrogen receptors. Such could clarify the diversity in decision-making between the two sexes. Structural neuroimaging investigations have also confirmed the presence of reduced grades of overall cortical thickness and increased cortical thickness decline in men and greater white-matter volume in women. Moreover, differences in neurotransmitter systems (i.e., adrenergic, serotonergic, cholinergic systems, and corticosterone, benzodiazepine, and cholecystokinin, factors largely associated with episodic memory) characterize the two sexes. Higher levels of serotonin are more typical of men than women and may impact disease conditions related to serotonin dysfunction [58]. These diversities can in turn determine variations in the learning process, as identified during stress conditions, as well as subsequent habituation (increased in males but restrained in females). Such variation has been associated with the diverse levels and profiles of sex hormones in the two sexes. Accordingly, interesting results have been found in studies on cognitive decline and neurodegenerative and psychiatric diseases conducted in human and animal models. They report that sex hormones alter the permeability of the BBB, which represents one of the key pathophysiological ND hallmarks. Such findings could illuminate these disease processes; however, further research is required for proving and supporting this relationship [58].

Sexual dimorphism also influences the immune system of both sexes, with typical and excessive responses of both innate and adaptive immunity versus pathogens and endogenous antigens in females than males. This impacts the outcomes of infections and the efficiency of vaccines [62], and simultaneously disposes females to a higher risk of autoimmune diseases, even if the mechanisms related to these modifications are not yet fully explained. Furthermore, the close relationship between variations in number and functions of immune cells, as well as in the levels of cytokines or other systemic immune mediators, and the biological consequence of sex chromosomes and sex hormones is well recognized [58]. Therefore, sex designates an important trigger of the physiological and pathological conditions of an organism, humans included, which acts via genetic, epigenetic, and hormonal regulations. This is object of study of sex/gender medicine, born in the late 1990s, to identify fluctuations between diseases and their determinants in the two sexes [58]. Variations in the mechanisms and pathways related to the pathophysiology of diseases have been observed between the two sexes, as well as in their clinical manifestation, prognosis, and outcomes. Accordingly, ongoing studies on such aspects of major diseases are being performed [58,63]. They can prove useful in establishing new criteria and guidelines for the two sexes, particularly women. Women are more challenging to diagnose, and the traditional diagnostic tests, created for men, have a lower sensitivity and specificity in quantifying the biomarkers in female blood samples. Thus, new protocols are imperative. However, clinicians and researchers have until now paid little attention to sex and gender in health planning and medical practice.

Therefore, numerous efforts are required to incorporate sex and gender in modern medical research, clinical trials, clinical practice, and medical societies and institutions. Finally, different gender variables need to be studied at all levels and by biomedical and pharmaceutical organizations for the development of distinctive biomarker panels and correct therapies for both sexes.

3. Changes in the MGB Axis and Neuroinflammation

Gut microbiota and the CNS are influenced by the MGB axis (see Figure 1), discovered in 2012 [64], and constituted by neuroanatomical brain structures and intestinal nerves, i.e., the vagus nerve, located in the intestinal wall [64]. The vagus nerve mediates a response of the descending branch, which in turn controls intestinal activities. In addition, the hypothalamic–pituitary–adrenal (HPA) axis represents another component of the MGB axis [65]. The HPA axis monitors the changes in the composition and functions of the gut microbiome. Thus, HPA dysfunctions result in MGB alterations related to the pathogenesis of neuropsychiatric diseases. Specifically, HPA activation results in the induction of inflammatory signaling pathways, releasing inflammatory mediators, such as tumor necrosis factor α (TNF-α), interferon γ (IFN-γ) and interleukin 6 (IL-6) [66]. In turn, these mediators contribute to damaging BBB integrity and the onset of brain diseases via systemic circulation and simultaneously to alter the gut mucosal barrier. Moreover, the inflammatory response induced via the HPA axis impacts the secretion of glucocorticoids [67], modulating gut function and production of proinflammatory factors [68]. This vicious cycle also evokes the activation of enteric immune cells, such as Th17 and NK cells, which infiltrate the brain, causing neuroinflammation [69]. Neuroinflammation, in turn, additionally contributes to modifying the gut microbial composition, and this further provokes activation of enteric immune cells and release of microbiota-derived metabolites, i.e., lipopolysaccharide (LPS). These exacerbate the bidirectional via of inflammatory signals, contributing to the onset of dysbiosis, a typical alteration in gut microbiota and the MGB axis. Dysbiosis has recently attracted increasing interest for its pathogenic role in immune-mediated diseases, including metabolic syndrome, gastrointestinal tract infections and inflammatory bowel disease, as well as autoimmune diseases such as systemic lupus erythematosus, systemic sclerosis, Sjogren's syndrome, antiphospholipid antibody syndrome, multiple sclerosis (MS) and myasthenia gravis (MG). MG is a typical neuromuscular autoimmune disease triggered by immune-mediated damage to the neuromuscular junction (NMJ), and with pathogenesis likely multifactorial [70]. Perturbations in human microbiota have been described to be related to MG pathogenesis and clinical course. MG cases compared with age-matched controls show a characteristic composition of the oral and gut microbiota, with a typical increase in *Streptococcus* and *Bacteroides* and a reduction in *Clostridia*, as well as a reduction in short-chain fatty acids. Moreover, it has been demonstrated that restoration of gut microbiota disorder after probiotic administration determines an improvement in symptoms in MG cases [7].

Such evidence highlights the double role of the MGB axis in maintaining host health and contributing to the typical alterations causing dysbiosis and MGB axis disorder, and to the onset of MG and some NDs, such as PD and AD. However, the related molecular and cellular mechanisms are not clear. Some studies also evidence that behavioral phenotypes can be transmitted from humans to animals via transplantation/translocation of the gut microbiota [7]. This emphasizes the role of MBG alterations in ND. However, further research is needed to confirm if the discoveries in animals may be also obtained in humans to identify all the relevant mechanisms by which the gut microbiota controls neuroinflammation and ND. Such studies could allow the development of new microbiota-based strategies for diagnosis, treatment, and clinical management of neuroinflammation and ND.

4. Autophagy

The term "autophagy" indicates a cell death process, which is physiologically regulated and evolutionarily conserved. It causes degradation of cytoplasmic proteins and other macromolecules within the lysosome in multicellular organisms [71]. It has been demonstrated to be involved in neuroinflammation and ND, which are characterized by protein aggregation and exacerbated autophagy [72–74]. However, the mechanisms involved are not clear [75–77]. Despite this, some drugs, including cocaine [78] or other toxic substances of exogenous or endogenous nature and pathogens have been tested to evoke autophagic cell death in astrocytes and in consequent pathogenesis of neurodegeneration [79]. Recently, a close relationship between glia maturation factor (GMF), autophagy-related proteins, and the NLRP3 inflammasome and a shift of microglia from M1 to M2 in AD patients has been detected [80,81]. However, further investigations are needed for understanding the role of autophagy in ND or neuroinflammation-associated disorders.

5. Ferroptosis in Neuroinflammation

Ferroptosis, discovered in 2012 by Brent R. Stockwell, is a new form of regulated cell death (RCD) described by the accumulation of lethal amounts of iron- and lipid-dependent reactive oxygen species [82]. Ferroptosis results in different morphological and biochemical characteristics from other conventional RCD forms [82]. It is evoked by severe peroxidation of membranes containing polyunsaturated fatty acids (PUFAs), and regulated by lipid, iron, and amino acid metabolism and signaling transduction. The critical phases involved in ferroptosis include the accumulation of intracellular free iron, glutathione depletion, and peroxidation of PUFA-rich membranes [82,83]. Recent evidence has documented ferroptosis as a crucial factor in the pathogenesis of several diseases, such as ND [82–85]. Accordingly, iron homeostasis, oxidative stress, and subsequent neuroinflammation have been described to contribute to the regulation of ferroptosis and neuronal health [82–85]. However, the precise molecular mechanisms underlying the involvement of ferroptosis in the pathological processes of neurodegeneration and its impact on neuronal dysfunction remain to be discovered. Nevertheless, ferroptosis has recently been reported to likely be regulated by ELA/APJ signaling of the apelinergic system [13], where ELA is a peptide hormone belonging to the adipokine group and a component of the apelinergic system, discovered in 2013–2014 [13]. This relationship, mediated by ELA/APJ signaling, might be a promising strategy for the treatment of NDs, such as stroke [13]. Accordingly, a recent study in mouse models of middle cerebral artery occlusion (MCAO) has demonstrated the protective role of the ELA–APJ axis in ischemic stroke after treatment with ELA-32 (widely quoted in [13]). A reduction in cerebral ischemic lesion and an improvement in neurobehavioral and cognitive deficits have been detected. Furthermore, ELA-32 administration has been revealed to ameliorate neuronal ferroptosis, iron deposition, mitochondrial damage, lipid peroxidation, and glutathione reduction. These results have emphasized the role of the ELA–APJ axis in attenuating neuronal ferroptosis after ischemic stroke (widely quoted in [13]). However, further data are needed to provide other/novel strategies to modulate the onset of neuroinflammation and ND, such as stroke.

6. The Close Link of Endothelial Dysfunction with Neuroinflammation and ND

Endothelial dysfunction represents another condition contributing to neuroinflammation and ND, which occurs with typical cellular and molecular mechanisms, including changes in the glycocalyx [86,87]. Such a close relationship of damaged endothelium with neuroinflammation is related to the relevance of the endothelium in the brain; it is a fundamental component of the neurovascular unit (NVU), composed of ECs arranged with neurons, glial cells, and other vascular elements [86,87]. NUV mediates diverse functions: maintenance of CNS homeostasis, physiological neurotransmission, and neuronal survival [62]. Furthermore, EC and glial cells, such as microglia cells, contribute to BBB integrity and provide both nutrients and oxygen. In systemic infections or in the presence of systemic inflammation, circulating toxins and inflammatory mediators infiltrate ECs and consequently the BBB. Accordingly, any alteration or disorder of the NVU also involves the BBB and causes neuroinflammation, which in turn contributes to evoking age-associated cognitive deficits and consequently ND onset [86–88]. Frequently, the altered clearance of amyloid-β peptide and its consequent accumulation in the brain constitute the typical trigger of NVU. In this case, the release of toxic small molecules and inflammatory products that cross the damaged BBB determines neuroinflammation [62]. However, cardiovascular disorders, including cerebrovascular diseases, i.e., macro-infarcts, lacuna, microbleeds, atherosclerosis, arteriolosclerosis, and cerebral amyloid angiopathy (CAA), have been documented to directly contribute to NVU dysfunction [62,86]. Among these, microvascular diseases have been demonstrated to affect NVU by determining alterations in the physiological process of brain oxygenation, as well as reduced blood flow and subsequent hypoxia [62,86–88]. Accordingly, chronic hypoxia–ischemia is accepted as a key trigger of chronic NVU damage and BBB dysfunction related to many NDs, such as stroke, MS, AD, and PD; however, there are increasing data linking BBB breakdown to physiological aging processes, specifically with vascular aging. This initially involves the hippocampus in subjects without cognitive impairment, and occurs more rapidity in old people and remarkably with concomitant MCI [86–88]. During normal aging, BBB dysfunction affects the CA1 region and the dentate gyrus, but not the CA3 region [86–88]. Moreover, hippocampal BBB distribution has been found to precede the onset of hippocampal atrophy [86–88]. Analysis of cerebrospinal fluids from MCI cases when compared with cognitively normal persons has evidenced a significant increase in pericyte damage biomarkers (i.e., platelet-derived growth factor receptor (PDGFR)-β) by implying an immediate role of pericytes, as opposed to other cell types, in BBB breakdown [86–88]. The involvement of BBB dysfunction in AD has been related to a reduced presence of tight junctions and an abnormal morphology in brain ECs. In addition, an altered diameter of blood vessels after typical tau deposition has been detected [86–88]. An anomalous angiogenesis in AD cases, likely due to altered function and quantity of trophic factors, has also been assessed. Patients with AD have expanded levels of VEGF, both in serum and temporal cortex and hypothalamus, and decreased expression of both VEGFR-1 and VEGFR-2 [86–88]. Some studies have pointed to VEGF itself being the cause of such decreases in levels of the two receptors. Indeed, VEGF mediates such effects through a ligand-mediated endocytosis mechanism [86–88]. Thus, VEGF in AD has a role of antagonist versus its receptors, resulting in an altered angiogenesis. Moreover, in vitro Aβ accumulation has been demonstrated to be able to reduce the mRNA levels of VEGFR-1 and VEGFR-2, resulting in increased VEGF reactivity [86–88].

Likewise, ALS patients show alterations in NVU and angiogenic factors, including VEGF. This finds confirmation in studies conducted in both SOD-1-mutant mice and ALS patients. Reduced levels of tight junctions, such as ZO-1, and BBB breakdown, which precede motor neuron death, have been detected in such studies. This suggests a key role of vascular damage as an early pathogenic ALS mechanism [62,86–89], data also validated by high levels of metalloproteinase (MMP)-2 and MMP-9 in peripheral blood samples from ALS cases and by the results obtained by Nicaise and colleagues on variations in the composition of vascular NVU elements in the early ALS stage [62]. The SOD-1 mouse model has also evidenced a blood–spinal cord barrier (BSCB) dysfunction, characterized by ex-erythrocyte extravasation, neurotoxic hemoglobin accumulation, and NUV injury via iron-dependent oxidative stress [62]. Studies in G93A SOD1 mice have demonstrated alterations in the NVU that are not only structural but also functional, as confirmed by a downregulation of Glut-1 and CD146 expression early and late in the disease [62]. Compared to healthy controls, ALS patients have also been demonstrated to have elevated levels of VEGF, particularly VEGF-A, in the blood and CSF [62], possibly due to a compensatory mechanism. Investigations into SOD-1-mutant mice have also revealed that VEGF-A also exercises neuroprotective effects by decreasing MN cell death via activation of the PI3K-Akt pathway [62], and this may result in a delay of disease onset [62,86–89]. However, other studies have proven the presence of reduced levels of VEGF and its receptor VEGFR in ALS cases and in subjects homozygous for certain haplotypes, i.e., three polymorphisms in their genes (−2578 C/A, −1154 G/A, and −634 G/C) [62]. The reason has been attributed to the destabilization of VEGF mRNA induced by SOD1 protein [62].

This evidence globally suggests that the maintenance of BBB and NVU integrity, as of entire cardiovascular system, could ameliorate the health of the cerebrovascular system and represent the best avenue for the development of potential strategies for improving blood flow at the cerebral microvascular level by protecting the BBB and NVU. Preserving the integrity, permeability, and function of the BBB and NVU could stop or delay the progression of neuroinflammation and ND. To achieve this goal, it is imperative to identify all the pathways involved in the pathophysiology of these diseases, and particularly those related to BBB and NVU dysfunction. Surely, this objective can be realized by performing multiple omics investigations, offering the opportunity of acquiring major, relevant, and new data. Accordingly, such studies are encouraged.

7. miRNAs and Epigenetic Factors in Neuroinflammation and ND

Recently, the modulation of genes related to neuroinflammation has been considered as a means to mitigate it. MicroRNAs and other epigenetic factors, universal regulators of differentiation, activation, and polarization of all the cells of human body, including immune and neuronal cells, appear to be directly responsible for neuroinflammatory processes. Recent investigations demonstrate different expression levels of miRNA and epigenetic factors in microglia, both in normal and inflamed CNSs, suggesting their role in brain health and neuroinflammation-associated disorders [90]. Cases of epilepsy and neuroinflammation in the hippocamp of patients with sclerosis have shown low levels of mature micro-RNAs in human temporal lobes [91,92]. Among these, microRNA-155 appears to negatively regulate BBB function in chronic neuroinflammation and neurodegeneration [93]. miR-195 inhibits autophagy after peripheral nerve damage [94]. In contrast, microRNA-188-3p, in an upregulated state, constrains the neuroinflammation and recovers memory in AD patients [95]. miR-137 attenuates beta-induced neurotoxicity in Neuro2a cells [96]. miR-124 expression modifies promoter DNA methylation and microglial functions [97]. Notably, microRNA-30e controlled neuroinflammation via NLRP3 in an MPTP-induced PD model [98]. MicroRNA-129-5p exacerbates neuroinflammation and BBB injury [99]. Similarly, miR-17-92 triggers the differentiation of neurons during neuroinflammatory conditions [100]. MicroRNA-139 favors AD pathogenesis via cannabinoid receptors [101]. Thus, microRNAs constitute good therapeutic targets to produce novel anti-neuroinflammatory AD treatments [90].

Moreover, other epigenetic factors related to post-transcriptional RNA changes modulate mRNA coding properties, stability, and translatability, expanding the genome's coding capacity. They appear to influence neuroinflammation. Among these, A-to-I RNA editing, m6A RNA methylation, and alternative splicing (AS) impact the neuronal cell life cycle, induce neuron death mechanisms, and contribute significantly to neuroinflammation and age-related neurodegeneration [17]. A-to-I RNA editing is a post-transcriptional mechanism modulating double-stranded (ds) RNA structures via the catalytic activity of adenosine-deaminase acting on RNA (ADAR) enzymes. It consists in the deamination of specific adenosine (A) into inosine (I) by altering both coding and non-coding transcripts [17]. Three ADAR enzymes are expressed in human cells–ADAR1, ADAR2 and ADAR3–and have high expression and activity in the brain in terms of regulating neurodevelopment, brain function, and physiological brain aging (widely quoted in [17]). Consequently, the brain appears to be susceptible to ADAR activity and RNA-editing dysregulation, which potentially initiate CNS disorders, such as glioblastoma, epilepsy, and ND (widely quoted in [17]).

Altered expression of N6-methyladenosine (m6A), a dynamic and reversible post-transcriptional alteration adding a methyl group to the N6 position in selected adenosines of each type of RNA [17], has been documented in aging mouse and human brains. In terms of ND, unusual m6A alterations have been identified in AD, PD, and ALS.

ND patients, including mainly AD, PD, ALS, frontotemporal dementia (FTD) and familial dysautonomia (FD) cases, have AS alterations [17].

The close relationship of post-transcriptional RNA modifications with brain aging and neurodegeneration emphasizes the possibility to reduce or inhibit these processes; antisense oligonucleotides (ASOs) can modify their expression. ASOs appear to eliminate causative splicing defects in PD, AD, FTD and ALS (widely quoted in [17]).

The growing evidence on the contribution and serious impact of A-to-I RNA editing, m6A RNA methylation, and alternative splicing on brain aging process, neuroinflammation, and ND points to the need for further investigations on these processes and how they may impact each other so as to control them simultaneously.

8. Transcriptional Factors and Related Pathways: Focus on NF-kB (Nuclear Factor Kappa-Light-Chain Enhancer of Activated B Cells) and Related Pathways

Other modulating factors of neuroinflammation are transcriptional factors, able to activate an inflammatory network in the brain and all linked to the NF-κB pathway, an ancient signaling pathway specialized in host defense [102]. The NF-κB pathway is a cytoplasmic molecular complex of diverse proteins comprising the Rel family proteins RelA/p65, c-Rel and RelB and NF-κB components-p50/p105 and p52/p100, and is commonly inhibited by binding to IκB proteins (i.e., IκBα, IκBβ, IκBγ, IκBδ, IκBε, IκBζ and Bcl3) via the action of many signaling pathways and negative feedback loops regulating diverse mechanisms at various levels of the signaling cascades. Immune insults and external and internal danger signals, such as oxidative and genotoxic stress and tissue injury, constitute its activators [102]. In addition, Toll-like receptors (TLRs) and inflammasome [103–105], as well as several upstream kinase cascades via canonical or non-canonical pathways, can activate the NF-κB pathway. IKKα/β and NIK are the most important upstream kinases. IKKγ is generally referred to as a nuclear factor-kappa B essential modulator (NEMO), an important regulatory component of the IKK complex linked upstream to genotoxic signals and IL-1 and TNF receptor-mediated signaling [103–105]. NF-κB complex activation, playing the crucial role of a pleiotropic mediator of gene expression, determines its translocation into the nucleus and the expression of target genes, encoding various molecules, such as proinflammatory cytokines, chemokines, adhesion molecules, eicosanoids, growth factors, metalloproteinases, nitric oxide, etc. [102]. NF-κB signaling has been reported to be one of the major pathways stimulating neuroinflammation [102,106].

Recent studies have evidenced the beneficial effects of dietary supplementation with anti-inflammatory compounds on cognitive decline, neuroinflammation and oxidative stress by acting on the NF-kB pathway in AD-like animal models [107,108]. Curcumin, krill oil, chicoric acid, plasmalogens, lycopene, tryptophan-related dipeptides, hesperidin, and selenium peptides have been tested, despite their heterogeneity, and have shown helpful actions on cognitive deficits and LPS-induced neuroinflammatory responses in rodents by affecting the NF-κB pathway [106–108]. Overall, dietary interventions could represent positive factors in countering AD, or other ND, by acting on neuroprotection and immune regulation. For example, treatment with metformin, an antidiabetic drug, has demonstrated anti-inflammatory effects via many mechanisms, revealing its potential as a therapy for neuroinflammation.

However, as evidenced in such reviews, the mechanisms involved in neuroinflammation are various and complex: numerous molecules are combined in a network and consequently can modify each other. For example, metformin significantly prevents nuclear translocation of p65, but pretreatment with compound C, an AMPK inhibitor, eliminates this effect, while silencing HMGB1 abolishes NF-κB activation. SIRT1 deacetylates FoxO, increasing its transcriptional activity. mTOR in dendritic cells regulates FoxO1 through AKT. Interactions between the various molecules need to be further explored to clarify their specific mechanisms and provide more guidance for the treatment of neuroinflammation [109].

Based on the evidence described above, mTOR and AKT pathways, as well as JAK-STAT, and PPARγ, and Notch pathways, constitute other crucial pathways in neuroinflammation [9,110–115]. They represent highly conserved signaling hubs that coordinate neuronal activity and brain development and participate in neuroinflammation. Accordingly, hyperactivation of JAK/STAT and mTOR and inhibition of PPARγ and AKT signaling have been associated with various neurological complications, including neuroinflammation, apoptosis, and oxidative stress [112–115]. Remarkably, target modulators have also been described to act during acute and chronic neurological deficits. For example, natural products, such as osthole, an important ingredient of traditional Chinese medicinal plants often found in various plants of the Apiaceae family, have been shown to target these pathways [116]. Osthole induces neurogenesis and neuronal function via the stimulation of Notch, BDNF/Trk, and P13k/Akt signaling pathways. This upregulates the expression of various proteins, such as BDNF, TrkB, CREB, Nrf-2, P13k, and Akt, and inhibits MAPK/NFκB-mediated transcription of genes involved in the production of inflammatory cytokines and the NLRP-3 inflammasome. Thus, modulation of Notch, BDNF/Trk, MAPK/NF-κB, and P13k/Akt signaling pathways by osthole confers protection against neuroinflammation and ND [116].

Te evidence described above suggests the neuroprotective potential of several compounds and natural products as possible therapeutic agents for neuroinflammation and NDs. However, a limitation of some of these substances is their low bioavailability and solubility in water. Furthermore, the use of innovative nanotechnology or the incorporation of a more polar group would be advantageous to increase the bioactivity and physicochemical properties of such compounds or natural products, such as osthole. To this end, liposomes, microspheres, nanoparticles, transferosomes, ectosomes, lipid-based systems, etc. have been developed for the modified delivery of various herbal drugs. For example, osthole-loaded nanoemulsion has been reported to effectively target the brain and have beneficial effects in the treatment of AD. Therefore, the development of potential nanocarriers such as liposomes, microspheres, and nanoemulsions could improve the bioavailability of such compounds [107,108]. However, further studies are needed to evaluate the real therapeutic effect of such compounds on neuroinflammation.

9. Circadian Cycle and Neuroinflammation

Another determinant of neuroinflammation is alteration in circadian cycle/rhythm, a fundamental process of life developed during the long-term evolution of organisms. It has diverse functions, maintaining the proliferation, migration, and activation of cells, and particularly of immune cells [117,118]. Circadian rhythm disorders impact the onset and development of neuroinflammation by activation of glial cells and peripheral immune responses [18,118,119]. Animal models exposed to nightshifts or night light have been confirmed to have significant levels of activated microglia and proinflammatory cytokines in brain [120,121]. Sleep deprivation has also been demonstrated to trigger the transcriptional factor NF-κB and intensify the release of IL-1β and TNF-α in the hippocampus, resulting in neuronal injury [122]. Studies have also revealed high mRNA levels of IL-1β and TNF-α in brain tissue of experimental animals, which evoke significant alterations in circadian rhythm, responsible for modifications in the sleep–wake cycle [123]. Inhibition of such cytokines has resulted to reduce spontaneous non-rapid eye movement sleep in experimental animals [124,125], confirming that proinflammatory cytokines induce effects on the circadian cycle and neuroimmune function. In addition, circadian system disorders influence microglial activation and their phenotypes [120–122]. Accordingly, under conditions of light exposure, diverse investigations in rats report an increased inflammatory activity of microglia, accompanied by significant rises of TNF-α, IL-1β, and IL-6 [120–122,126].

A critical role of circadian cycle/rhythm has been evidenced in the regulation of the peripheral immune system [127], including innate and adaptive cells. They have their own molecular clock and display significant rhythmic differences during recruitment and activation processes [128]. A regulation of the bone marrow chemokine CXCL12 on the hypothalamic sympathetic–parasympathetic nervous system in a circadian manner has also been detected. It determines periodic fluctuations in CXCL12 levels and CXCR4 receptor activation to sustain daily rhythmic changes in the number of neutrophil cells in the bone marrow blood reserve [129,130]. In macrophages, the circadian cycle impacts their pattern-recognition receptor signaling pathways, inflammatory mediators, and phagocytic activity [131]. Krüppel-like factor 4 (KLF4), whose expression is time-regulated, appears to regulate the macrophage phenotype and rhythmic expression of inflammatory factors [132]. In addition, the REV-ERBα clock gene has been demonstrated to modulate the expression of the PI3K/Akt signaling pathway and the regulation of the diurnal rhythm of macrophage polarization [133]. Thus, REV-ERBα also represents a potential target for regulating circadian rhythms and inflammatory response. Similarly, adaptive immune cells also display rhythmicity, with immune responses that differ significantly during the different hours of the day. For example, more CD8+ T cells are produced during the day than at night in response to antigen immunization, and the rhythmic response dissolves by knocking out the Bmal1 gene in T cells, further validating the relevance of circadian rhythms in modulating adaptive immune responses [134].

10. Chronic Low-Grade Inflammation and Neurodegenerative Diseases

Currently, the precise nature and temporal characteristics of the relationship between neuroinflammation and ND remain largely unknown. Clinical and preclinical studies have described how systemic chronic inflammation (SCI) should be considered a potential driver of the onset of the neurodegenerative process associated with cognitive impairment [135,136]. Several studies have proposed the concept of chronic low-grade inflammation as potentially causal in the etiopathogenesis of dementia and other ARDs of the elderly individual, and the term "inflammaging" has been coined for this phenomenon [137,138]. Specifically, inflammaging refers to the presence of chronic low-grade systemic inflammation that occurs during aging in the absence of overt infection (the so-called sterile inflammation). Clinical and epidemiological studies have shown that this process is a relevant risk factor for morbidity and mortality in the elderly [137,138]. In particular, the presence of SCI leads to an increased risk of metabolic diseases (e.g., hypertension, diabetes, dyslipidemia) and osteoporosis, cancer, and cardiovascular, neurodegenerative, and autoimmune diseases [135].

SCI implies the involvement of several cytokines and transcription factors that regulate chronic inflammation at the tissue and causal levels for different ARDs. Among the cytokines, IL-6 is probably the one most associated with a robust chronic inflammatory response that characterizes different ARDs [139]; other inflammatory cytokines that participate in the inflammatory process during ARD are IL-1β and TNF-α [139,140]. In turn, cytokines interact with specific tissue surface receptors, regulating the inflammatory cascade by regulating transcriptional processes. The two main protein transcription factors associated with SCI are STAT (signal transducer and activator of transcription) and NF-κB [102]. These proteins regulate a series of genes that code for the formation of inflammatory cytokines.

Over the last decade, the role of low-grade SCI in periodontal disease (PeD) has been suggested as a potential risk factor for overall dementia and particularly AD. Several authors have described the presence of significantly elevated antibody levels toward specific oral cavity opportunistic pathogens causing PeD in subjects with AD but also MCI compared with control subjects without cognitive impairment [141]. Regarding specific oral pathogens, the one most implicated in the link between dementia and PeD appears to be porphyromonas gingivalis [142], but significantly elevated levels of oral microbial load of other pathogens such as fusobacterium nucleatum and treponema denticola have been described in subjects with AD and MCI compared with control subjects [142]. Data from a recent national US retrospective cohort study showed that periodontal pathogens increase the risk of AD incidence and mortality [143]. In addition, data from a recent meta-analysis showed that the risk of cognitive disorder in individuals with PeD increases as the severity of PeD increases, and this risk appears to be greater in the female sex [144]. There are at least two main mechanisms by which PeD can cause cognitive disorders. The first involves the presence of an increased cerebral inflammatory state caused by the SCI process originating from oral pathogens; the second involves a direct action of periodontal bacteria on the CNS that cross the BBB and cause its breakdown with subsequent, potential triggering of the preexisting neurodegenerative process [141–144].

In addition to increased risk of dementia, some studies have suggested that PeD may increase the risk of PD [145]; however, data from a recent meta-analysis revealed no association between PeD and increased risk of PD [145]. In conclusion, PeD is associated with an increased risk of overall dementia, AD, and MCI, and this appears to be due to low-grade SCI sustained by the oral pathogens that cause PeD. However, prospective data on large population cohorts are needed to confirm the role of PeD as a risk factor for AD, dementia, and possibly other neurodegenerative diseases. If confirmed, such data will have major implications for the treatment and prevention of cognitive disorders.

11. Conclusions

With this review, we have provided an overview of the new mechanisms associated with the relationship between neuroinflammation and subsequent onset of ND (see Figure 2A,B). The latter offers the possibility of hypothesizing and developing new treatments and identifying diagnostic and prognostic biomarker profiles for neuroinflammation and ND. They could include the assessment of transmigration and activation levels of monocytes, as well as the levels, activation, and quantification of clonotypic cells and their mediators, and the evaluation of expression of NF-kB and other transcriptional factors (i.e., ERG factor) and the related pathways [146,147]. The concomitant assessment of microRNAs and epigenetic factors involved in the regulation of these mechanisms could be additionally helpful. Furthermore, the targeting of autophagy and ferroptosis is gaining more and more interest, as it contributes to the modulation of neuroinflammation and the onset of ND, as well as to endothelium-related BBB dysfunction. Considering all the findings to date, the complex pathophysiology and pathogenesis of neuroinflammation and ND might appear clearer, as well as all pathways and cells. This could facilitate the identification of biomarkers and targets and, consequently, the management of NDs. Therefore, further efforts and investigations are needed. Advances in omics methodologies, artificial intelligence, machine learning, advanced biological techniques, metagenomics, and meta-transcriptomics are currently important in neuroscientific research and could be used to achieve this goal. Similarly, large-scale randomized controlled trials are needed. Such studies would pave the way for next-generation treatment strategies capable of modulating neuroinflammation during ND.

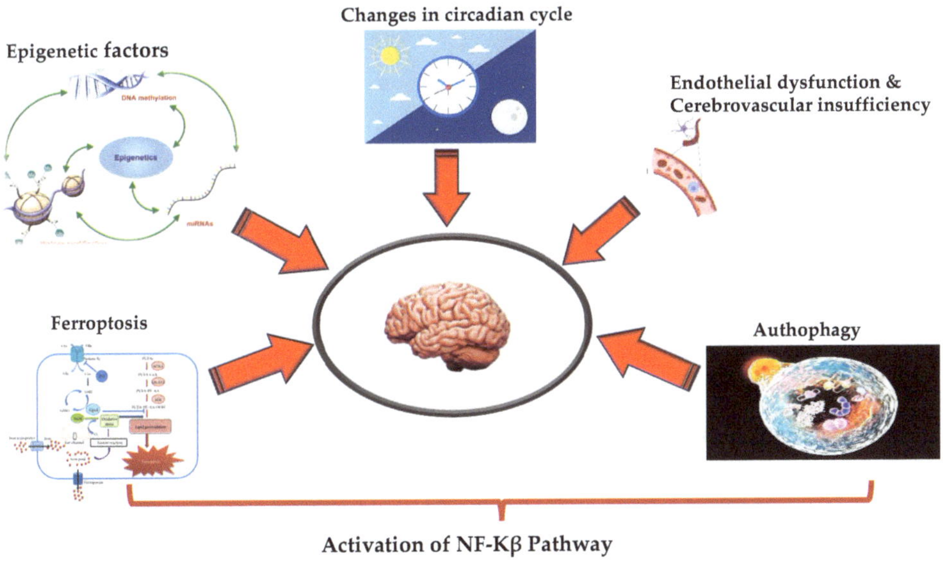

Figure 2. *Cont.*

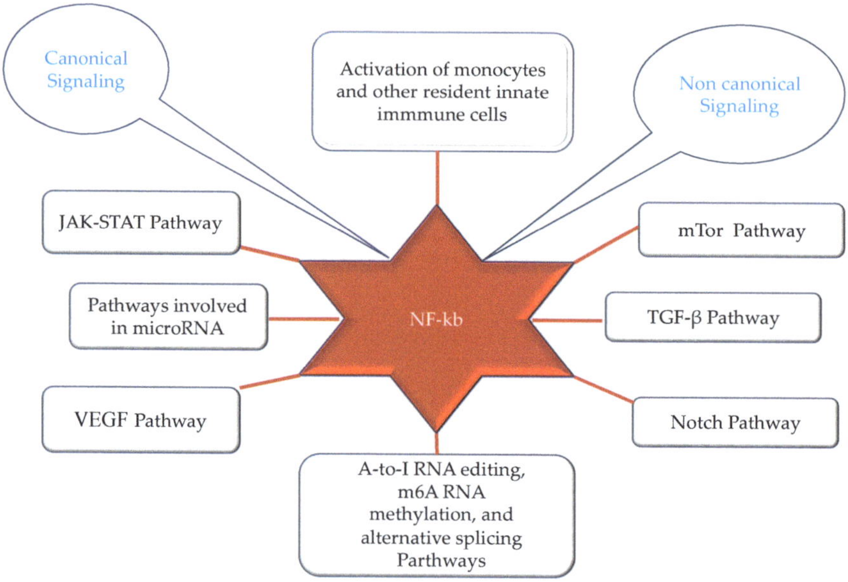

Figure 2. (**A,B**) (by Biorender software): Model describing the novel mechanisms involved in neuroinflammation and its relation to the onset of ND. In (**A**), it illustrates how the ferroptosis, autophagy, epigenetic factors, changes in circadian rhythm, and endothelial dysfunction associated with cerebrovascular insufficiency determine all the activation of NF-kB pathway through canonical or non-canonical signaling. In (**B**), it shows how activation of the NF-kB pathway through canonical or non-canonical signaling can activate a network of different signaling pathways, all related to the onset of neuroinflammation and the consequent onset of ND.

Author Contributions: Conceptualization, C.R.B.; literature search, C.R.B. and R.M.; data curation, C.R.B. and R.M.; writing—original draft preparation, C.R.B. and R.M.; writing—review and editing C.R.B. and R.M.; figure production, C.R.B.; supervision, C.R.B. and R.M. All authors have read and agreed to the published version of the manuscript.

Funding: This research received no external funding.

Conflicts of Interest: The authors declare no conflict of interest.

References

1. Disabato, D.J.; Quan, N.; Godbout, J.P. Neuroinflammation: The devil is in the details. *J. Neurochem.* **2016**, *139* (Suppl. 2), 136–153. [CrossRef] [PubMed]
2. Gilhus, N.E.; Deuschl, G. Neuroinflammation—A common thread in neurological disorders. *Nat. Rev. Neurol.* **2019**, *15*, 429–430. [CrossRef] [PubMed]
3. Gao, C.; Jiang, J.; Tan, Y.; Chen, S. Microglia in neurodegenerative diseases: Mechanism and potential therapeutic targets. *Signal Transduct. Target. Ther.* **2023**, *8*, 359. [CrossRef] [PubMed]
4. Ní Chasaide, C.; Lynch, M.A. The role of the immune system in driving neuroinflammation. *Brain Neurosci. Adv.* **2020**, *4*, 2398212819901082. [CrossRef] [PubMed]
5. Becher, B.; Spath, S.; Goverman, J. Cytokine networks in neuroinflammation. *Nat. Rev. Immunol.* **2017**, *17*, 49–59. [CrossRef] [PubMed]
6. Carata, E.; Muci, M.; Di Giulio, S.; Mariano, S.; Panzarini, E. Looking to the Future of the Role of Macrophages and Extracellular Vesicles in Neuroinflammation in ALS. *Int. J. Mol. Sci.* **2023**, *24*, 11251. [CrossRef] [PubMed]
7. Schirò, G.; Iacono, S.; Balistreri, C.R. The Role of Human Microbiota in Myasthenia Gravis: A Narrative Review. *Neurol. Int.* **2023**, *15*, 392–404. [CrossRef]
8. Celorrio, M.; Shumilov, K.; Friess, S.H. Gut microbial regulation of innate and adaptive immunity after traumatic brain injury. *Neural Regen. Res.* **2024**, *19*, 272–276. [CrossRef]

9. Sanghai, N.; Tranmer, G.K. Biochemical and Molecular Pathways in Neurodegenerative Diseases: An Integrated View. *Cells* **2023**, *12*, 2318. [CrossRef]
10. Li, L.; Guo, L.; Gao, R.; Yao, M.; Qu, X.; Sun, G.; Fu, Q.; Hu, C.; Han, G. Ferroptosis: A new regulatory mechanism in neuropathic pain. *Front. Aging Neurosci.* **2023**, *15*, 1206851. [CrossRef]
11. Massaccesi, L.; Balistreri, C.R. Biomarkers of Oxidative Stress in Acute and Chronic Diseases. *Antioxidants* **2022**, *11*, 1766. [CrossRef] [PubMed]
12. Zhang, J.B.; Jia, X.; Cao, Q.; Chen, Y.T.; Tong, J.; Lu, G.D.; Li, D.J.; Han, T.; Zhuang, C.L.; Wang, P. Ferroptosis-Regulated Cell Death as a Therapeutic Strategy for Neurodegenerative Diseases: Current Status and Future Prospects. *ACS Chem. Neurosci.* **2023**, *14*, 2995–3012. [CrossRef] [PubMed]
13. Monastero, R.; Magro, D.; Venezia, M.; Pisano, C.; Balistreri, C.R. A promising therapeutic peptide and preventive/diagnostic biomarker for age-related diseases: The Elabela/Apela/Toddler peptide. *Ageing Res. Rev.* **2023**, *91*, 102076. [CrossRef] [PubMed]
14. Schirò, G.; Iacono, S.; Ragonese, P.; Aridon, P.; Salemi, G.; Balistreri, C.R. A Brief Overview on BDNF-Trk Pathway in the Nervous System: A Potential Biomarker or Possible Target in Treatment of Multiple Sclerosis? *Front. Neurol.* **2022**, *13*, 917527. [CrossRef] [PubMed]
15. Brites, D.; Fernandes, A. Neuroinflammation and Depression: Microglia Activation, Extracellular Microvesicles and microRNA Dysregulation. *Front. Cell. Neurosci.* **2015**, *9*, 476. [CrossRef] [PubMed]
16. Tregub, P.P.; Ibrahimli, I.; Averchuk, A.S.; Salmina, A.B.; Litvitskiy, P.F.; Manasova, Z.S.; Popova, I.A. The Role of microRNAs in Epigenetic Regulation of Signaling Pathways in Neurological Pathologies. *Int. J. Mol. Sci.* **2023**, *24*, 12899. [CrossRef] [PubMed]
17. Tassinari, V.; La Rosa, P.; Guida, E.; Colopi, A.; Caratelli, S.; De Paolis, F.; Gallo, A.; Cenciarelli, C.; Sconocchia, G.; Dolci, S.; et al. Contribution of A-to-I RNA editing, M6A RNA Methylation, and Alternative Splicing to physiological brain aging and neurodegenerative diseases. *Mech. Ageing Dev.* **2023**, *212*, 111807. [CrossRef]
18. Xu, X.; Wang, J.; Chen, G. Circadian cycle and neuroinflammation. *Open Life Sci.* **2023**, *18*, 20220712. [CrossRef]
19. Nair, A.L.; Groenendijk, L.; Overdevest, R.; Fowke, T.M.; Annida, R.; Mocellin, O.; de Vries, H.E.; Wevers, N.R. Human BBB-on-a-chip reveals barrier disruption, endothelial inflammation, and T cell migration under neuroinflammatory conditions. *Front. Mol. Neurosci.* **2023**, *16*, 1250123. [CrossRef]
20. Bruno, M.; Bonomi, C.G.; Ricci, F.; Di Donna, M.G.; Mercuri, N.B.; Koch, G.; Martorana, A.; Motta, C. Blood-brain barrier permeability is associated with different neuroinflammatory profiles in Alzheimer's disease. *Eur. J. Neurol.* **2023**, *31*, e16095, *advance online publication*. [CrossRef]
21. Lauritsen, J.; Romero-Ramos, M. The systemic immune response in Parkinson's disease: Focus on the peripheral immune component. *Trends Neurosci.* **2023**, *46*, 863–878. [CrossRef] [PubMed]
22. Miró-Mur, F.; Pérez-de-Puig, I.; Ferrer-Ferrer, M.; Urra, X.; Justicia, C.; Chamorro, A.; Planas, A.M. Immature monocytes recruited to the ischemic mouse brain differentiate into macrophages with features of alternative activation. *Brain Behav. Immun.* **2016**, *53*, 18–33. [CrossRef] [PubMed]
23. Jones, K.A.; Maltby, S.; Plank, M.W.; Kluge, M.; Nilsson, M.; Foster, P.S.; Walker, F.R. Peripheral immune cells infiltrate into sites of secondary neurodegeneration after ischemic stroke. *Brain Behav. Immun.* **2018**, *67*, 299–307. [CrossRef]
24. Kanazawa, M.; Ninomiya, I.; Hatakeyama, M.; Takahashi, T.; Shimohata, T. Microglia and Monocytes/Macrophages Polarization Reveal Novel Therapeutic Mechanism against Stroke. *Int. J. Mol. Sci.* **2017**, *18*, 2135. [CrossRef] [PubMed]
25. Ritzel, R.M.; Li, Y.; Jiao, Y.; Doran, S.J.; Khan, N.; Henry, R.J.; Brunner, K.; Loane, D.J.; Faden, A.I.; Szeto, G.L.; et al. The brain-bone marrow axis and its implications for chronic traumatic brain injury. *Res. Sq.* **2023**, rs.3.rs-3356007. [CrossRef]
26. Côté, M.; Poirier, A.A.; Aubé, B.; Jobin, C.; Lacroix, S.; Soulet, D. Partial depletion of the proinflammatory monocyte population is neuroprotective in the myenteric plexus but not in the basal ganglia in a MPTP mouse model of Parkinson's disease. *Brain Behav. Immun.* **2015**, *46*, 154–167. [CrossRef]
27. Williams, D.W.; Veenstra, M.; Gaskill, P.J.; Morgello, S.; Calderon, T.M.; Berman, J.W. Monocytes mediate HIV neuropathogenesis: Mechanisms that contribute to HIV associated neurocognitive disorders. *Curr. HIV Res.* **2014**, *12*, 85–96. [CrossRef]
28. Molnarfi, N.; Benkhoucha, M.; Bjarnadottir, K.; Juillard, C.; Lalive, P.H. Interferonbeta induces hepatocyte growth factor in monocytes of multiple sclerosis patients. *PLoS ONE* **2012**, *7*, e49882. [CrossRef]
29. Man, S.; Tucky, B.; Cotleur, A.; Drazba, J.; Takeshita, Y.; Ransohoff, R.M. CXCL12- induced monocyte-endothelial interactions promote lymphocyte transmigration across an in vitro blood-brain barrier. *Sci. Transl. Med.* **2012**, *4*, 119ra114. [CrossRef]
30. Schmitt, C.; Strazielle, N.; Ghersi-Egea, J.F. Brain leukocyte infiltration initiated by peripheral inflammation or experimental autoimmune encephalomyelitis occurs through pathways connected to the CSF-filled compartments of the forebrain and midbrain. *J. Neuroinflammation* **2012**, *9*, 187. [CrossRef]
31. Vojdani, A.; Lambert, J. The role of Th17 in neuroimmune disorders: Target for CAM therapy. Part II. *Evid. Based Complement. Altern. Med.* **2011**, *2011*, 984965. [CrossRef] [PubMed]
32. Kadiu, I.; Glanzer, J.G.; Kipnis, J.; Gendelman, H.E.; Thomas, M.P. Mononuclear phagocytes in the pathogenesis of neurodegenerative diseases. *Neurotox. Res.* **2005**, *8*, 25–50. [CrossRef] [PubMed]
33. Cashman, J.R.; Gagliardi, S.; Lanier, M.; Ghirmai, S.; Abel, K.J.; Fiala, M. Curcumins promote monocytic gene expression related to beta-amyloid and superoxide dismutase clearance. *Neurodegener. Dis.* **2012**, *10*, 274–276. [CrossRef] [PubMed]
34. Kohman, R.A. Aging microglia: Relevance to cognition and neural plasticity. *Meth. Mol. Biol.* **2012**, *934*, 193–218.

35. Bottger, D.; Ullrich, C.; Humpel, C. Monocytes deliver bioactive nerve growth factor through a brain capillary endothelial cell-monolayer in vitro and counteract degeneration of cholinergic neurons. *Brain Res.* **2010**, *1312*, 108–119. [CrossRef] [PubMed]
36. Karlmark, K.R.; Tacke, F.; Dunay, I.R. Monocytes in health and disease—Minireview. *Eur. J. Microbiol. Immunol. Bp* **2012**, *2*, 97–102. [CrossRef] [PubMed]
37. Schirò, G.; Di Stefano, V.; Iacono, S.; Lupica, A.; Brighina, F.; Monastero, R.; Balistreri, C.R. Advances on Cellular Clonotypic Immunity in Amyotrophic Lateral Sclerosis. *Brain Sci.* **2022**, *12*, 1412. [CrossRef] [PubMed]
38. Jorfi, M.; Park, J.; Hall, C.K.; Lin, C.J.; Chen, M.; von Maydell, D.; Kruskop, J.M.; Kang, B.; Choi, Y.; Prokopenko, D.; et al. Infiltrating CD8$^+$ T cells exacerbate Alzheimer's disease pathology in a 3D human neuroimmune axis model. *Nat. Neurosci.* **2023**, *26*, 1489–1504. [CrossRef]
39. Kimura, K.; Nishigori, R.; Hamatani, M.; Sawamura, M.; Ashida, S.; Fujii, C.; Takata, M.; Lin, Y.; Sato, W.; Okamoto, T.; et al. Resident Memory-like CD8$^+$ T Cells Are Involved in Chronic Inflammatory and Neurodegenerative Diseases in the CNS. *Neurol. R Neuroimmunol. Neuroinflammation* **2023**, *11*, e200172. [CrossRef]
40. Harms, A.S.; Yang, Y.T.; Tansey, M.G. Central and peripheral innate and adaptive immunity in Parkinson's disease. *Sci. Transl. Med.* **2023**, *15*, eadk3225. [CrossRef]
41. Jia, Z.; Guo, M.; Ge, X.; Chen, F.; Lei, P. IL-33/ST2 Axis: A Potential Therapeutic Target in Neurodegenerative Diseases. *Biomolecules* **2023**, *13*, 1494. [CrossRef] [PubMed]
42. Lutshumba, J.; Wilcock, D.M.; Monson, N.L.; Stowe, A.M. Sex-based differences in effector cells of the adaptive immune system during Alzheimer's disease and related dementias. *Neurobiol. Dis.* **2023**, *184*, 106202. [CrossRef] [PubMed]
43. Zeng, J.; Liu, J.; Qu, Q.; Zhao, X.; Zhang, J. JKAP, Th1 cells, and Th17 cells are dysregulated and inter-correlated, among them JKAP and Th17 cells relate to cognitive impairment progression in Alzheimer's disease patients. *Ir. J. Med. Sci.* **2022**, *191*, 1855–1861. [CrossRef] [PubMed]
44. Shi, Y.; Wei, B.; Li, L.; Wang, B.; Sun, M. Th17 cells and inflammation in neurological disorders: Possible mechanisms of action. *Front. Immunol.* **2022**, *13*, 932152. [CrossRef] [PubMed]
45. Oberstein, T.J.; Taha, L.; Spitzer, P.; Hellstern, J.; Herrmann, M.; Kornhuber, J.; Maler, J.M. Imbalance of Circulating T_h17 and Regulatory T Cells in Alzheimer's Disease: A Case Control Study. *Front. Immunol.* **2018**, *9*, 1213. [CrossRef]
46. Jafarzadeh, A.; Sheikhi, A.; Jafarzadeh, Z.; Nemati, M. Differential roles of regulatory T cells in Alzheimer's disease. *Cell. Immunol.* **2023**, *393–394*, 104778. [CrossRef]
47. Fu, J.; Duan, J.; Mo, J.; Xiao, H.; Huang, Y.; Chen, W.; Xiang, S.; Yang, F.; Chen, Y.; Xu, S. Mild Cognitive Impairment Patients Have Higher Regulatory T-Cell Proportions Compared With Alzheimer's Disease-Related Dementia Patients. *Front. Aging Neurosci.* **2021**, *12*, 624304. [CrossRef]
48. Yeapuri, P.; Machhi, J.; Lu, Y.; Abdelmoaty, M.M.; Kadry, R.; Patel, M.; Bhattarai, S.; Lu, E.; Namminga, K.L.; Olson, K.E.; et al. Amyloid-β specific regulatory T cells attenuate Alzheimer's disease pathobiology in APP/PS1 mice. *Mol. Neurodegener.* **2023**, *18*, 97. [CrossRef]
49. Boillée, S.; Yamanaka, K.; Lobsiger, C.S.; Copeland, N.G.; Jenkins, N.A.; Kassiotis, G.; Kollias, G.; Cleveland, D.W. Onset and Progression in Inherited ALS Determined by Motor Neurons and Microglia. *Science* **2006**, *312*, 1389–1392. [CrossRef]
50. Hooten, K.G.; Beers, D.R.; Zhao, W.; Appel, S.H. Protective and Toxic Neuroinflammation in Amyotrophic Lateral Sclerosis. *Neurotherapeutics* **2015**, *12*, 364–375. [CrossRef]
51. Seksenyan, A.; Ron-Harel, N.; Azoulay, D.; Cahalon, L.; Cardon, M.; Rogeri, P.; Ko, M.K.; Weil, M.; Bulvik, S.; Rechavi, G.; et al. Thymic involution, a co-morbidity factor in amyotrophic lateral sclerosis. *J. Cell. Mol. Med.* **2009**, *14*, 2470–2482. [CrossRef] [PubMed]
52. Ehrhart, J.; Smith, A.J.; Kuzmin-Nichols, N.; Zesiewicz, T.A.; Jahan, I.; Shytle, R.D.; Kim, S.-H.; Sanberg, C.D.; Vu, T.H.; Gooch, C.L.; et al. Humoral factors in ALS patients during disease progression. *J. Neuroinflamm.* **2015**, *12*, 127. [CrossRef] [PubMed]
53. Zhao, W.; Beers, D.R.; Liao, B.; Henkel, J.S.; Appel, S.H. Regulatory T lymphocytes from ALS mice suppress microglia and effector T lymphocytes through different cytokine-mediated mechanisms. *Neurobiol. Dis.* **2012**, *48*, 418–428. [CrossRef] [PubMed]
54. Yang, Z.; He, L.; Ren, M.; Lu, Y.; Meng, H.; Yin, D.; Chen, S.; Zhou, Q. Paraneoplastic Amyotrophic Lateral Sclerosis: Case Series and Literature Review. *Brain Sci.* **2022**, *12*, 1053. [CrossRef] [PubMed]
55. Plantone, D.; Pardini, M.; Locci, S.; Nobili, F.; De Stefano, N. B Lymphocytes in Alzheimer's Disease-A Comprehensive Review. *J. Alzheimer's Dis. JAD* **2022**, *88*, 1241–1262. [CrossRef] [PubMed]
56. Ashford, M.T.; Zhu, D.; Bride, J.; McLean, E.; Aaronson, A.; Conti, C.; Cypress, C.; Griffin, P.; Ross, R.; Duncan, T.; et al. Understanding Online Registry Facilitators and Barriers Experienced by Black Brain Health Registry Participants: The Community Engaged Digital Alzheimer's Research (CEDAR) Study. *J. Prev. Alzheimer's Dis.* **2023**, *10*, 551–561. [CrossRef] [PubMed]
57. Asher, S.; Priefer, R. Alzheimer's disease failed clinical trials. *Life Sci.* **2022**, *306*, 120861. [CrossRef]
58. Balistreri, C.R. New frontiers in ageing and longevity: Sex and gender medicine. *Mech. Ageing Dev.* **2023**, *214*, 111850. [CrossRef]
59. Garofalo, S.; Cocozza, G.; Bernardini, G.; Savage, J.; Raspa, M.; Aronica, E.; Tremblay, M.E.; Ransohoff, R.M.; Santoni, A.; Limatola, C. Blocking immune cell infiltration of the central nervous system to tame Neuroinflammation in Amyotrophic lateral sclerosis. *Brain Behav. Immun.* **2022**, *105*, 1–14. [CrossRef]
60. Rekik, A.; Aissi, M.; Rekik, I.; Mhiri, M.; Frih, M.A. Brain atrophy patterns in multiple sclerosis patients treated with natalizumab and its clinical correlates. *Brain Behav.* **2022**, *12*, e2573. [CrossRef]

61. Laaker, C.; Baenen, C.; Kovács, K.G.; Sandor, M.; Fabry, Z. Immune cells as messengers from the CNS to the periphery: The role of the meningeal lymphatic system in immune cell migration from the CNS. *Front. Immunol.* **2023**, *14*, 1233908. [CrossRef] [PubMed]
62. Schirò, G.; Balistreri, C.R. The close link between brain vascular pathological conditions and neurodegenerative diseases: Focus on some examples and potential treatments. *Vasc. Pharmacol.* **2022**, *142*, 106951. [CrossRef] [PubMed]
63. Nicoletti, A.; Baschi, R.; Cicero, C.E.; Iacono, S.; Re, V.L.; Luca, A.; Schirò, G.; Monastero, R.; Gender Neurology Study Group of the Italian Society of Neurology. Sex and gender differences in Alzheimer's disease, Parkinson's disease, and Amyotrophic Lateral Sclerosis: A narrative review. *Mech. Ageing Dev.* **2023**, *212*, 111821. [CrossRef] [PubMed]
64. Han, W.; Tellez, L.A.; Perkins, M.H.; Perez, I.O.; Qu, T.; Ferreira, J.; Ferreira, T.L.; Quinn, D.; Liu, Z.W.; Gao, X.B.; et al. A Neural Circuit for Gut-Induced Reward. *Cell* **2018**, *175*, 665–678.e23. [CrossRef] [PubMed]
65. Agirman, G.; Yu, K.B.; Hsiao, E.Y. Signaling Inflammation across the Gut-Brain Axis. *Science* **2021**, *374*, 1087–1092. [CrossRef] [PubMed]
66. McCarville, J.L.; Chen, G.Y.; Cuevas, V.D.; Troha, K.; Ayres, J.S. Microbiota Metabolites in Health and Disease. *Annu. Rev. Immunol.* **2020**, *38*, 147–170. [CrossRef] [PubMed]
67. Luo, Y.; Zeng, B.; Zeng, L.; Du, X.; Li, B.; Huo, R.; Liu, L.; Wang, H.; Dong, M.; Pan, J.; et al. Gut microbiota regulates mouse behaviors through glucocorticoid receptor pathway genes in the hippocampus. *Transl. Psychiatry* **2018**, *8*, 187. [CrossRef]
68. O'Riordan, K.J.; Collins, M.K.; Moloney, G.M.; Knox, E.G.; Aburto, M.R.; Fülling, C.; Morley, S.J.; Clarke, G.; Schellekens, H.; Cryan, J.F. Short chain fatty acids: Microbial metabolites for gut-brain axis signalling. *Mol. Cell. Endocrinol.* **2022**, *546*, 111572. [CrossRef]
69. Thibaut, M.M.; Bindels, L.B. Crosstalk between bile acid-activated receptors and microbiome in entero-hepatic inflammation. *Trends Mol. Med.* **2022**, *28*, 223–236. [CrossRef]
70. Neumann, B.; Angstwurm, K.; Mergenthaler, P.; Kohler, S.; Schönenberger, S.; Bösel, J.; Neumann, U.; Vidal, A.; Huttner, H.B.; Gerner, S.T.; et al. Myasthenic crisis demanding mechanical ventilation: A multicenter analysis of 250 cases. *Neurology* **2020**, *94*, e299–e313. [CrossRef]
71. Wang, H.; Li, X.; Zhang, Q.; Fu, C.; Jiang, W.; Xue, J.; Liu, S.; Meng, Q.; Ai, L.; Zhi, X.; et al. Autophagy in Disease Onset and Progression. *Aging Dis.* **2023**. *advance online publication*. [CrossRef]
72. Li, L.; Li, T.; Qu, X.; Sun, G.; Fu, Q.; Han, G. Stress/cell death pathways, neuroinflammation, and neuropathic pain. *Immunol. Rev.* **2023**. *advance online publication*. [CrossRef] [PubMed]
73. Qu, L.; Li, Y.; Liu, F.; Fang, Y.; He, J.; Ma, J.; Xu, T.; Wang, L.; Lei, P.; Dong, H.; et al. Microbiota-Gut-Brain Axis Dysregulation in Alzheimer's Disease: Multi-Pathway Effects and Therapeutic Potential. *Aging Dis.* **2023**. *advance online publication*. [CrossRef] [PubMed]
74. Chen, Y.; Chen, J.; Xing, Z.; Peng, C.; Li, D. Autophagy in Neuroinflammation: A Focus on Epigenetic Regulation. *Aging Dis.* **2023**. *advance online publication*. [CrossRef] [PubMed]
75. Zhang, X.W.; Zhu, X.X.; Tang, D.S.; Lu, J.H. Targeting autophagy in Alzheimer's disease: Animal models and mechanisms. *Zool. Res.* **2023**, *44*, 1132–1145. [CrossRef] [PubMed]
76. Nechushtai, L.; Frenkel, D.; Pinkas-Kramarski, R. Autophagy in Parkinson's Disease. *Biomolecules* **2023**, *13*, 1435. [CrossRef] [PubMed]
77. Azad, A.; Gökmen, Ü.R.; Uysal, H.; Köksoy, S.; Bilge, U.; Manguoğlu, A.E. Autophagy dysregulation plays a crucial role in regulatory T-cell loss and neuroinflammation in amyotrophic lateral sclerosis (ALS). *Amyotroph. Lateral Scler. Front. Degener.* **2023**, 1–9, *advance online publication*. [CrossRef] [PubMed]
78. Sil, S.; Niu, F.; Tom, E.; Liao, K.; Periyasamy, P.; Buch, S. Cocaine Mediated Neuroinflammation: Role of Dysregulated Autophagy in Pericytes. *Mol. Neurobiol.* **2019**, *56*, 3576–3590. [CrossRef]
79. Singh, S.; Thangaraj, A.; Chivero, E.T.; Guo, M.L.; Periyasamy, P.; Buch, S. Role of Dysregulated Autophagy in HIV Tat, Cocaine, and cART Mediated NLRP3 Activation in Microglia. *J. Neuroimmune Pharmacol.* **2023**, *18*, 327–347. [CrossRef]
80. Elkholy, A.; Estfanous, R.S.; Alrobaian, M.; Borg, H.M.; Kabel, A.M. Repositioning itraconazole for amelioration of bleomycin-induced pulmonary fibrosis: Targeting HMGB1/TLR4 Axis, NLRP3 inflammasome/NF-κB signaling, and autophagy. *Life Sci.* **2023**, *313*, 121288. [CrossRef]
81. Yang, S.; Zhang, X.; Zhang, H.; Lin, X.; Chen, X.; Zhang, Y.; Lin, X.; Huang, L.; Zhuge, Q. Dimethyl itaconate inhibits LPS-induced microglia inflammation and inflammasome-mediated pyroptosis via inducing autophagy and regulating the Nrf-2/HO-1 signaling pathway. *Mol. Med. Rep.* **2021**, *24*, 672. [CrossRef] [PubMed]
82. Li, S.; Huang, P.; Lai, F.; Zhang, T.; Guan, J.; Wan, H.; He, Y. Mechanisms of Ferritinophagy and Ferroptosis in Diseases. *Mol. Neurobiol.* **2023**. *advance online publication*. [CrossRef] [PubMed]
83. Wang, Y.; Li, H.; He, Q.; Zou, R.; Cai, J.; Zhang, L. Ferroptosis: Underlying mechanisms and involvement in neurodegenerative diseases. *Apoptosis Int. J. Program. Cell Death* **2023**. *advance online publication*. [CrossRef] [PubMed]
84. Chavoshinezhad, S.; Beirami, E.; Izadpanah, E.; Feligioni, M.; Hassanzadeh, K. Molecular mechanism and potential therapeutic targets of necroptosis and ferroptosis in Alzheimer's disease. *Biomed. Pharmacother. Biomed. Pharmacother.* **2023**, *168*, 115656. [CrossRef] [PubMed]
85. Dar, N.J.; John, U.; Bano, N.; Khan, S.; Bhat, S.A. Oxytosis/Ferroptosis in Neurodegeneration: The Underlying Role of Master Regulator Glutathione Peroxidase 4 (GPX4). *Mol. Neurobiol.* **2023**. *advance online publication*. [CrossRef]

86. Balistreri, C.R. Vascular ageing and the related complications in the brain: New insights on related mechanisms and their translational applications. *Mech. Ageing Dev.* **2021**, *196*, 111469. [CrossRef]

87. Allegra, A.; Giarratana, R.M.; Scola, L.; Balistreri, C.R. The close link between the fetal programming imprinting and neurodegeneration in adulthood: The key role of "hemogenic endothelium" programming. *Mech. Ageing Dev.* **2021**, *195*, 111461. [CrossRef]

88. Balistreri, C.R. Promising Strategies for Preserving Adult Endothelium Health and Reversing Its Dysfunction: From Liquid Biopsy to New Omics Technologies and Noninvasive Circulating Biomarkers. *Int. J. Mol. Sci.* **2022**, *23*, 7548. [CrossRef]

89. Olivieri, F.; Pompilio, G.; Balistreri, C.R. Endothelial progenitor cells in ageing. *Mech. Ageing Dev.* **2016**, *159*, 1–3. [CrossRef]

90. Ho, P.T.B.; Clark, I.M.; Le, L.T.T. MicroRNA-Based Diagnosis and Therapy. *Int. J. Mol. Sci.* **2022**, *23*, 7167. [CrossRef]

91. Ferrante, M.; Conti, G.O. Environment and Neurodegenerative Diseases: An Update on miRNA Role. *MicroRNA* **2017**, *6*, 157–165. [CrossRef] [PubMed]

92. Zhang, J.; Li, A.; Gu, R.; Tong, Y.; Cheng, J. Role and regulatory mechanism of microRNA mediated neuroinflammation in neuronal system diseases. *Front. Immunol.* **2023**, *14*, 1238930. [CrossRef]

93. Rastegar-Moghaddam, S.H.; Ebrahimzadeh-Bideskan, A.; Shahba, S.; Malvandi, A.M.; Mohammadipour, A. Roles of the miR-155 in Neuroinflammation and Neurological Disorders: A Potent Biological and Therapeutic Target. *Cell. Mol. Neurobiol.* **2023**, *43*, 455–467. [CrossRef] [PubMed]

94. Ren, Y.; Li, H.; Xie, W.; Wei, N.; Liu, M. MicroRNA-195 triggers neuroinflammation in Parkinson's disease in a Rho-associated kinase 1-dependent manner. *Mol. Med. Rep.* **2019**, *19*, 5153–5161. [CrossRef]

95. Zhang, J.; Hu, M.; Teng, Z.; Tang, Y.P.; Chen, C. Synaptic and cognitive improvements by inhibition of 2-AG metabolism are through upregulation of microRNA-188-3p in a mouse model of Alzheimer's disease. *J. Neurosci.* **2014**, *34*, 14919–14933. [CrossRef] [PubMed]

96. He, D.; Tan, J.; Zhang, J. miR-137 attenuates Aβ-induced neurotoxicity through inactivation of NF-κB pathway by targeting TNFAIP1 in Neuro2a cells. *Biochem. Biophys. Res. Commun.* **2017**, *490*, 941–947. [CrossRef] [PubMed]

97. Yang, J.; Gong, Z.; Dong, J.; Bi, H.; Wang, B.; Du, K.; Zhang, C.; Chen, L. lncRNA XIST inhibition promotes M2 polarization of microglial and aggravates the spinal cord injury via regulating miR-124-3p/IRF1 axis. *Heliyon* **2023**, *9*, e17852. [CrossRef] [PubMed]

98. Li, D.; Yang, H.; Ma, J.; Luo, S.; Chen, S.; Gu, Q. MicroRNA-30e regulates neuroinflammation in MPTP model of Parkinson's disease by targeting Nlrp3. *Hum. Cell* **2018**, *31*, 106–115. [CrossRef]

99. Singh, V.; Kushwaha, S.; Ansari, J.A.; Gangopadhyay, S.; Mishra, S.K.; Dey, R.K.; Giri, A.K.; Patnaik, S.; Ghosh, D. MicroRNA-129-5p-regulated microglial expression of the surface receptor CD200R1 controls neuroinflammation. *J. Biol. Chem.* **2022**, *298*, 101521. [CrossRef]

100. Nasirishargh, A.; Kumar, P.; Ramasubramanian, L.; Clark, K.; Hao, D.; Lazar, S.V.; Wang, A. Exosomal microRNAs from mesenchymal stem/stromal cells: Biology and applications in neuroprotection. *World J. Stem Cells* **2021**, *13*, 776–794. [CrossRef]

101. Tang, Y.; Bao, J.S.; Su, J.H.; Huang, W. MicroRNA-139 modulates Alzheimer's-associated pathogenesis in SAMP8 mice by targeting cannabinoid receptor type 2. *Genet. Mol. Res. GMR* **2017**, *16*, gmr16019166. [CrossRef] [PubMed]

102. Balistreri, C.R.; Candore, G.; Accardi, G.; Colonna-Romano, G.; Lio, D. NF-κB pathway activators as potential ageing biomarkers: Targets for new therapeutic strategies. *Immun. Ageing* **2013**, *10*, 24. [CrossRef] [PubMed]

103. Balistreri, C.R.; Colonna-Romano, G.; Lio, D.; Candore, G.; Caruso, C. TLR4 polymorphisms and ageing: Implications for the pathophysiology of age-related diseases. *J. Clin. Immunol.* **2009**, *29*, 406–415. [CrossRef] [PubMed]

104. Yang, J.; Wise, L.; Fukuchi, K.I. TLR4 Cross-Talk With NLRP3 Inflammasome and Complement Signaling Pathways in Alzheimer's Disease. *Front. Immunol.* **2020**, *11*, 724. [CrossRef] [PubMed]

105. Gurram, P.C.; Manandhar, S.; Satarker, S.; Mudgal, J.; Arora, D.; Nampoothiri, M. Dopaminergic Signaling as a Plausible Modulator of Astrocytic Toll-Like Receptor 4: A Crosstalk between Neuroinflammation and Cognition. *CNS Neurol. Disord. Drug Targets* **2023**, *22*, 539–557. [CrossRef]

106. Sivamaruthi, B.S.; Raghani, N.; Chorawala, M.; Bhattacharya, S.; Prajapati, B.G.; Elossaily, G.M.; Chaiyasut, C. NF-κB Pathway and Its Inhibitors: A Promising Frontier in the Management of Alzheimer's Disease. *Biomedicines* **2023**, *11*, 2587. [CrossRef]

107. Cui, D.; Chen, Y.; Ye, B.; Guo, W.; Wang, D.; He, J. Natural products for the treatment of neurodegenerative diseases. *Phytomedicine Int. J. Phytother. Phytopharm.* **2023**, *121*, 155101. [CrossRef]

108. Decandia, D.; Gelfo, F.; Landolfo, E.; Balsamo, F.; Petrosini, L.; Cutuli, D. Dietary Protection against Cognitive Impairment, Neuroinflammation and Oxidative Stress in Alzheimer's Disease Animal Models of Lipopolysaccharide-Induced Inflammation. *Int. J. Mol. Sci.* **2023**, *24*, 5921. [CrossRef]

109. Balistreri, C.R. Anti-Inflamm-Ageing and/or Anti-Age-Related Disease Emerging Treatments: A Historical Alchemy or Revolutionary Effective Procedures? *Mediat. Inflamm.* **2018**, *2018*, 3705389. [CrossRef]

110. Dias-Carvalho, A.; Sá, S.I.; Carvalho, F.; Fernandes, E.; Costa, V.M. Inflammation as common link to progressive neurological diseases. *Arch. Toxicol.* **2023**. *advance online publication*. [CrossRef]

111. Koole, L.; Martinez-Martinez, P.; Amelsvoort, T.V.; Evelo, C.T.; Ehrhart, F. Interactive neuroinflammation pathways and transcriptomics-based identification of drugs and chemical compounds for schizophrenia. *World J. Biol. Psychiatry* **2023**, 1–14, *advance online publication*. [CrossRef] [PubMed]

112. Lashgari, N.A.; Roudsari, N.M.; Shamsnia, H.S.; Shayan, M.; Momtaz, S.; Abdolghaffari, A.H. TLR/mTOR inflammatory signaling pathway: Novel insight for the treatment of schizophrenia. *Can. J. Physiol. Pharmacol.* **2023**. *advance online publication*. [CrossRef]
113. Sarapultsev, A.; Gusev, E.; Komelkova, M.; Utepova, I.; Luo, S.; Hu, D. JAK-STAT signaling in inflammation and stress-related diseases: Implications for therapeutic interventions. *Mol. Biomed.* **2023**, *4*, 40. [CrossRef] [PubMed]
114. Islam, F.; Roy, S.; Zehravi, M.; Paul, S.; Sutradhar, H.; Yaidikar, L.; Kumar, B.R.; Dogiparthi, L.K.; Prema, S.; Nainu, F.; et al. Polyphenols Targeting MAP Kinase Signaling Pathway in Neurological Diseases: Understanding Molecular Mechanisms and Therapeutic Targets. *Mol. Neurobiol.* **2023**. *advance online publication*. [CrossRef] [PubMed]
115. Zheng, Y.; Zhang, X.; Zhang, R.; Wang, Z.; Gan, J.; Gao, Q.; Yang, L.; Xu, P.; Jiang, X. Inflammatory signaling pathways in the treatment of Alzheimer's disease with inhibitors, natural products and metabolites (Review). *Int. J. Mol. Med.* **2023**, *52*, 111. [CrossRef] [PubMed]
116. Singh, L.; Bhatti, R. Signaling Pathways Involved in the Neuroprotective Effect of Osthole: Evidence and Mechanisms. *Mol. Neurobiol.* **2023**. *advance online publication*. [CrossRef]
117. Mohawk, J.A.; Green, C.B.; Takahashi, J.S. Central and peripheral circadian clocks in mammals. *Annu. Rev. Neurosci.* **2012**, *35*, 445–462. [CrossRef] [PubMed]
118. Chen, R.; Routh, B.N.; Gaudet, A.D.; Fonken, L.K. Circadian Regulation of the Neuroimmune Environment Across the Lifespan: From Brain Development to Aging. *J. Biol. Rhythm.* **2023**, *38*, 419–446. [CrossRef] [PubMed]
119. Ahmad, F.; Sachdeva, P.; Sarkar, J.; Izhaar, R. Circadian dysfunction and Alzheimer's disease—An updated review. *Aging Med.* **2022**, *6*, 71–81. [CrossRef]
120. Meng, D.; Yang, M.; Zhang, H.; Zhang, L.; Song, H.; Liu, Y.; Zeng, Y.; Yang, B.; Wang, X.; Chen, Y.; et al. Microglia activation mediates circadian rhythm disruption-induced cognitive impairment in mice. *J. Neuroimmunol.* **2023**, *379*, 578102. [CrossRef]
121. Srinivasan, M.; Walker, C. Circadian Clock, Glucocorticoids and NF-κB Signaling in Neuroinflammation- Implicating Glucocorticoid Induced Leucine Zipper as a Molecular Link. *ASN Neuro* **2022**, *14*, 17590914221120190. [CrossRef]
122. Aho, V.; Ollila, H.M.; Rantanen, V.; Kronholm, E.; Surakka, I.; van Leeuwen, W.M.; Lehto, M.; Matikainen, S.; Ripatti, S.; Härmä, M.; et al. Partial sleep restriction activates immune response-related gene expression pathways: Experimental and epidemiological studies in humans. *PLoS ONE* **2013**, *8*, e77184. [CrossRef] [PubMed]
123. Taishi, P.; Chen, Z.; Obál, F., Jr.; Hansen, M.K.; Zhang, J.; Fang, J.; Krueger, J.M. Sleep-associated changes in interleukin-1beta mRNA in the brain. *J. Interferon Cytokine Res.* **1998**, *18*, 793–798. [CrossRef] [PubMed]
124. Szentirmai, É.; Kapás, L. Sleep and body temperature in TNFα knockout mice: The effects of sleep deprivation, β3-AR stimulation and exogenous TNFα. *Brain Behav. Immun.* **2019**, *81*, 260–271. [CrossRef] [PubMed]
125. Nguyen, J.; Gibbons, C.M.; Dykstra-Aiello, C.; Ellingsen, R.; Koh, K.M.S.; Taishi, P.; Krueger, J.M. Interleukin-1 receptor accessory proteins are required for normal homeostatic responses to sleep deprivation. *J. Appl. Physiol.* **2019**, *127*, 770–780. [CrossRef]
126. Choudhury, M.E.; Miyanishi, K.; Takeda, H.; Islam, A.; Matsuoka, N.; Kubo, M.; Matsumoto, S.; Kunieda, T.; Nomoto, M.; Yano, H.; et al. Phagocytic elimination of synapses by microglia during sleep. *Glia* **2020**, *68*, 44–59. [CrossRef] [PubMed]
127. Klarica, M.; Radoš, M.; Orešković, D. The Movement of Cerebrospinal Fluid and Its Relationship with Substances Behavior in Cerebrospinal and Interstitial Fluid. *Neuroscience* **2019**, *414*, 28–48. [CrossRef] [PubMed]
128. Hand, L.E.; Gray, K.J.; Dickson, S.H.; Simpkins, D.A.; Ray, D.W.; Konkel, J.E.; Hepworth, M.R.; Gibbs, J.E. Regulatory T cells confer a circadian signature on inflammatory arthritis. *Nat. Commun.* **2020**, *11*, 1658. [CrossRef]
129. García-García, A.; Korn, C.; García-Fernández, M.; Domingues, O.; Villadiego, J.; Martín-Pérez, D.; Isern, J.; Bejarano-García, J.A.; Zimmer, J.; Pérez-Simón, J.A.; et al. Dual cholinergic signals regulate daily migration of hematopoietic stem cells and leukocytes. *Blood* **2019**, *133*, 224–236. [CrossRef]
130. Adrover, J.M.; Del Fresno, C.; Crainiciuc, G.; Cuartero, M.I.; Casanova-Acebes, M.; Weiss, L.A.; Huerga-Encabo, H.; Silvestre-Roig, C.; Rossaint, J.; Cossío, I.; et al. A Neutrophil Timer Coordinates Immune Defense and Vascular Protection. *Immunity* **2019**, *50*, 390–402.e10. [CrossRef]
131. O'Siorain, J.R.; Curtis, A.M. Circadian Control of Redox Reactions in the Macrophage Inflammatory Response. *Antioxid. Redox Signal.* **2022**, *37*, 664–678. [CrossRef]
132. Blacher, E.; Tsai, C.; Litichevskiy, L.; Shipony, Z.; Iweka, C.A.; Schneider, K.M.; Chuluun, B.; Heller, H.C.; Menon, V.; Thaiss, C.A.; et al. Aging disrupts circadian gene regulation and function in macrophages. *Nat. Immunol.* **2022**, *23*, 229–236. [CrossRef] [PubMed]
133. Cui, L.; Jin, X.; Xu, F.; Wang, S.; Liu, L.; Li, X.; Lin, H.; Du, M. Circadian rhythm-associated Rev-erbα modulates polarization of decidual macrophage via the PI3K/Akt signaling pathway. *Am. J. Reprod. Immunol.* **2021**, *86*, e13436. [CrossRef] [PubMed]
134. Nobis, C.C.; Dubeau Laramée, G.; Kervezee, L.; De Sousa, D.M.; Labrecque, N.; Cermakian, N. The circadian clock of CD8 T cells modulates their early response to vaccination and the rhythmicity of related signaling pathways. *Proc. Natl. Acad. Sci. USA* **2019**, *116*, 20077–20086. [CrossRef] [PubMed]
135. Furman, D.; Campisi, J.; Verdin, E.; Carrera-Bastos, P.; Targ, S.; Franceschi, C.; Ferrucci, L.; Gilroy, D.W.; Fasano, A.; Miller, G.W.; et al. Chronic inflammation in the etiology of disease across the life span. *Nat. Med.* **2019**, *25*, 1822–1832. [CrossRef] [PubMed]
136. Monastero, R.; Caruso, C.; Vasto, S. Alzheimer's disease and infections, where we stand and where we go. *Immun. Ageing* **2014**, *11*, 26. [CrossRef] [PubMed]
137. Franceschi, C.; Garagnani, P.; Parini, P.; Giuliani, C.; Santoro, A. Inflammaging: A new immune-metabolic viewpoint for age-related diseases. *Nat. Rev. Endocrinol.* **2018**, *14*, 576–590. [CrossRef] [PubMed]

138. Proctor, M.J.; McMillan, D.C.; Horgan, P.G.; Fletcher, C.D.; Talwar, D.; Morrison, D.S. Systemic inflammation predicts all-cause mortality: A glasgow inflammation outcome study. *PLoS ONE* **2015**, *10*, e0116206. [CrossRef]

139. Maggio, M.; Guralnik, J.M.; Longo, D.L.; Ferrucci, L. Interleukin-6 in aging and chronic disease: A magnificent pathway. *J. Gerontol. Ser. A Biol. Sci. Med. Sci.* **2006**, *61*, 575–584. [CrossRef]

140. Feghali, C.A.; Wright, T.M. Cytokines in acute and chronic inflammation. *Front. Biosci. A J. Virtual Libr.* **1997**, *2*, d12–d26. [CrossRef]

141. Parra-Torres, V.; Melgar-Rodríguez, S.; Muñoz-Manríquez, C.; Sanhueza, B.; Cafferata, E.A.; Paula-Lima, A.C.; Díaz-Zúñiga, J. Periodontal bacteria in the brain-Implication for Alzheimer's disease: A systematic review. *Oral Dis.* **2023**, *29*, 21–28. [CrossRef]

142. Liu, S.; Dashper, S.G.; Zhao, R. Association Between Oral Bacteria and Alzheimer's Disease: A Systematic Review and Meta-Analysis. *J. Alzheimer's Dis. JAD* **2023**, *91*, 129–150. [CrossRef] [PubMed]

143. Beydoun, M.A.; Beydoun, H.A.; Hossain, S.; El-Hajj, Z.W.; Weiss, J.; Zonderman, A.B. Clinical and Bacterial Markers of Periodontitis and Their Association with Incident All-Cause and Alzheimer's Disease Dementia in a Large National Survey. *J. Alzheimer's Dis. JAD* **2020**, *75*, 157–172. [CrossRef] [PubMed]

144. Larvin, H.; Gao, C.; Kang, J.; Aggarwal, V.R.; Pavitt, S.; Wu, J. The impact of study factors in the association of periodontal disease and cognitive disorders: Systematic review and meta-analysis. *Age Ageing* **2023**, *52*, afad015. [CrossRef] [PubMed]

145. Chen, Y.; Jin, Y.; Li, K.; Qiu, H.; Jiang, Z.; Zhu, J.; Chen, S.; Xie, W.; Chen, G.; Yang, D. Is There an Association Between Parkinson's Disease and Periodontitis? A Systematic Review and Meta-Analysis. *J. Park. Dis.* **2023**, *13*, 1107–1125. [CrossRef]

146. Yazdanpanahi, N.; Etemadifar, M.; Kardi, M.T.; Shams, E.; Shahbazi, S. Investigation of ERG Gene Expression in Iranian Patients with Multiple Sclerosis. *Immunol. Investig.* **2018**, *47*, 351–359. [CrossRef]

147. Pisano, C.; Terriaca, S.; Scioli, M.G.; Nardi, P.; Altieri, C.; Orlandi, A.; Ruvolo, G.; Balistreri, C.R. The Endothelial Transcription Factor ERG Mediates a Differential Role in the Aneurysmatic Ascending Aorta with Bicuspid or Tricuspid Aorta Valve: A Preliminary Study. *Int. J. Mol. Sci.* **2022**, *23*, 10848. [CrossRef]

Disclaimer/Publisher's Note: The statements, opinions and data contained in all publications are solely those of the individual author(s) and contributor(s) and not of MDPI and/or the editor(s). MDPI and/or the editor(s) disclaim responsibility for any injury to people or property resulting from any ideas, methods, instructions or products referred to in the content.

Review

Investigation of the Relationship between Apolipoprotein E Alleles and Serum Lipids in Alzheimer's Disease: A Meta-Analysis

Huaxue Xu [1], Jiajia Fu [1], Risna Begam Mohammed Nazar [1], Jing Yang [1], Sihui Chen [1], Yan Huang [2], Ting Bao [2] and Xueping Chen [1,*]

1 Department of Neurology, West China Hospital, Sichuan University, Chengdu 610041, China; xuxuhuaxue@sina.com (H.X.); fujiajia0421@foxmail.com (J.F.); begamrisna@gmail.com (R.B.M.N.); yo_screw1900@163.com (J.Y.); chensihuiyyds@163.com (S.C.)
2 Management Center, West China Hospital, Sichuan University, Chengdu 610041, China; huangyanhy513@163.com (Y.H.); baoting199@163.com (T.B.)
* Correspondence: chenxueping0606@sina.com; Fax: +86-028-85423550

Abstract: Prior studies have yielded mixed findings concerning the association between apolipoprotein E(*APOE*)-ε4 and serum lipids in patients with Alzheimer's disease (AD) and healthy individuals. Some studies suggested a relationship between *APOEε4* and serum lipids in patients with AD and healthy individuals, whereas others proposed that the *APOEε4* allele affects lipids only in patients with AD. Our study aimed to investigate whether *APOE* alleles have a distinct impact on lipids in AD. We conducted a comprehensive search of the PubMed and Embase databases for all related studies that investigate *APOE* and serum lipids of AD from the inception to 30 May 2022. Elevated total cholesterol (TC) and low-density lipoprotein (LDL) levels were found in *APOEε4* allele carriers compared with non-carriers. No significant differences were found for high-density lipoprotein (HDL) and triglyceride (TG) levels in *APOEε4* allele carriers compared to non-carriers. Notably, elevated TC and LDL levels showed considerable heterogeneity between patients with AD and healthy controls. A network meta-analysis did not find a distinct effect of carrying one or two *APOEε4* alleles on lipid profiles. Higher TC and LDL levels were found in *APOEε4* allele carriers compared with non-carriers, and the difference was more significant in patients with AD than in healthy controls.

Keywords: apolipoprotein E; serum lipids; Alzheimer's disease; meta-analysis

Citation: Xu, H.; Fu, J.; Mohammed Nazar, R.B.; Yang, J.; Chen, S.; Huang, Y.; Bao, T.; Chen, X. Investigation of the Relationship between Apolipoprotein E Alleles and Serum Lipids in Alzheimer's Disease: A Meta-Analysis. *Brain Sci.* **2023**, *13*, 1554. https://doi.org/10.3390/brainsci13111554

Academic Editors: Andrew Clarkson and Ashu Johri

Received: 5 October 2023
Revised: 30 October 2023
Accepted: 2 November 2023
Published: 6 November 2023

Copyright: © 2023 by the authors. Licensee MDPI, Basel, Switzerland. This article is an open access article distributed under the terms and conditions of the Creative Commons Attribution (CC BY) license (https://creativecommons.org/licenses/by/4.0/).

1. Introduction

Alzheimer's disease (AD) presents as a prevalent progressive neurodegenerative disease characterized by an insidious onset, progressive memory decline, cognitive impairment, and a spectrum of behavioral and psychological symptoms [1]. The development of AD appears to be a result of the complex interplay of genetic and environmental factors [2], hence rendering effective treatment of AD a formidable challenge [3]. The multifactorial etiology of this global health challenge has driven many research endeavors to unravel the complex web of causative elements of AD. Among these factors are apolipoprotein E (*APOE*) and its allelic variants, specifically the *APOE ε4* allele, which have emerged as being noteworthy.

The human APOE gene is encoded on chromosome 19, and it has three allelic variants: ε2, ε3, and ε4 [4]. Notably, the individuals carrying the *APOEε4* allele exhibit a high risk of sporadic AD [5]. Individuals with a single *APOEε4* allele have a 3.2 times higher risk of developing AD, whereas, in those with two *APOEε4* alleles, the risk of developing AD is increased by 8 to 10 folds [6]. This can be attributed to the influence of the *APOEε4* allele on amyloid-β (Aβ), either by reducing its clearance or by increasing its production in the brain [7].

In neuroimaging investigations of APOE polymorphism in healthy individuals, there has been a predominant focus on examining gray matter alterations in middle or late life, particularly in brain regions associated with significant AD pathological findings. Even in individuals showing no clinical symptoms, documentation has shown a reduction in the gray matter within the hippocampal and frontotemporal regions in *APOEε4* allele carriers compared with non-carriers [8].

Moreover, the human *APOE* allele encodes a polyclonal lipoprotein integral to metabolic processes, including cholesterol transport [9]. Although *APOE* alleles have a certain impact on lipid profiles [10–14], current research results are inconsistent [11,14]. Some studies have identified elevated levels of low-density lipoprotein (LDL) and total cholesterol (TC) in *APOEε4* allele carriers (*APOEε4* allele-C) compared with non-carriers (*APOEε4* allele-N), whereas others [10,12] have reported the opposite. Furthermore, some studies [13,15] have reported that significant differences exist in high-density lipoprotein (HDL) levels between carriers and non-carriers of the *APOEε4* allele. However, such distinctions were not observed in other studies [12,13]. Intriguingly, no systematic analyses have focused on the differences in lipid profiles between single *APOEε4* allele carriers and *APOEε4* homozygous individuals concerning lipid profiles.

Most researchers believe that lipid metabolism is very important in the pathophysiological mechanism of AD [16]. Notably, the latest meta-analysis summarized the disparities in lipid profiles between individuals with AD and healthy controls [17]. Since the *APOEε4* allele affects both lipid metabolism and the pathophysiology of AD, it has been hypothesized that the special relationship between the *APOEε4* allele and lipid metabolism is unique in AD. Some studies have found a relationship between the *APOEε4* allele and lipid profiles in patients with AD and healthy control populations [18], whereas others have discerned this association exclusively within the AD population [13].

Additionally, most meta-analyses summarized the differences in lipids between patients with AD and healthy controls, but there has been no relevant summary analysis that has explored whether the unique relationship between the *APOEε4* allele and lipids differs between patients with AD and healthy controls. Therefore, we systematically compared the lipid profiles between carriers and non-carriers of the *APOE4* allele among patients with AD and healthy controls and investigated whether the effect of *APOE* on lipids is unique in AD. We hypothesized that the *APOEε4* allele might cause the development of AD by influencing lipid metabolism.

2. Materials and Methods

2.1. Search Strategy

Two independent investigators searched the PubMed, Embase, Web of Science, and Chinese databases on 30 May 2022. The following medical subject heading (MeSH) terms and topic terms were used as the search terms: "Lipid", "Cholesterol", "Triglycerides", "Alzheimer's disease", "Alzheimer Dementia", "Apoprotein E", and "APOE".

2.2. Inclusion and Exclusion Criteria

The inclusion criteria were as follows: (1) all articles that reported the results of *APOE* alleles and were grouped participants according to whether they carried the *APOEε4* allele and/or different *APOE* alleles; (2) articles reporting data as mean ± standard derivation (SD); (3) studies that analyzed patients with AD patients or healthy controls as the study populations; (4) studies that included patients diagnosed with AD; and (5) studies that included healthy controls with normal cognitive function and no neurological disease.

The exclusion criteria were as follows: reviews, conference papers, letters, comments, editorials, case reports, and abstracts without an available full text.

2.3. Data Extraction and Quality Evaluation

FJJ and YJ conducted the preliminary screening of titles and abstracts and then screened potentially relevant full texts according to the inclusion criteria. A third investigator verified all the data. From each study, we collected the following data: the sample size, publication year, and participant characteristics (age, number of participants, sex ratio, country, and Mini-Mental State Examination scores). Relevant information was extracted independently by two investigators and verified by a third investigator. The Newcastle–Ottawa Quality Assessment Scale (NOS Scale) was used to assess the quality of the included studies [19]. The total score on this scale is 9, and a score of ≥ 6 is acceptable.

2.4. Statistical Analysis

We performed a meta-analysis using Stata, version 15.0 software (StataCorp LLC., College Station, TX, USA) and used the standardized mean difference (SMD) to obtain aggregate effects. The random-effects model was used if there was significant heterogeneity between the included studies (the Cochrane Q test result and I^2 statistic: $I^2 > 50\%$ or $p < 0.1$). The z-test was used to determine the overall effect. We assessed heterogeneity using sensitivity, meta-regression, and subgroup analyses and evaluated the publication bias using Begg's and Egger's tests. The standardized effect size was compared between multiple groups using network meta-analysis, and related indicators of each group were compared using the cumulative ranking curve (SUCRA).

3. Results
3.1. Study Selection and Characteristics

The flow chart illustrates the systematic search and selection process (Figure 1); 17 studies were included in the final analysis [10,12–15,18,20–30] (Table 1). These selected studies, which were carefully evaluated for their relevance and contribution to our research objectives, are shown in Table 1. Eight of these studies grouped the participants on the basis of their APOEε4 allele status. Among them, four studies exclusively focused on individuals with AD, whereas the other four studies examined both patients with AD and healthy controls. APOE allele classification was further extended in six studies, which divided participants into three specific groups: *APOEε2* allele carriers, *APOEε3/3* carriers, and *APOEε4* allele carriers. Of these, two studies exclusively focused on the AD population, and the remaining four encompassed both AD and healthy control populations. The participants were divided into six subgroups based on their *APOE* alleles status across a total of five studies. Among them, one study was exclusively dedicated to the AD population, one study was exclusively dedicated to the control population, and three studies grouped both the AD and control populations. It is important to note that some studies did not analyze all pertinent variables, including but not limited to TC, triglycerides (TG), HDL, and LDL levels. These variances are essential to consider when interpreting the collective findings. The NOS Scale scores are shown in Table S1.

Table 1. Details of the original studies included in the meta-analysis for *APOE* and AD.

Author-Year	Country	n (AD)	n (CON)	Sex (Male%) (AD)	Sex (Male%) (CON)	Age (AD)	Age (CON)	*APOE* (*n*)	Lipid Profiles (mmol/L)
Fernandes, 1999 [20]	Portugal	74	35	43.2	48.6	68.24 ± 9.02	64.97 ± 10.42	AD: *APOEε4*+(18), *APOEε4*−(27) CON: *APOEε4*+(4), *APOEε4*−(24) AD: ε2/ε2(0), ε2/ε3(3), ε2/ε4(0), ε3/ε3(24), ε3/ε4(13), ε4/ε4(5) CON: ε2/ε2(0), ε2/ε3(3), ε2/ε4(0), ε3/ε3(21), ε3/ε4(4), ε4/ε4(0)	TC, TG
Wehra, 2000 [21]	Poland	26	39	30.8	38.5	70.6 ± 7.3	70.0 ± 8.3	AD: *APOEε4*+(16), *APOEε4*−(10)	TC, TG, HDL, LDL
Sheng, 2000 [22]	China	39	40	54.8				AD: ε2+(2), ε3/ε3(20), ε4+(17)	TC, TG, HDL, LDL
Isbir, 2001 [12]	Turkey	35	29	25.7	70	73.91 ± 7.35	73.62 ± 13.63	AD: *APOEε4*+(7), *APOEε4*−(28)	TC, TG, HDL, LDL
Jingbin, 2002 [23]	China	109	98	41.3	54.1	3.7 ± 7.1	9.2 ± 6.5	AD: ε2/ε2(0), ε2/ε3(8), ε2/ε4(0), ε3/ε3(46), ε3/ε4(37), ε4/ε4(18) CON: ε2/ε2(1), ε2/ε3(14), ε2/ε4(1), ε3/ε3(73), ε3/ε4(8), ε4/ε4(1)	TC, TG
Xiangyu, 2002 [24]	China	48	84	64.6	73.8	73 ± 8	61 ± 1	AD: ε2+(10), ε3/ε3(23), ε4+(15) CON: ε2+(10), ε3/ε3(67), ε4+(5)	TC, TG
Al-Shammari, 2004 [25]	Kuwait		106		94.3		40.5 ± 4.7	CON: ε2/ε2(2), ε2/ε3(7), ε2/ε4(1), ε3/ε3(78), ε3/ε4(18), ε4/ε4(0)	TC, TG, HDL, LDL

Table 1. *Cont.*

Author-Year	Country	n (AD)	n (CON)	Sex (Male%) (AD)	Sex (Male%) (CON)	Age (AD)	Age (CON)	APOE (n)	Lipid Profiles (mmol/L)
Raygani, 2006 [13]	Iran	94	111	43.6	36.9	74.2 ± 10	72 ± 11.4	AD: APOEε4+(34), APOEε4−(60) CON: APOEε4+(14), APOEε4−(97)	TC, TG, HDL, LDL
Hall, 2006 [10]	India	29	1046					AD: APOEε4+(14), APOEε4−(15) CON: APOEε4+(416), APOEε4−(630)	TC, TG, HDL, LDL
Sabbagh, 2006 [15]	America	142				52–96		AD: APOEε4+(86), APOEε4−(60) AD: ε2/ε2(0), ε2/ε3(10), ε2/ε4(0), ε3/ε3(50), ε3/ε4(65), ε4/ε4(17)	TC, TG, HDL, LDL
Dongmei, 2008 [26]	China	77	158	59.7	55.7	3.3 ± 4.6	3.8 ± 5.0	AD: ε2+(4), ε3/ε3(57), ε4+(16)	TC, TG, HDL, LDL
Singh, 2012 [27]	India	70	75	50–85				AD: ε2/ε2(0), ε2/ε3(4), ε2/ε4(2), ε3/ε3(23), ε3/ε4(40), ε4/ε4(1) CON: ε2/ε2(0), ε2/ε3(9), ε2/ε4(1), ε3/ε3(55), ε3/ε4(10), ε4/ε4(0)	TC, TG, HDL, LDL
Tieqiang, 2012 [28]	China	100	102	37	41.2	77.5 ± 57.3	77.0 ± 6.3	AD: ε2+(15), ε3/ε3(54), ε4+(31) CON: ε2+(18), ε3/ε3(70), ε4+(14)	TC, TG, HDL
Jie, 2013 [29]	China	157	155			71.7 ± 10.9	72.1 ± 11.5	AD: ε2+(20), ε3/ε3(85), ε4+(52) CON: ε2+(24), ε3/ε3(106), ε4+(25)	TC
Shafagoj, 2018 [14]	Jordan	38	33					AD: APOEε4+(11), APOEε4−(27)	TC, TG, HDL, LDL

Table 1. *Cont.*

Author-Year	Country	*n* (AD)	*n* (CON)	Sex (Male%) (AD)	Sex (Male%) (CON)	Age (AD)	Age (CON)	*APOE* (*n*)	Lipid Profiles (mmol/L)
Mengzhen, 2018 [30]	China	47	35	31.9	31.4	69.96 ± 8.66	68.57 ± 8.64	AD: ε2+(7), ε3/ε3(26), ε4+(14) CON: ε2+(7), ε3/ε3(25), ε4+(3)	TC, TG, HDL, LDL
Wang, 2020 [31]	China	63	33	47.6	66.7	66.3 ± 9.6	66.0 ± 8.7	AD: *APOEε4*+(28), *APOEε4*−(35) CON: *APOEε4*+(8), *APOEε4*−(25)	TC, TG, HDL, LDL

Note: APOE: apolipoprotein E; AD: Alzheimer's disease; TC: total cholesterol (TC); TG: triglycerides; HDL: high-density lipoprotein; LDL: low-density lipoprotein; CON: healthy control population; GHP: general healthy population; *APOEε4*+: carriers of the apolipoprotein *Eε4* allele; *APOEε4*−: non-carriers of the apolipoprotein *Eε4* allele; ε2+: carriers of the apolipoprotein *Eε2* allele (the apolipoprotein *Eε2*/apolipoprotein *Eε4* was classified into the ε2+ group); ε4+: carriers of the apolipoprotein *Eε2* allele; ε2/ε2: apolipoprotein *Eε2*/apolipoprotein *Eε2*; ε2/ε3: apolipoprotein *Eε2*/apolipoprotein *Eε3*; ε2/ε4: apolipoprotein *Eε2*/apolipoprotein *Eε4*; ε3/ε3: apolipoprotein *Eε3*/apolipoprotein *Eε3*; ε3/ε4: apolipoprotein *Eε3*/apolipoprotein *Eε4*; ε4/ε4:apolipoprotein *Eε4*/apolipoprotein *Eε4*.

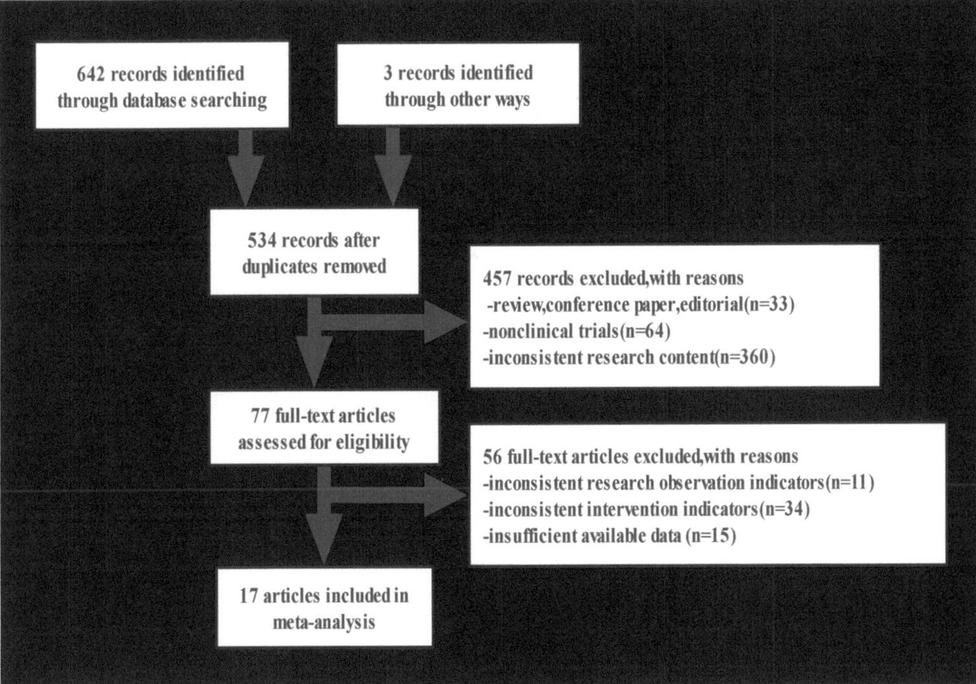

Figure 1. The literature screening flow chart.

3.2. Data Extraction and Study Population

We extracted data from eight articles focusing on TC and TG levels that included 652 individuals carrying *APOEε4* allele-C and 1038 individuals with *APOEε4* allele-N. Additionally, we collected information from seven articles regarding HDL and LDL levels that included 630 individuals carrying *APOEε4* allele-C and 987 individuals with *APOEε4* allele-N.

3.3. Overall Effect, Heterogeneity, Publication Bias, and Subgroup Analysis

To elucidate the overall effect across the studies, we used a random effect model to address the variances arising from differences present in the included studies. Noteworthy differences in TC and LDL levels were observed when comparing the APOEε4 allele-C and APOEε4 allele-N groups. Specifically, individuals in the *APOEε4* allele-C group showed higher TC and LDL levels than those in the *APOEε4* allele-N group (TC: SMD = 0.62 [0.2, 1.04], p = 0.004; LDL: SMD = 0.63 [0.9, 1.08], p = 0.005). However, studies indicated no difference in TG and HDL levels between those groups (TG: SMD = 0.08 [−0.19, 0.41], p = 0.108; HDL: SMD = −0.08 [−0.45, 0.28], p = 0.655) (Figures 2 and S1–S4).

Sensitivity analysis was conducted to identify the causes of heterogeneity and we were able to identify a clear cause of heterogeneity (Figures S5–S8). The funnel plot and bias test showed no significant publication bias (Figures S9–12, Tables S2–S5).

However, subgroup analysis showed great heterogeneity in TC and LDL levels between the *APOEε4* allele-C and *APOEε4* allele-N groups among the AD and healthy control populations (TC: p = 0.042; LDL: p = 0.001). However, no heterogeneity was shown in HDL and TG levels (TG: p = 0.794; HDL: p = 0.823) (Figures 2 and S13–S16).

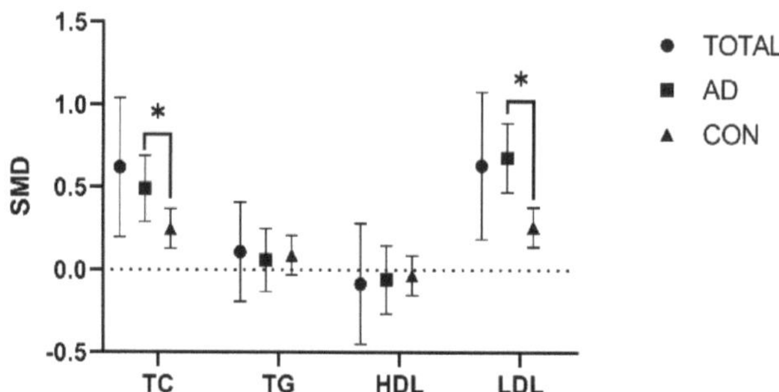

Figure 2. Comparison of the effect of the *APOEε4* allele on lipids in the AD and healthy control populations. The data are presented as the standardized mean difference (SMD) and 95% confidence interval. TOTAL: both the Alzheimer's disease and healthy control populations; AD: Alzheimer's disease population; CON: healthy control population; TC: total cholesterol; TG: triglycerides; HDL: high-density lipoprotein; LDL: low-density lipoprotein. *: $p < 0.05$.

Notably, the AD and healthy control populations had elevated TC and LDL levels in the APOEε4 allele-C group compared with the APOEε4 allele-N group, but the degree of elevation was lower in the AD population than in the healthy control population (AD population: TC: SMD = 0.49 [0.29, 0.69], $p = 0.000$; LDL: SMD = 0.68 [0.47, 0.89], $p = 0.000$) (healthy control population: TC: SMD = 0.25 [0.13, 0.37], $p = 0.000$; LDL: SMD = 0.26 [0.14,0.38], $p = 0.000$) (Figures 2 and S13–S16).

3.4. Comparison of the Lipids in APOEε3/ε3, APOEε2 Allele, and APOEε4 Allele Carriers

The network meta-analysis was performed to compare TC, TG, and LDL levels in individuals carrying different *APOE* alleles; *APOEε4* allele carriers had the highest SUCRA value, followed by *APOEε3/3* and *APOEε2* allele carriers, respectively (Tables S6–S8). However, regarding HDL levels, *APOEε4* allele carriers had the lowest SUCRA value, followed by *APOEε2* allele carriers, whereas *APOEε3/3* allele carriers had the highest SUCRA value (Table S9).

3.5. Comparison of the Lipids between Six Groups of APOE Alleles

The network meta-analysis of six distinct groups formed by *APOE* alleles showed variations in the SUCRA values of TC levels as follows: *APOEε3/ε4 > APOEε4/ε4 > APOEε3/ε3 > APOEε2/ε2 > APOEε2/ε4 > APOEε2/ε3* (Table S10).

Similarly, SUCRA values of TG levels were as follows: *APOEε2/ε2 > APOEε3/ε4 > APOEε3/ε3 > APOEε4/ε4 > APOEε2/ε3 > APOEε2/ε4* (Table S11).

Owing to the limited availability of multiple data sets, sequencing comparisons of HDL and LD levels between these six groups could not be performed.

4. Discussion

4.1. Main Findings

We hypothesized that the *APOEε4* allele might cause the development of AD by influencing lipid metabolism. Studies have found that high levels of serum cholesterol are positively associated with an increased risk of dementia, and the prevalence of AD is reduced in patients taking cholesterol-lowering drugs [32]. A Mendelian randomization study of AD metabolism and risk confirmed the causal role of LDL, cholesterol, and serum total cholesterol in the high-risk of AD [33]. Some studies have found an association between blood lipids and Alzheimer's disease, proving that blood lipids can be used as biomarkers for the early diagnosis of Alzheimer's disease. It can also help predict the stage of prognosis and disease severity, and further studies are needed to find out the exact mechanisms behind these changes [34]. This study focused on the relationship between *APOE* alleles and serum lipid profiles, specifically TC, LDL, TG, and HDL levels in individuals with AD compared to healthy controls. Through our meta-analysis, we found that individuals carrying the *APOEε4* allele showed increased TC and LDL levels compared with those without the *APOEε4* allele. There was a statistically significant difference in TC levels between the *APOEε4* allele-C and *APOEε4* allele-N groups. The *p*-value indicated that the difference did not occur by chance and is, therefore, statistically significant. *APOEε4* allele-C carriers had higher LDL levels than non-*APOEε4* allele-C carriers.

Notably, no significant statistical differences were found in TG and HDL levels between these groups. These data reinforce the absence of statistically significant differences in TG and HDL levels between individuals with *APOEε4* allele-C and those without. Further analysis showed differences in TC and LDL levels between *APOEε4* allele-C and *APOEε4* allele-N groups with significant heterogeneity when considering AD and healthy controlled populations separately. These data suggest that in AD populations, TC and LDL levels are higher in *APOEε4* allele-C carriers than in *APOEε4* allele-N carriers, but the degree of elevation is lower than that seen in the healthy control populations.

It is crucial to note that *APOEε4* acts as a main genetic risk factor for AD. Genome-wide association studies have shown that *APOEε4* is the strongest genetic risk factor for AD, irrespective of the age of onset [31].

4.2. APOE Functions in the Brain

The APOE gene encodes the APOE protein, which plays an important role in the transportation and metabolism of lipids [35]. APOE is responsible for the transportation of lipids and the maintenance of cholesterol homeostasis in the brain. It plays a crucial role in supplying neurons with cholesterol and facilitating the removal of excess cholesterol. It is also involved in other brain functions, such as promoting synaptic plasticity, transmitting signals, maintaining protein balance, modulating the immune system, and repairing after an injury [36].

4.3. APOE Isomers and Their Binding Specificity

Research has shown that the C-terminal domain of APOE is the key to lipoprotein binding and determines the specificity of APOE subtype lipidosis [37]. Specifically, *APOEε4* shows distinct characteristics, including poor lipidation compared with *APOEε2* and *APOEε3* alleles [38]. The *APOEε3* and *APOEε2* alleles prefer to bind to HDL, whereas the *APOEε4* allele prefers to bind to very low-density lipoprotein (VLDL) [39]. This variation in lipoprotein association is determined by differences in the interactions of the carboxyl-terminal domains among the isoforms, leading to *APOEε2* and *APOEε3* binding to smaller more phospholipid-enriched HDL, and *APOEε4* binding to larger triglyceride-rich VLDL [40].

4.4. Lipid-Binding Effects of APOEε 4 and Cholesterol Efflux

The lipid-binding features of *APOEε4* have substantial effects on the efflux of cholesterol and the metabolism of amyloid-beta (Aβ). The functional attributes of *APOE*, including receptor binding capabilities, molecular stability, and overall functionality, are conditional based on its lipidation status [41]. In vitro model studies have shown a pivotal role of lipidation in preventing self-aggregation of *APOE* [42]. Given the considerable influence of lipidation on many roles of *APOE*, it has been proposed as a potential therapeutic treatment for AD. Hence, there is the possibility to correct, as well as prevent, certain outcomes associated with neurodegeneration. The benefit of increasing lipidation and reducing lipid-free availability may offer greater advantages to the individuals who carry the *APOEε4* allele, which accounts for a larger percentage of both AD populations and healthy control populations [43].

Another complementary study observed that pharmacologically promoting cholesterol efflux can increase myelination in vitro and in vivo and improve cognition in APOE4/e-TR mice. This finding indicates a link between cholesterol dysregulation and myelination in *APOEε4* carriers, which may impact the onset and severity of cognitive decline in AD.

Interventions such as pharmacological treatments, lifestyle, and dietary modifications aiming at restoring cholesterol equilibrium and myeline volume might help to increase the cognitive reserves in *APOEε4* carriers [44].

This proposal to augment APOE lipidation as a therapeutic approach shows the increasing understanding of the complex connection between lipid metabolism, APOE genetics, and AD pathogenesis. Further investigations are required to determine the practicality and effectiveness of using this approach in the clinical setting as a means to develop successful therapeutic interventions for AD.

Furthermore, the *APOEε4* allele has a strong lipid-binding affinity and a low recovery capacity, leading to impaired cholesterol efflux, culminating in an increased accumulation of cholesterol in cell membranes [45]. The distribution of elevated cholesterol levels on the plasma membrane of neurons correlated with increases in the metabolism of Aβ precursor protein (APP), which results in increased Aβ production [46]. In addition to neurons, astrocytes and microglia are also affected by impaired cholesterol efflux. In these cells, less cholesterol efflux reduces Aβ degradation, which may increase aggregation of Aβ into plaques [47].

4.5. HDL and Cholesterol Metabolism in APOE Non-Carriers

Longitudinal studies have shown that individuals with AD who are non-carriers of the *APOEε4* allele have elevated HDL levels. This elevation is associated with impaired cholesterol metabolism and impaired function, possibly resulting from reduced lipid availability in neuronal membranes [48]. Furthermore, in *APOEε4* allele non-carriers of AD-stratified populations, the enzyme 3-hydroxy-3-methylglutaryl-CoA synthetase was significantly associated with sporadic AD. This suggests potential cholesterol metabolic dysfunction in patients with AD who do not carry the *APOEε4* allele [49].

4.6. Heterogeneity and Implications in Clinical Practice

Subgroup analysis based on different populations yielded findings showing significant inter-group heterogeneity in patients with AD and the healthy controls, especially since the influence of the *APOEε4* allele on TC and LDL levels appears to be more pronounced in patients with AD than in the healthy control population. One important consideration is that TC and LDL in peripheral blood rarely enter the central nervous system (CNS). These lipids typically do not cross the blood–brain barrier in substantial amounts to cause harm to CNS function. Therefore, any effect of *APOE* alleles on peripheral TC and LDL levels may differ from their potential roles in the CNS. This raises an important question as to whether the influence of *APOE* alleles on peripheral lipid levels is related to the central pathological mechanism of AD. The exact nature of this relationship remains unclear, so it is an important area that warrants more comprehensive investigations. Therefore, more attention should be paid to AD in clinical practice and future studies, especially the lipid levels of patients with AD carrying the *APOEε4* allele.

Given the high degree of heterogeneity in this meta-analysis, we acknowledge that the exact cause of this variability has not been definitively identified despite performing sensitivity meta-regression and other analyses. We tried to exclude influential studies and found that heterogeneity could not be significantly reduced after re-analysis. This heterogeneity could be due to a combination of various factors, including different study populations, methodologies, and patient characteristics, such as age, sex, genetic background, medication use, ethnicity, and race.

4.7. Sex-Based Analysis

Sex-based analysis can provide more insight into how sex-specific hormonal factors interact with *APOE* alleles to modulate lipid profiles and AD risk differently in men and women. There are significant differences between males and females in the regulation of fatty acid metabolism. Premenopausal women tend to have higher levels of polyunsaturated fatty acids than men [31], which may be due to higher estrogen levels affecting lipid metabolism in premenopausal women [50]. Additionally, women in premenopausal, menopausal transition states have alterations in various body fats, which are also related to changes in their estrogen concentrations [51]. Decreased estrogen levels in postmenopausal women can affect lipid metabolism, which increases the risk of cognitive decline [52].

Females with one copy of the *APOEε4* allele had about four times the risk of AD, whereas males with one copy of the *APOEε4* allele had only twice the risk [53]. It is unknown whether there are differences in lipid metabolism between different *APOEε4* allele groups with different sexes. A study conducted in 2022 showed that within the AD population, both sexes showed high levels of TC and LDL compared with the control group. Notably, among female patients with AD, TC and LDL levels were significantly higher in *APOEε4* allele carriers than in non-carriers. In contrast, the presence of the *APOEε2* allele was linked to reduced TC levels in male patients with AD compared with non-carriers. This particular influence was not evident among male controls, female controls, or female AD populations. However, further prospective studies are required to confirm these findings [54].

In our study, owing to insufficient data, it was not possible to conduct subgroup analysis based on sex, age, and medication use to explore various causes of heterogeneity. It is worth further exploring the sex-based differences in lipid metabolism between different *APOEε4* allele groups and how these differences can influence AD risk.

4.8. Dual APOE4 and Lipid Profiles

In addition, this study used network meta-analyses to explore the effect of both single and dual *APOEε4* alleles on lipid profiles. Interestingly, the presence of dual *APOEε4* alleles did not increase the degree of influence on lipid profiles compared with a single APOEε4 allele. This finding negates the notion that having a higher genetic predisposition (possessing two *APOEε4* alleles) leads to more lipid-related impacts in AD.

4.9. Comparisons with Other Studies and What This Study Added to the Existing Knowledge

In contrast to previous meta-analyses that primarily examined the differences in lipids between AD and healthy controls, this study took a more focused approach. We investigated the difference in lipid levels between those carrying *APOEε4* allele-C and *APOEε4* allele-N within the context of AD. Thus, we were able to evaluate the specific influence of the *APOEε4* allele on lipids in AD, which adds novel knowledge to improve understanding of the complex interplay between genetics and lipid metabolism in AD pathogenesis.

4.10. Study Strengths and Limitations

This study is the first comprehensive analysis of the distinctive relationship between the *APOEε4* allele and lipids in patients with AD and healthy controls. The influence of the presence of the *APOEε4* allele on blood lipids, and the differences between single and dual *APOEε4* allele lipids, were analyzed using MeSH terms in meta-analysis, which is the strength of this study. However, this study has some limitations. First, since the data on age and the sex ratio of the *APOEε4* allele carriers and the non-carriers were insufficient, we could not conduct a deeper subgroup analysis stratified by age and sex. Second, despite our best efforts to contact the respective authors, some articles had incomplete data.

5. Conclusions

This meta-analysis showed that *APOEε4* allele-C carriers had higher TC and LDL levels than *APOEε4* allele-N carriers, and the difference was significant between patients with AD and healthy participants. The dual *APOEε4* allele may not have an increased effect on the lipid profiles. The effect of dyslipidemia and interventions on lipids levels in AD, especially in *APOEε4* allele carriers, should be extensively studied in the future. Currently, there are no therapies targeting APOE for AD treatment. These studies offer new insights for potential future AD treatments and provide a basis for precision medicine.

Supplementary Materials: The following supporting information can be downloaded: https://www.mdpi.com/article/10.3390/brainsci13111554/s1, Figure S1: random effect forest map of TC; Figure S2: random effect forest map of LDL; Figure S3: random effect forest map of TG; Figure S4: random effect forest map of HDL; Figure S5: sensitivity analysis for TC; Figure S6: sensitivity analysis for LDL; Figure S7: sensitivity analysis for TG; Figure S8: sensitivity analysis for HDL; Figure S9: funnel plot of TC; Figure S10: funnel plot of LDL; Figure S11: funnel plot of TG; Figure S12: funnel plot of HDL; Figure S13: fixed effect subgroup analysis of TC; Figure S14: fixed effect subgroup analysis of LDL; Figure S15: fixed effect subgroup analysis of TG; Figure S16: fixed effect subgroup analysis of HDL; Table S1: quality assessment of NOS; Table S2: Begg bias test and Egger's test of TC; Table S3: Begg bias test and Egger's test of LDL; Table S4: Begg bias test and Egger's test of TG; Table S5: Begg bias test and Egger's test of HDL; Table S6: SUCRA ranking of ApoE4 allele carrying, ApoE3/3, and ApoE4 allele carrying (TC); Table S7: SUCRA ranking of ApoE4 allele carrying, ApoE3/3, and ApoE4 allele carrying (TG); Table S8: SUCRA ranking of ApoE4 allele carrying, ApoE3/3, and ApoE4 allele carrying (LDL); Table S9: SUCRA ranking of ApoE4 allele carrying, ApoE3/3, and ApoE4 allele carrying (HDL); Table S10: SUCRA ranking of 6 genotypes (TC); Table S11: SUCRA ranking of 6 genotypes (TG).

Author Contributions: H.X., J.F., R.B.M.N. and J.Y. contributed to the literature search, data extraction, and manuscript preparation. S.C., Y.H. and T.B. verified all the data. H.X., J.F. and X.C. conceived this study and revised this article. All authors have read and agreed to the published version of the manuscript.

Funding: This research was funded by a grant from the Cadres Health Care Project in Sichuan Province (grant number: 2023–111).

Institutional Review Board Statement: Not applicable.

Informed Consent Statement: Not applicable.

Data Availability Statement: The original material presented in this study is included as Supplementary Material.

Acknowledgments: The authors thank all those who contributed to the maintenance of the database.

Conflicts of Interest: The authors declare no conflict of interest.

Abbreviations

APOE: apolipoprotein E; AD: Alzheimer's disease; TC: total cholesterol; LDL: low-density lipoprotein; TG: triglycerides; HDL: high-density lipoprotein; *APOEε4* allele-C: *APOEε4* allele carriers; *APOEε4* allele-N: *APOEε4* allele non-carriers; Aβ: amyloid-β; NOS Scale: the Newcastle–Ottawa Quality Assessment Scale; SMD: standardized mean difference; CI: confidence interval; SUCRA: the cumulative ranking curve; APP: Aβ precursor protein; VLDL: very low-density lipoprotein; CNS: the central nervous system.

References

1. Breijyeh, Z.; Karaman, R. Comprehensive review on Alzheimer's disease: Causes and treatment. *Molecules* **2020**, *25*, 5789. [CrossRef]
2. Poirier, J.; Miron, J.; Picard, C.; Gormley, P.; Théroux, L.; Breitner, J.; Dea, D. Apolipoprotein E and lipid homeostasis in the etiology and treatment of sporadic Alzheimer's disease. *Neurobiol. Aging* **2014**, *35* (Suppl. S2), S3–S10. [CrossRef]
3. Gremer, L.; Schölzel, D.; Schenk, C.; Reinartz, E.; Labahn, J.; Ravelli, R.B.G.; Tusche, M.; Lopez-Iglesias, C.; Hoyer, W.; Heise, H.; et al. Fibril structure of amyloid-β(1–42) by cryo-electron microscopy. *Science* **2017**, *358*, 116–119. [CrossRef] [PubMed]
4. Filippini, N.; MacIntosh, B.J.; Hough, M.G.; Goodwin, G.M.; Frisoni, G.B.; Smith, S.M.; Matthews, P.M.; Beckmann, C.F.; Mackay, C.E. Distinct patterns of brain activity in young carriers of the APOE-epsilon4 allele. *Proc. Natl. Acad. Sci. USA* **2009**, *106*, 7209–7214. [CrossRef]
5. Corder, E.H.; Saunders, A.M.; Strittmatter, W.J.; Schmechel, D.E.; Gaskell, P.C.; Small, G.W.; Roses, A.D.; Haines, J.L.; Pericak-Vance, M.A. Gene dose of apolipoprotein E type 4 allele and the risk of Alzheimer's disease in late onset families. *Science* **1993**, *261*, 921–923. [CrossRef] [PubMed]

6. Farrer, L.A.; Cupples, L.A.; Haines, J.L.; Hyman, B.; Kukull, W.A.; Mayeux, R.; Myers, R.H.; Pericak-Vance, M.A.; Risch, N.; van Duijn, C.M. Effects of age, sex, and ethnicity on the association between apolipoprotein E genotype and Alzheimer disease. A meta-analysis. APOE and Alzheimer disease meta analysis consortium. *JAMA* **1997**, *278*, 1349–1356. [CrossRef]
7. Stalmans, P.; Parys-Vanginderdeuren, R.; De Vos, R.; Feron, E.J. ICG staining of the inner limiting membrane facilitates its removal during surgery for macular holes and puckers. *Bull. Soc. Belge Ophtalmol.* **2001**, *281*, 21–26.
8. Wishart, H.A.; Saykin, A.J.; McAllister, T.W.; Rabin, L.A.; McDonald, B.C.; Flashman, L.A.; Roth, R.M.; Mamourian, A.C.; Tsongalis, G.J.; Rhodes, C.H. Regional brain atrophy in cognitively intact adults with a single APOE epsilon4 allele. *Neurology* **2006**, *67*, 1221–1224. [CrossRef] [PubMed]
9. Tarasoff-Conway, J.M.; Carare, R.O.; Osorio, R.S.; Glodzik, L.; Butler, T.; Fieremans, E.; Axel, L.; Rusinek, H.; Nicholson, C.; Zlokovic, B.V.; et al. Clearance systems in the brain-implications for Alzheimer disease. *Nat. Rev. Neurol.* **2015**, *11*, 457–470. [CrossRef]
10. Hall, K.; Murrell, J.; Ogunniyi, A.; Deeg, M.; Baiyewu, O.; Gao, S.; Gureje, O.; Dickens, J.; Evans, R.; Smith-Gamble, V.; et al. Cholesterol, APOE genotype, and Alzheimer disease: An epidemiologic study of Nigerian Yoruba. *Neurology* **2006**, *66*, 223–227. [CrossRef] [PubMed]
11. Hoshino, T.; Kamino, K.; Matsumoto, M. Gene dose effect of the APOE-epsilon4 allele on plasma HDL cholesterol level in patients with Alzheimer's disease. *Neurobiol. Aging* **2002**, *23*, 41–45. [CrossRef]
12. Isbir, T.; Agaçhan, B.; Yilmaz, H.; Aydin, M.; Kara, I.; Eker, E.; Eker, D. Apolipoprotein-E gene polymorphism and lipid profiles in Alzheimer's disease. *Am. J. Alzheimer's Dis. Other Dement.* **2001**, *16*, 77–81. [CrossRef] [PubMed]
13. Raygani, A.V.; Rahimi, Z.; Kharazi, H.; Tavilani, H.; Pourmotabbed, T. Association between apolipoprotein E polymorphism and serum lipid and apolipoprotein levels with Alzheimer's disease. *Neurosci. Lett.* **2006**, *408*, 68–72. [CrossRef]
14. Shafagoj, Y.A.; Naffa, R.G.; El-Khateeb, M.S.; Abdulla, Y.L.; Al-Qaddoumi, A.A.; Khatib, F.A.; Al-Motassem, Y.F.; Al-Khateeb, E.M. APOE Gene polymorphism among Jordanian Alzheimer's patients with relation to lipid profile. *Neurosciences* **2018**, *23*, 29–34. [CrossRef] [PubMed]
15. Sabbagh, M.N.; Sandhu, S.; Kolody, H.; Lahti, T.; Silverberg, N.B.; Sparks, D.L. Studies on the effect of the apolipoprotein E genotype on the lipid profile in Alzheimer's disease. *Curr. Alzheimer Res.* **2006**, *3*, 157–160. [CrossRef] [PubMed]
16. Anstey, K.J.; Lipnicki, D.M.; Low, L.F. Cholesterol as a risk factor for dementia and cognitive decline: A systematic review of prospective studies with meta-analysis. *Am. J. Geriatr. Psychiatry* **2008**, *16*, 343–354. [CrossRef]
17. Tang, Q.; Wang, F.; Yang, J.; Peng, H.; Li, Y.; Li, B.; Wang, S. Revealing a novel landscape of the association between blood lipid levels and Alzheimer's disease: A meta-analysis of a case-control study. *Front. Aging Neurosci.* **2019**, *11*, 370. [CrossRef]
18. Wang, P.; Zhang, H.; Wang, Y.; Zhang, M.; Zhou, Y. Plasma cholesterol in Alzheimer's disease and frontotemporal dementia. *Transl. Neurosci.* **2020**, *11*, 116–123. [CrossRef]
19. Tugwell, G.W.P.; Shea, B.; O'Connell, D.; Welch, J.P.V.; Losos, M. The Newcastle-Ottawa Scale (NOS) for Assessing the Quality of Nonrandomised Studies in Meta-Analyses. 2021. Available online: https://www.ohri.ca/programs/clinical_epidemiology/oxford.asp (accessed on 5 January 2022).
20. Fernandes, M.A.; Proença, M.T.; Nogueira, A.J.; Oliveira, L.M.; Santiago, B.; Santana, I.; Oliveira, C.R. Effects of apolipoprotein E genotype on blood lipid composition and membrane platelet fluidity in Alzheimer's disease. *Biochim. Biophys. Acta* **1999**, *1454*, 89–96. [CrossRef] [PubMed]
21. Wehr, H.; Parnowski, T.; Puzyński, S.; Bednarska-Makaruk, M.; Bisko, M.; Kotapka-Minc, S.; Rodo, M.; Wołkowska, M. Apolipoprotein E genotype and lipid and lipoprotein levels in dementia. *Dement. Geriatr. Cogn. Disord.* **2000**, *11*, 70–73. [CrossRef]
22. Sheng, B. *Preliminary Studies on Genes Associated with Alzheimer's Disease*; Jilin University: Jilin, China, 2000.
23. Cui, J.; Wang, J.; Guo, L. Effects of apolipoprotein E genotype on plasma apolipoprotein E, total cholesterol and triglyceride levels in patients with Alzheimer's disease. *Shanghai Med.* **2002**, *25*, 444–446. [CrossRef]
24. Zeng, X.; Qin, B.; Guo, H. Effect of apolipoprotein E gene on lipid metabolism in patients with Alzheimer's disease. *Chin. J. Psychiatry* **2002**, *3*. [CrossRef]
25. Al-Shammari, S.; Fatania, H.; Al-Radwan, R.; Akanji, A.O. Apolipoprotein E polymorphism and lipoprotein levels in a Gulf Arab population in Kuwait: A pilot study. *Ann. Saudi Med.* **2004**, *24*, 361–364. [CrossRef] [PubMed]
26. Li, D.; Wang, Y.; Yang, J. Study on the association between APOE gene polymorphism and Alzheimer's disease. *J. Mod. Lab. Med.* **2008**, *23*, 20–23. [CrossRef]
27. Singh, N.K.; Chhillar, N.; Banerjee, B.D.; Bala, K.; Mukherjee, A.K.; Mustafa, M.D.; Mitrabasu. Gene-environment interaction in Alzheimer's disease. *Am. J. Alzheimer's Dis. Other Demen.* **2012**, *27*, 496–503. [CrossRef]
28. Dai, T. *Relationship between APOE Gene Polymorphism and Cognitive Function in Sporadic Alzheimer's Disease*; Central South University: Shenzhen, China, 2012.
29. Liang, J.; Kabinver; Ke, Y. Study on the relationship between lipid levels and apolipoprotein E gene polymorphism in patients with sporadic Alzheimer's disease in Uygurs and Han nationalities. *Chin. J. Clin. Health Care* **2013**, *16*, 565–568.
30. Bian, M. Study on Apolipoprotein E Gene Polymorphism and Lipid Level in Alzheimer's Disease Patients. Master's Thesis, Dalian Medical University, Dalian, China, 2018.

31. Lim, W.L.F.; Huynh, K.; Chatterjee, P.; Martins, I.; Jayawardana, K.S.; Giles, C.; Mellett, N.A.; Laws, S.M.; Bush, A.I.; Rowe, C.C.; et al. Relationships between plasma lipids species, gender, risk factors, and Alzheimer's disease. *J. Alzheimer's Dis.* **2020**, *76*, 303–315. [CrossRef]

32. Loera-Valencia, R.; Goikolea, J.; Parrado-Fernandez, C.; Merino-Serrais, P.; Maioli, S. Alterations in cholesterol metabolism as a risk factor for developing Alzheimer's disease: Potential novel targets for treatment. *J. Steroid Biochem. Mol. Biol.* **2019**, *190*, 104–114. [CrossRef] [PubMed]

33. Huang, S.-Y.; Yang, Y.-X.; Zhang, Y.-R.; Kuo, K.; Li, H.-Q.; Shen, X.-N.; Chen, S.-D.; Chen, K.-L.; Dong, Q.; Tan, L.; et al. Investigating Causal Relations Between Circulating Metabolites and Alzheimer's Disease: A Mendelian Randomization Study. *J. Alzheimer's Dis.* **2022**, *87*, 463–477. [CrossRef]

34. Agarwal, M.; Khan, S. Plasma Lipids as Biomarkers for Alzheimer's Disease: A Systematic Review. *Cureus* **2020**, *12*, e12008. [CrossRef]

35. Strittmatter, W.J.; Saunders, A.M.; Schmechel, D.; Pericak-Vance, M.; Enghild, J.; Salvesen, G.S.; Roses, A.D. Apolipoprotein E: High-avidity binding to beta-amyloid and increased frequency of type 4 allele in late-onset familial Alzheimer disease. *Proc. Natl. Acad. Sci. USA* **1993**, *90*, 1977–1981. [CrossRef]

36. Liu, C.C.; Liu, C.C.; Kanekiyo, T.; Xu, H.; Bu, G. Apolipoprotein E and Alzheimer disease: Risk, mechanisms and therapy. *Nat. Rev. Neurol.* **2013**, *9*, 106–118. [CrossRef] [PubMed]

37. Hu, J.; Liu, C.C.; Chen, X.F.; Zhang, Y.W.; Xu, H.; Bu, G. Opposing effects of viral mediated brain expression of apolipoprotein E2 (apoE2) and apoE4 on apoE lipidation and Aβ metabolism in apoE4-targeted replacement mice. *Mol. Neurodegener.* **2015**, *10*, 6. [CrossRef]

38. Kanekiyo, T.; Xu, H.; Bu, G. ApoE and Aβ in Alzheimer's disease: Accidental encounters or partners? *Neuron* **2014**, *81*, 740–754. [CrossRef]

39. Nguyen, D.; Dhanasekaran, P.; Nickel, M.; Nakatani, R.; Saito, H.; Phillips, M.C.; Lund-Katz, S. Molecular basis for the differences in lipid and lipoprotein binding properties of human apolipoproteins E3 and E4. *Biochemistry* **2010**, *49*, 10881–10889. [CrossRef] [PubMed]

40. Morrow, J.A.; Segall, M.L.; Lund-Katz, S.; Phillips, M.C.; Knapp, M.; Rupp, B.; Weisgraber, K.H. Differences in stability among the human apolipoprotein E isoforms determined by the amino-terminal domain. *Biochemistry* **2000**, *39*, 11657–11666. [CrossRef] [PubMed]

41. Mahley, R.W.; Rall, S.C. Apolipoprotein E: Far more than a lipid transport protein. *Annu. Rev. Genomics Hum. Genet.* **2000**, *1*, 507–537. [CrossRef] [PubMed]

42. Hubin, E.; Verghese, P.B.; van Nuland, N.; Broersen, K. Apolipoprotein E associated with reconstituted high-density lipoprotein-like particles is protected from aggregation. *FEBS Lett.* **2019**, *593*, 1144–1153. [CrossRef]

43. Fernández-Calle, R.; Konings, S.C.; Frontiñán-Rubio, J.; García-Revilla, J.; Camprubí-Ferrer, L.; Svensson, M.; Martinson, I.; Boza-Serrano, A.; Venero, J.L.; Nielsen, H.M.; et al. APOE in the bullseye of neurodegenerative diseases: Impact of the APOE genotype in Alzheimer's disease pathology and brain diseases. *Mol. Neurodegener.* **2022**, *17*, 62. [CrossRef]

44. Blanchard, J.W.; Akay, L.A.; Davila-Velderrain, J.; von Maydell, D.; Mathys, H.; Davidson, S.M.; Effenberger, A.; Chen, C.Y.; Maner-Smith, K.; Hajjar, I.; et al. APOE4 impairs myelination via cholesterol dysregulation in oligodendrocytes. *Nature* **2022**, *611*, 769–779. [CrossRef]

45. Yassine, H.N.; Finch, C.E. APOE alleles and diet in brain aging and Alzheimer's disease. *Front. Aging Neurosci.* **2020**, *12*, 150. [CrossRef] [PubMed]

46. Cui, W.; Sun, Y.; Wang, Z.; Xu, C.; Xu, L.; Wang, F.; Chen, Z.; Peng, Y.; Li, R. Activation of liver X receptor decreases BACE1 expression and activity by reducing membrane cholesterol levels. *Neurochem. Res.* **2011**, *36*, 1910–1921. [CrossRef]

47. Prasad, H.; Rao, R. Amyloid clearance defect in ApoE4 astrocytes is reversed by epigenetic correction of endosomal pH. *Proc. Natl. Acad. Sci. USA* **2018**, *115*, E6640–E6649. [CrossRef] [PubMed]

48. Sacks, D.; Baxter, B.; Campbell, B.C.V.; Carpenter, J.S.; Cognard, C.; Dippel, D.; Eesa, M.; Fischer, U.; Hausegger, K.; Hirsch, J.A.; et al. Multisociety Consensus Quality Improvement Revised Consensus Statement for Endovascular Therapy of Acute Ischemic Stroke. *Int. J. Stroke* **2018**, *13*, 612–632. [CrossRef] [PubMed]

49. Leduc, V.; De Beaumont, L.; Théroux, L.; Dea, D.; Aisen, P.; Petersen, R.C.; Alzheimer's Disease Neuroimaging Initiative; Dufour, R.; Poirier, J. HMGCR is a genetic modifier for risk, age of onset and MCI conversion to Alzheimer's disease in a three cohorts study. *Mol. Psychiatry* **2015**, *20*, 867–873. [CrossRef]

50. Palmisano, B.T.; Zhu, L.; Eckel, R.H.; Stafford, J.M. Sex differences in lipid and lipoprotein metabolism. *Mol. Metab.* **2018**, *15*, 45–55. [CrossRef] [PubMed]

51. Ko, S.H.; Kim, H.S. Menopause-associated lipid metabolic disorders and foods beneficial for postmenopausal women. *Nutrients* **2020**, *12*, 202. [CrossRef] [PubMed]

52. Ancelin, M.L.; Ripoche, E.; Dupuy, A.M.; Samieri, C.; Rouaud, O.; Berr, C.; Carrière, I.; Ritchie, K. Gender-specific associations between lipids and cognitive decline in the elderly. *Eur. Neuropsychopharmacol.* **2014**, *24*, 1056–1066. [CrossRef]

53. Altmann, A.; Tian, L.; Henderson, V.W.; Greicius, M.D.; Alzheimer's Disease Neuroimaging Initiative Investigators. Sex modifies the APOE-related risk of developing Alzheimer disease. *Ann. Neurol.* **2014**, *75*, 563–573. [CrossRef]
54. Fu, J.; Huang, Y.; Bao, T.; Ou, R.; Wei, Q.; Chen, Y.; Yang, J.; Chen, X.; Shang, H. Effects of sex on the relationship between apolipoprotein E gene and serum lipid profiles in Alzheimer's disease. *Front. Aging Neurosci.* **2022**, *14*, 844066. [CrossRef]

Disclaimer/Publisher's Note: The statements, opinions and data contained in all publications are solely those of the individual author(s) and contributor(s) and not of MDPI and/or the editor(s). MDPI and/or the editor(s) disclaim responsibility for any injury to people or property resulting from any ideas, methods, instructions or products referred to in the content.

Review

Thirty Risk Factors for Alzheimer's Disease Unified by a Common Neuroimmune–Neuroinflammation Mechanism

Donald F. Weaver

Krembil Research Institute, University Health Network, Departments of Medicine, Chemistry, Pharmaceutical Sciences, University of Toronto, Toronto, ON M5T 0S8, Canada; donald.weaver@uhnresearch.ca

Abstract: One of the major obstacles confronting the formulation of a mechanistic understanding for Alzheimer's disease (AD) is its immense complexity—a complexity that traverses the full structural and phenomenological spectrum, including molecular, macromolecular, cellular, neurological and behavioural processes. This complexity is reflected by the equally complex diversity of risk factors associated with AD. However, more than merely mirroring disease complexity, risk factors also provide fundamental insights into the aetiology and pathogenesis of AD as a neurodegenerative disorder since they are central to disease initiation and subsequent propagation. Based on a systematic literature assessment, this review identified 30 risk factors for AD and then extended the analysis to further identify neuroinflammation as a unifying mechanism present in all 30 risk factors. Although other mechanisms (e.g., vasculopathy, proteopathy) were present in multiple risk factors, dysfunction of the neuroimmune–neuroinflammation axis was uniquely central to all 30 identified risk factors. Though the nature of the neuroinflammatory involvement varied, the activation of microglia and the release of pro-inflammatory cytokines were a common pathway shared by all risk factors. This observation provides further evidence for the importance of immunopathic mechanisms in the aetiopathogenesis of AD.

Keywords: Alzheimer's disease; dementia; neurodegeneration; neuroinflammation; neuroimmune; microglia; cytokine

Citation: Weaver, D.F. Thirty Risk Factors for Alzheimer's Disease Unified by a Common Neuroimmune–Neuroinflammation Mechanism. *Brain Sci.* **2024**, *14*, 41. https://doi.org/10.3390/brainsci 14010041

Academic Editors: Junhui Wang, Hongxing Wang and Jing Sun

Received: 29 November 2023
Revised: 27 December 2023
Accepted: 30 December 2023
Published: 31 December 2023

Copyright: © 2023 by the author. Licensee MDPI, Basel, Switzerland. This article is an open access article distributed under the terms and conditions of the Creative Commons Attribution (CC BY) license (https://creativecommons.org/licenses/by/4.0/).

1. Introduction

The brain is the human body's most complex and convoluted organ, and neurodegenerative disorders such as Alzheimer's disease (AD) are arguably amongst the most complex diseases of the brain. One of the major hurdles encountered when formulating a mechanistic understanding with which to facilitate management strategies for AD is its immense complexity—a complexity that traverses the full structural and phenomenological spectrum, including molecular, macromolecular, cellular, behavioural and neurological processes [1,2]. AD risk factors are excellent examples of this immense complexity; these risk factors include such bewilderingly diverse conditions as medical diseases (diabetes), psychiatric disorders (depression), personal injuries (head trauma), societal factors (social isolation) and environmental issues (air pollution).

To identify a harmonizing mechanistic explanation with which to unify the many and varied risk factors for AD, a comprehensive literature review was initially completed (in PubMed, Web of Science, Scopus and Google Scholar databases including publications dating up to November 2023) and identified 30 "risk" factors for AD, employing a broad definition of "risk factor": some are modifiable risk factors connected in a causative manner with AD (e.g., smoking, alcohol abuse, obesity); others are concomitant disorders occurring as co-morbidities (e.g., glaucoma; people with glaucoma are at risk for also developing AD); others are bidirectional risk factors (e.g., chronic pain causes neuroinflammation, which is a risk factor for AD, yet the neuroinflammation is a positive feedback risk factor for continuing pain). This comprehensive list of 30 risk factors includes the 12 modifiable risk factors identified in the 2020 Lancet commission (air pollution, alcohol abuse, brain injury, depression, diabetes, hearing impairment/deafness, hypertension, lower educational level, obesity, physical inactivity/sedentary lifestyle, smoking and social isolation) [3]. Beyond these 12 *Lancet* commission risk factors, 18 additional factors have been added, which include well-recognized risk factors that are non-modifiable (e.g., age, sex), risk factors that are modifiable but not at the personal level (e.g., climate change), concomitant co-morbidities as risks (e.g., glaucoma, migraine) and other newer factors for which convincing data are emerging but they remain less conclusive (e.g., oral hygiene, allergies).

Next, all literature sources discussing the 30 identified risk factors were searched for common terms providing mechanistic explanations. The term uniting all 30 risk factors was "neuroinflammation", where neuroinflammation is defined as a functional process of the brain's innate immune system following activation by diverse external (physical trauma, toxin (microbiological, chemical)) and/or internal (ischaemia) challenges, and manifesting as integrated cellular (microglial) and molecular (especially cytokine: e.g., Interleukin (IL)-1β, IL-6 and Tumour Necrosis Factor (TNF)-α) alterations within the brain [4,5]. Since many studies provide data strongly implicating neuroinflammation as a significant contributor and culprit in the aetiopathogenesis of AD, a shared neuroimmune–neuroinflammation mechanism clearly emerges as a unifying thread providing harmonization within the rich tapestry of diverse risk factors associated with AD.

Herein, an overview of the neuroimmune–neuroinflammation axis as related to AD is presented followed by a consideration of the 30 risk factors for AD in conjunction with a description of their neuroinflammatory mechanisms (Figure 1).

Neuroimmune–Neuroinflammatory Contributions to Alzheimer's Disease

Traditionally, AD has been regarded as a proteopathy (i.e., protein-based disorder) arising from the misfolding and oligomerization of β-amyloid (Aβ) and tau. Regrettably, however, this conceptualization has failed to yield a definitive curative therapy, thereby necessitating the need to explore other mechanistic approaches, including immunopathy, gliopathy, mitochondriopathy, membranopathy, synaptotoxicity, metal dyshomeostasis and oxidative damage, reflecting the biochemical complexity and heterogeneity of AD. Of these mechanisms, immunopathy is emerging as a lead contender [6–10].

Not surprisingly, an immunopathic mechanistic explanation of AD is likewise complex and involves a host of cellular (microglia) and molecular (cytokine) participants. Emerging data indicate that the homeostatic balance between pro-inflammatory and anti-inflammatory processes becomes disordered over the time duration of the disease, ultimately tilting towards a neuropathic pro-inflammatory milieu and manifesting with increased concentrations of activated microglia and pro-inflammatory cytokines (IL-1β, IL-6, TNFα) [6–8]. During the initial pre-symptomatic phases of the disease, immune processes are neuroprotective with microglia-mediated phagocytosis of cytotoxic Aβ aggregates. However, as the disease progresses, such neuroprotective effects are supplanted by neuro-toxic effects with elevated pro-inflammatory processes. These neurotoxic pro-inflammatory effects occur within the context of both innate immunity and adaptive immunity, with deleterious neuroinflammation arising primarily from the actions of prolonged innate immunity activity. Neuroinflammation involves reactive, pro-inflammatory microglia and astrocytic phenotypes, which paradoxically enhance Aβ oligomerization and promote tau hyperphosphorylation, complement activation and the catabolism of neurotransmitters and brain biomolecules into neurotoxic metabolites—changes which both initiate and/or propagate neurodegeneration, heralding cognitive reduction and dementia in susceptible (usually geriatric) adults. Since the neurotoxicity of excessive pro-inflammatory processes occurs not only at the level of innate immunity, via neuroinflammation, but also at the level of adaptive immunity, the neuropathological mechanisms of neuronal death involve both auto-inflammatory and autoimmune mechanisms. Additional support for the ae-tiopathogenic role of the neuroimmune–neuroinflammation axis AD comes from genetic studies: genome-wide association studies (GWASs) reveal that multiple polymorphisms associated with AD occur in genes that regulate innate immune function (e.g., CD33, CLU, CR1, TREM-2), which encode proteins that regulate complement activation and cellular phagocytic activities [9]. Thus, although inflammation, in general, is a non-specific response to many different types of injury, within the specific context of AD, the neuroimmune–neuroinflammation axis is a key contributor to disease pathogenesis and progression; accordingly, factors that affect the biochemistry or histology of this axis emerge as risk factors for AD [10].

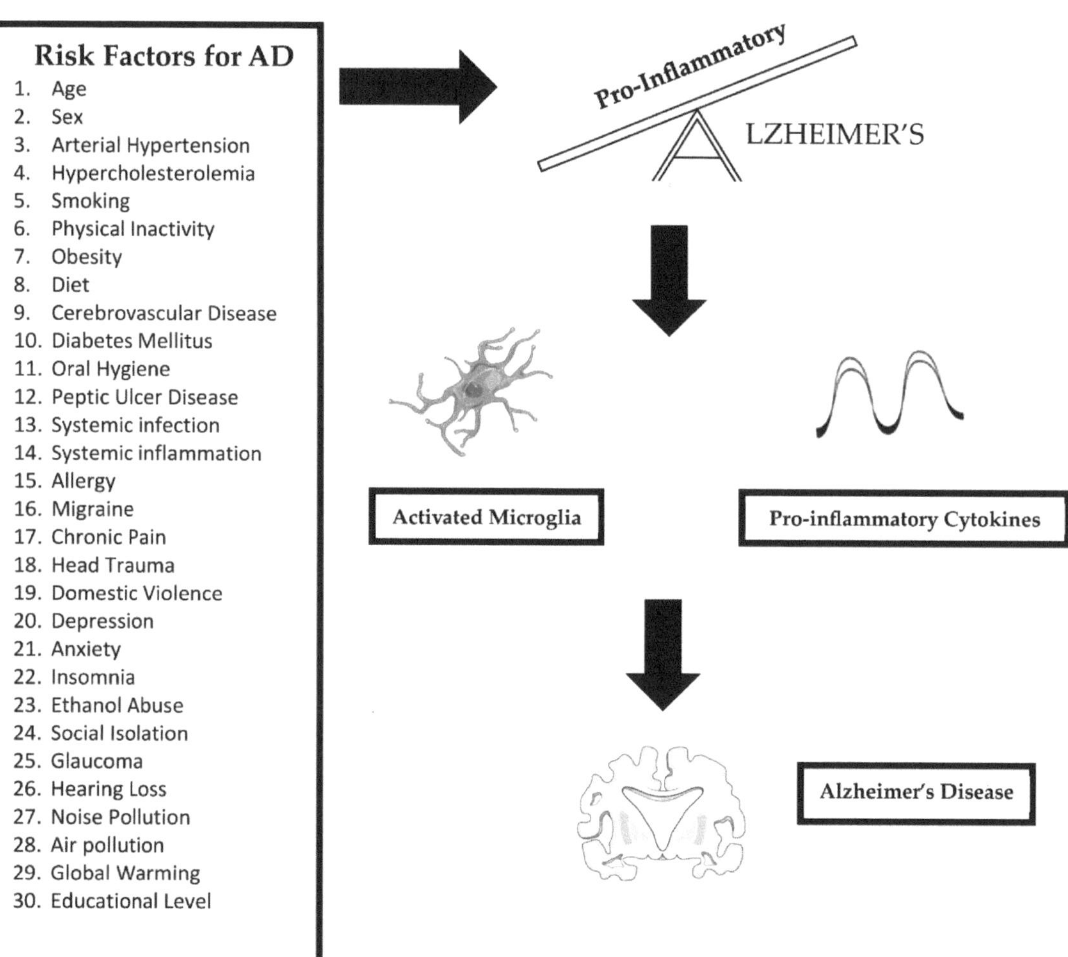

Figure 1. Thirty risk factors for Alzheimer's disease: traditionally AD was regarded as a proteopathic disease arising from protein misfolding and aggregation; however, immunopathy also contributes to AD particularly as an excessive pro-inflammatory innate immune response. The 30 very diverse risk factors for AD identified in this review are uniquely unified by their common ability to elicit neuroinflammation, manifesting as microglial activation and pro-inflammatory cytokine (IL-1β, IL-6, TNFα) release, ultimately causing neuronal loss and brain atrophy thereby contributing to the pathogenesis of the disease. These risk factors cause an imbalance in immune homeostasis triggering excessive pro-inflammatory activities which are neurotoxic.

2. Thirty Risk Factors

2.1. Age

Although AD is not a normal long-term outcome of aging, age is regarded as the best established risk factor for the disease. The number of people living with AD doubles every five years after age 65 years; 40% of people aged 90 years and older have AD [11]. In the preponderance of people diagnosed with AD, symptoms onset after they reach their mid-60 s in age or even later. When the disease manifests clinically before age 65, it is regarded as uncommon.

The links between aging and AD are many and complex; however, neuroinflammation is a key component of this link, with aging being associated with neuroinflammation and neuroinflammation being associated with AD. D'Avila et al. established that aged mice exhibit dystrophic pro-inflammatory microglia in the entorhinal cortex and hippocampus within the medial temporal lobe [12]. Aged mice also release higher levels of pro-inflammatory (IL-1β and IL-6) cytokines in the brain and higher levels of NADPH oxidase 2 (Nox2) expression compared to younger animals [13].

In humans, aging and a chronic inflammatory state frequently co-exist in the periphery and in the brain. Aging impairs functional interactions between the brain and the immune system; microglia and astrocytes, functioning in their capacity as innate immune cells, become more pro-inflammatory during aging [14]. This age-associated increase in innate immune reactivity heralds an augmented inflammatory cytokine brain response after activation of the innate immune system during the initiation and progression of AD, leading to more severe long-lasting behavioural and cognitive deficits.

2.2. Sex

After age, sex is the other most commonly cited risk factor for AD. Women are more likely to develop dementia over the course of their life (even after greater longevity is considered); twice as many women have AD compared to men. A Swedish study by Beam et al. followed 16,926 people and noted that commencing at age 80 years, women are more likely to be identified as having AD than men at corresponding ages [15]. An analogous Taiwanese study by Liu et al. concluded that the likelihood of developing AD throughout a seven-year time duration was greater in women compared to men [16]. Finally, a meta-analysis by Niu et al. studying the European incidence of AD calculated that, annually, 13 women out of 1000 developed AD, compared to only 7 men [17].

Immune-mediated neuroinflammatory responses are different between men and women. Women are more susceptible to inflammatory pathological consequences than men via neuroimmune alterations, including microglial activation, pro-inflammatory cytokine expression and dysinformational synaptic plasticity [18]. In a study involving injecting volunteers with immunogenic lipopolysaccharides (LPSs), Engler et al. ascertained that women undergo a significantly enhanced pro-inflammatory response, with higher circulating levels of TNFα and IL-6; conversely, the LPS-triggered rise in anti-inflammatory IL-10 was significantly greater in men [19]. Finally, women constitute >80% of all diagnoses of autoimmunity, particularly as demonstrated by differences in the incidence for Sjögren syndrome, systemic lupus erythematosus, Hashimoto thyroiditis, Graves' disease, scleroderma and myasthenia gravis [20]; Meier-Stephenson et al. argued that AD is an autoimmune disease. Such sex-based neuroimmune differences provide a possible explanation for the corresponding sex differences in the incidence and prevalence of AD [21].

2.3. Arterial Hypertension

Hypertension is a well-documented and accepted risk factor for AD. Multiple studies have concluded the existence of a correlation between cognitive decline and systemic arterial hypertension in different age cohorts [22,23]. Systemic arterial hypertension, particularly midlife high blood pressure (BP), has been related to a higher risk of dementia, including AD. In the middle years of life (age 40–64 years), there is a positively correlated relationship between BP elevation and cognitive dysfunction in AD, whilst in elderly populations (age $\geq$ 65 years), this relationship is more controversial, with hypotension being deleterious to intellectual function.

Not surprisingly, the link between hypertension and AD is multifactorial, with vascular factors playing a major contributing role. However, neuroinflammation is another major mechanism linking hypertension and AD [24]. Animal studies have established that prolonged BP elevation culminates in neurotoxic glial activation and increased cerebral inflammatory mediators, particularly pro-inflammatory cytokines, such as TNFα and IL-1β. Solé-Guardia et al. observed that individuals experiencing chronic hypertension had an enhanced neuroinflammatory response, manifesting as augmented microglial activation and astrogliosis and more apparent perivascular inflammation compared to non-hypertensives [25]. Carnevale et al. showed that hypertension induced microglial activation, and interleukin IL-1β upregulation triggers neuroinflammation before Aβ deposition [26].

2.4. Hypercholesterolemia

Dysregulated cholesterol biosynthesis and metabolism constitute a risk factor for AD and multiple other diseases. In vivo and human-based investigations have concluded that a high-cholesterol diet (HCD) induces A. In rats and mice, HCD produces significant cognitive decline and AD-like disease [27,28]; in Japanese white rabbits on an HCD, alterations in brain structure and function analogous to those of human AD were noted [29]. Epidemiological investigations have also suggested a relationship between hypercholesterolemia and AD [30]. Xu et al. suggested that high cholesterol levels were associated with increased AD pathology severity, and that the mechanism for this enhanced pathology is not entirely mediated by cerebrovascular conditions [31]. Thus, mounting evidence indicates that excessive cholesterol accumulates in AD, driving AD-associated pathological changes, and that hypercholesterolemia promotes AD development as a risk factor, especially with elevated cholesterol levels in the middle years of life.

As with hypertension, the link between hypercholesterolemia and AD is multifactorial, with vascular factors playing a major contributing role. However, neuroinflammation is another major mechanism linking hypercholesterolemia and AD [32]. For example, Thirumangalakudi et al. demonstrated that hyperlipidemic mice showed increased expression of pro-inflammatory microglia and cytokines/mediators, including TNFα, IL-1β, IL-6, NOS2 (Nitric oxide synthase 2) and COX2 (Cyclooxygenase-2) [33]. Chen et al. also showed, in mice, that a high-cholesterol diet enhanced pro-inflammatory NLRP3 (NLR family pyrin domain containing 3) inflammasome activation and IL-1β expression [34].

2.5. Smoking

Based on a comprehensive review, Durazzo et al. concluded that smoking tobacco products gives rise to a significantly intensified risk for AD and dementia [35]. Cigarette smoke/smoking is associated with AD neuropathology in both preclinical models and human studies. Jeong et al. showed that smoking discontinuation resulted in a reduced risk of dementia [36].

The negative consequences of smoking are numerous, providing multiple mechanisms by which smoking contributes to pathology. However, immune-based inflammation is a significant contributor to this pathology. Alrouji et al. concluded that smoking inflicts complex immunological effects, which include enhancements in inflammatory responses (activated microglia with increased pro-inflammatory cytokine responses) with a concomitant lessening of immune defences, causing an increased vulnerability to the deleterious effects of a chronic ongoing pro-inflammatory environment [37]. In a case–control study, Liu et al. found that cigarette smoking was associated with elevated concentrations of at-risk biomarkers for AD, as indicated by higher neuroinflammation biomarkers in the cerebrospinal fluid of participants in the active smoking group [38].

2.6. Physical Inactivity

Based on a comprehensive literature review, Meng et al. concluded that physical inactivity is one of the most readily addressable and avoidable risk factors for AD and that improved physical activity levels are linked to a diminished risk of AD development [39]. Physical exercise is also helpful in improving behavioural and psychiatric symptomatic indicators of AD, notably via better cognitive function. Chen et al., likewise, concluded that physical exercise is important in the prevention of AD, providing non-pharmacological treatment options [40].

The case correlating physical inactivity with AD via a neuroinflammatory mechanism is strong. Recently, Wang et al. demonstrated that exercise ameliorates AD by directly and indirectly regulating brain immune responses and promoting hippocampal neurogenesis [41]. Similarly, Seo et al. showed that neuroinflammation-mediated microglia activation with pro-inflammatory cytokine release is enhanced by physical inactivity and downregulated by exercise [42]. Svensson and co-workers likewise showed that exercise leads to the elevated biosynthesis and release of anti-inflammatory cytokines, and lower concentrations of pro-inflammatory cytokines and activated microglia [43].

2.7. Obesity

Obese people exhibit a higher risk of acquiring age-correlated cognitive reduction, mild cognitive impairment and AD [44,45]. An association between body mass index (BMI) and AD has been described, with multiple groups studying the relationship between elevated BMI and AD. Obesity is, thus, a recognized risk factor for AD [46–49].

Miller and Spencer suggested that neuroinflammation is the linkage that unites obesity with AD; obesity (and high fat consumption) culminates in systemic inflammation as well as elevated levels of circulating free fatty acids and inflammatory mediators. These circulating cytokines and activated immune cells reach the brain and initiate local neuroinflammation, including microglial proliferation and causing synaptic remodelling and neurodegeneration [50]. Similarly, Henn et al. also suggested that immune dysregulation, including inflammaging (e.g., age-related increase in the levels of cytotoxic pro-inflammatory biomarkers in blood and tissues) and immunosenescence (e.g., age-related reduction in the efficacy of immune system function), commonly occur prematurely as a consequence of obesity, promoting cognitive impairment and AD [51].

2.8. Dietary Factors

In recent years, numerous studies have confirmed that, especially with advancing age, diet affects cognitive capacities and ultimate susceptibility to AD. In a systematic search of randomized clinical trials, reviews and meta-analyses evaluating the connection between diet and AD, Xu Lou et al. examined 38 studies and concluded that a Western diet pattern is a risk factor for AD, whereas the Mediterranean diet, ketogenic diet and supplementation with omega-3 fatty acids and probiotics are potentially neuroprotective diets [52]. The Mediterranean diet, the related MIND diet (which incorporates constituents designed to lower blood pressure) and other healthful dietary regimens are associated with cognitive benefits in studies and a decreased probability of AD [53–56].

Kip and Parr-Brownlie noted that many dietary risks factors are linked to AD-promoting neuroinflammation, particularly high saturated and trans-fat intake; indeed, dietary modifications in mice can influence levels of pro-inflammatory microglia and cytokines [57]. Conversely, dietary restriction (DR) has been shown to diminish age-related pro-inflammatory activation of microglia, astrocytes and cytokines while prolonging lifespan in various organisms [58,59].

The microbiome also plays an essential role in the link between diet and AD. Dietary changes (either deleterious or beneficial) influence the microbiome composition, thereby altering the gut–brain homeostatic axis with the release of pro-inflammatory bacterial metabolites, which predispose people to AD progression [60].

2.9. Cerebrovascular Disease

Cerebrovascular disease, manifesting as cerebral atherosclerosis and arteriolosclerosis, is a risk factor associated with AD; thus, cerebral vasculopathy is a pervasive risk factor for both vascular dementia and AD [61]. A number of midlife vascular risk factors are significantly associated with AD—findings consistent with a role of vascular disease in the development of AD [62]. Stroke is a common pathology associated with AD among elderly individuals—a co-morbid relationship at its fullest when accompanied by a plethora of commonly acknowledged vascular risk factors [63]. Vascular risk factors associated with AD include the conventionally recognized risk factors (hypertension, cholesterolemia) which contribute to atherosclerotic vascular change, as well as amyloid angiopathy, in which amyloid deposits in the walls of small to medium cerebral blood vessels lead to microhaemorrhages with consequent neurologic deficits, which may include impairments in memory or cognition.

Beyond the obvious vascular contributions (ischaemia, hypoxemia) to dementia, neurotoxic brain inflammation (pro-inflammatory microglia and cytokines) accompanies the ischaemic conditions of cerebrovascular disease, thereby contributing to AD pathogenesis [64–66]. Jurcau and Simion showed that neuroinflammatory mechanisms significantly contribute to neuronal injury during cerebral ischemia, ultimately further increasing the extent of cerebral damage and neurological deficits in AD [67].

2.10. Diabetes Mellitus

Numerous studies have shown that people with diabetes, especially Type 2 Diabetes, are at higher risk for AD; indeed, AD has even been referred to as Type 3 Diabetes [68–70]. Among people with diabetes, the risk for AD is 65% higher than in non-diabetic controls. Conversely, but analogously, in people with AD, the prevalence of diabetes is higher than anticipated, approaching 35%. An even greater number of people with AD (46%) have glucose intolerance, which is often a metabolic predictor of diabetes. Even without overt clinically demonstrated diabetes, dysregulation of the glucose metabolism is associated with cognitive decline and AD risk.

Given the complexity of diabetes, the possible mechanistic links between diabetes and AD are multi-fold and include Aβ misfolding and oligomerization, tau hyperphosphorylation and aggregation, neuroinflammation, damaging pro-oxidative processes and dysfunctional mitochondria. Amongst these, Van Dyken and Lacoste argued that neuroinflammation is one of the key mechanistic connectors between diabetes and AD [71]. Similarly, based on a systematic review of in vitro, preclinical and clinical studies, Vargas-Soria et al. concluded that diabetes triggers specific responses that include the upregulation of activated microglia and secretion of a wide variety of pro-inflammatory cytokines and chemokines [72]. Pathways commonly activated by diabetic pathological changes include the NLRP3 inflammasome.

2.11. Oral Hygiene (Porphyromonas gingivalis)

Bacteria and their associated inflammatory molecules are able to travel from regions of mouth infections to the brain via the bloodstream [73]. Researchers in the School of Dentistry, University of Central Lancashire, initially drew attention to the link between oropharyngeal disease and AD, concluding that periodontitis/gingivitis is a risk factor for AD [74]. The mouth contains 700 bacterial species, including ones that cause periodontal gingivitis; *Porphyromonas gingivalis*, a Gram-negative, rod-shaped, pathogenic anaerobic bacterium from the phylum Bacteroidota, is the most common culprit of gum disease. Recent studies indicate that Aβ oligomerization and its associated neuroinflammatory responses may be triggered in response to this infection. *Porphyromonas gingivalis* and the gingipains enzyme which it produces have been identified in AD brains. Thus, periodontitis is an anatomically specific infection and risk factor for AD [75,76].

Neuroinflammatory processes constitute the connection between chronic, inflammatory disease of the oropharyngeal cavity and gums (periodontitis) and AD [77]. This neuroinflammation may occur through two basic processes: a. local (oral) and/or its associated systemic inflammation, triggering a neuroinflammatory reaction within the brain via the distribution of pro-inflammatory mediators; or b. direct entry of bacteria into the cranial space, eliciting a protective innate immune response manifesting as neuroinflammation. Also, pathogenic oropharyngeal bacteria release structurally diverse metabolites and inflammatory mediators into the bloodstream, ultimately crossing the brain–blood barrier (BBB); these bacteria can instigate alterations in gut microbiota, further enhancing inflammation and affecting brain function via the gut–brain axis. The fifth cranial (trigeminal) nerve has been proposed as an alternative route for connecting oral bacterial products to the brain. Whatever the mechanism, periodontitis/gingivitis leads to microglial activation and pro-inflammatory cytokine release in the brain, thereby triggering and promoting AD pathogenesis [78,79] (see Figure 2).

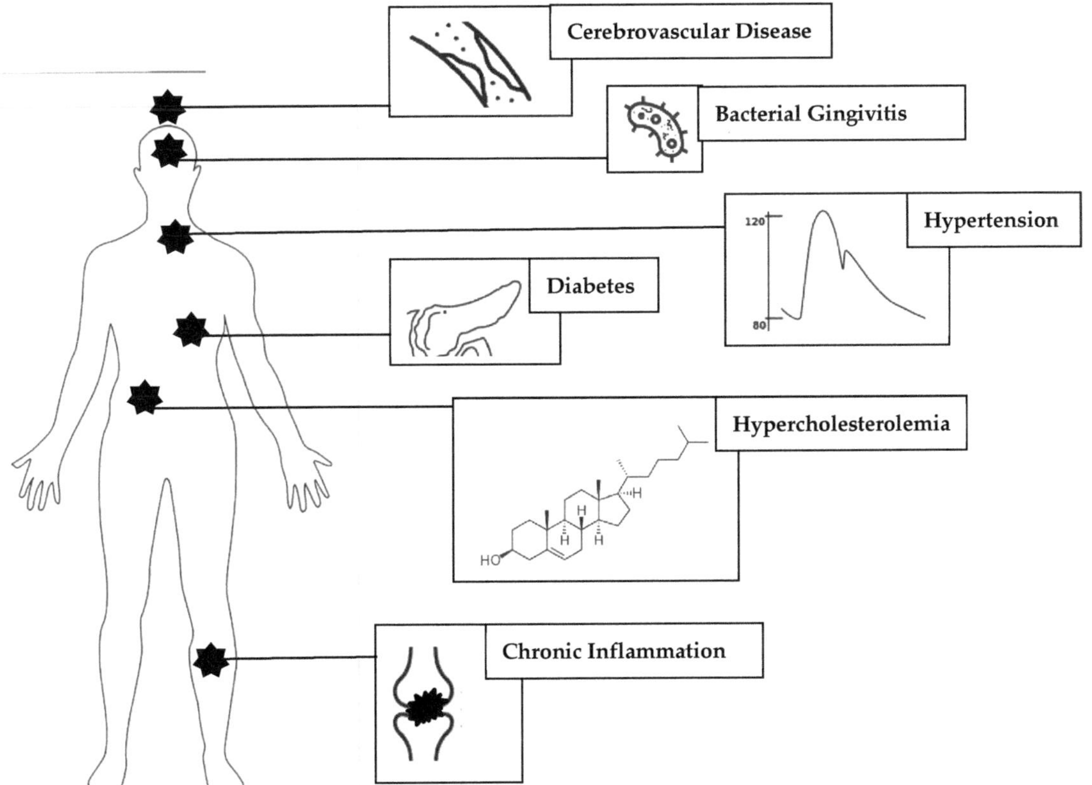

Figure 2. Risk factors for Alzheimer's disease: although AD is a disease of the central nervous system, the diverse risk factors that contribute to its initiation and progression are not confined to the brain and often are systemic disorders such as diabetes mellitus, arterial hypertension or chronic inflammation. Non-systemic localized factors, anatomically distinct from brain, such as chronic periodontitis/gingivitis dental inflammation, are also emerging risks. May of the risk factors are interconnected (e.g., vascular disease, hypertension, hypercholesterolemia and diabetes) and mutually self-sustaining; these factors also contribute to the pathogenesis of AD via a multi-factorial route, through not only neuroinflammatory processes, but also vascular dysfunction.

2.12. Peptic Ulcer Disease (Helicobacter pylori)

There is also an association between peptic ulcer disease and AD, analogous to the connection between oral bacteria and AD, but with the peptic ulcer bacterium (Helicobacter pylori) being further down the gastrointestinal tract [80,81]. Studies have shown that peptic ulcer disease increases the risk of AD via the mechanisms of systemic inflammation and altered gut microbiota [82]. In a population-based study, Chang et al. showed that the suppression of Helicobacter pylori yields decreased progression of dementia [83]. Thus, periodontitis and peptic ulcer disease are two anatomically specific infections implicated as risk factors for AD.

Noori et al. showed that Helicobacter pylori infection contributes to the expression of AD-associated risk factors and neuroinflammation, particularly enhanced concentrations of activated microglia and pro-inflammatory cytokines [84]. In a rat model of peptic ulcer disease, increased levels of circulating pro-inflammatory cytokines such as IL-1β were documented [85].

2.13. Systemic Infection

The relationship between systemic non-CNS infections and AD is complex, but a preponderance of evidence supports the supposition that systemic infection is a risk factor for AD [86]. Giridharan et al. showed that infection-induced systemic sepsis accelerates cognitive decline and neuropathology in an AD mouse model [87]. Based on a systematic review and meta-analysis of human studies, Lei et al. showed that surviving sepsis was linked to a greater risk of dementia (OR = 1.62, 95% CI = 1.23–2.15, I^2 = 96.4%, p = 0.001) and that septicaemia is associated with increased risk for dementia and AD [88]. Though many microorganisms have been implicated, Herpes simplex virus 1, Chlamydia pneumonia and Borrelia burgdorferi have been discussed as infectious agents, which are possible specific microbiological risk factors for AD. Conversely, systemic infection exacerbates pre-existing AD, accelerating cognitive decline and disease progression.

Systemic infections provoke a systemic inflammatory response, which, in turn, elicits neuroinflammation. In a prospective human pilot study, Holmes et al. demonstrated that cognitive function is negatively impacted for two months or longer following the resolution of a systemic infection and that elevated serum levels of IL-1β herald this cognitive impairment [89]. In a post-mortem study, Asby et al. provided evidence that systemic infection raises the levels of multiple cytokines (TNFα, IL-1β, IL-6, IL-8 and IL-15) in the brain [90].

2.14. Systemic Inflammation

Acute and chronic systemic inflammation is characterized by the systemic production of pro-inflammatory cytokines (e.g., TNFα) that play a role in immune to brain communication; systemic inflammation increases pro-inflammatory cytokine secretion in the brain, which, in turn, causes an increase in cognitive decline and disease progression in AD [91]. Walker et al. discussed how systemic pro-inflammatory cytokines can traverse the BBB and enter the brain to regionally promote a pro-inflammatory environment, via a process which also involves signalling through endothelial cells and/or activating the vagus nerve [92]. Systemic inflammation, thereby, induces phenotypically reactive pro-inflammatory microglia and astrocytes, which further can foster β-amyloid oligomerization, tau hyperphosphorylation and complement activation. Similarly, Xie et al., likewise, discussed how peripheral inflammation is a risk factor contributing to AD by means of neuroinflammation [93]. Finally, diseases typically associated with chronic systemic inflammation, such as rheumatoid arthritis, are regarded as risk factors for AD [94,95].

2.15. Allergies

Joh et al. studied 6,785,948 adults aged ≥40 years who participated in a national health examination without any history of dementia before baseline; during 8.1 years of follow-up, 260,705 dementia cases (195,739 AD) were identified, and three allergic diseases (asthma, atopic dermatitis, allergic rhinitis) were positively associated with dementia risk [96]. Compared with individuals without allergies, those with all three allergic diseases had a substantially increased risk of AD (multivariable hazard ratios 1.46; 95% CI 1.25–1.70). Bożek et al. also noted a similar correlation between allergies and AD [97]. Conversely, allergies can exacerbate existing health issues for older adults with AD.

Not surprisingly, there is a relationship between allergies and inflammation [98]. Kabata and Artis described how allergies affect a variety of cytokines, inflammatory mediators and neuropeptides to yield an enhanced neuroinflammatory response [99]. Similarly, Mirotti et al. extensively reviewed the relationship between allergies and brain inflammation, particularly microglial activation and pro-inflammatory cytokine release [100].

2.16. Migraine Headache

In a nationwide (South Korea) cohort study, Kim et al. showed that the overall incidence of AD was greater in people with a chronic migraine history than in non-migraineurs (8.0 per 1000 person-years vs. 4.1 per 1000 person-years) [101]. Similarly, in a population-based cohort study involving 88,390 participants, Hurh et al. concluded that migraine is associated with an increased risk of subsequent AD [102]. Multiple other epidemiological studies support the observation that migraine is a risk factor for AD [103,104].

Migraine is a neuroinflammatory disorder [105], with evidence of neuroinflammation in vascular and perivascular spaces. The implications of co-existing migrainous neurogenic inflammation and neuroinflammation in the histochemical pathophysiology of migraine have been repeatedly demonstrated in preclinical models involving dural vessels and trigeminal endings within the trigemino-vascular system. Neuroinflammatory pathways, especially those invoking inflammasome protein involvement, are regarded as clinical biomarkers and promising druggable targets for migraine [106].

2.17. Chronic Pain

In a France-wide propensity-matched cohort group, Bornier et al. noted that among 64,496 people, the incidence of AD was higher in the chronic pain population than in a control group (1.13% vs. 0.95%, $p < 0.001$); chronic pain increases the risk of AD [107]. Supportively, in a systematic review, Innes and Sambamoorthi documented the possible involvement of chronic pain to cognitive impairment and subsequent dementia including AD [108]. Also, Cao et al. provided evidence that supports a risk factor link between chronic pain and AD [109].

In mechanistic considerations, Vergne-Salle and Bertin discussed how sensory peripheral nerve fibres conveying pain messages are able to mediate peripheral sensitization processes, which, in turn, are linked to the elaboration of inflammation molecules; these afferent nerve fibres trigger neurotransmitter release in spinal cord dorsal root ganglia and dorsal horns, thereby activating microglia and producing pro-inflammatory cytokines and chemokines throughout the CNS [110]. Moreover, as with many of these risk factors, the relationship is bidirectional, self-sustaining and mutually triggering, as evidenced by the fact that neuroinflammation enhances chronic pain perception [111].

2.18. Head Trauma

Young adults who experience moderate to severe head trauma have a greater-than-two-fold enhanced risk for developing AD or a related dementia later in life [112]. In a study based on a population-based prospective historical cohort design, Plassman et al. showed that both moderate head injury (hazard ratio (HR) = 2.32; CI = 1.04 to 5.17) and severe head injury (HR = 4.51; CI = 1.77 to 11.47) were associated with an increased risk of AD [113]. Thus, there is evidence for an association between a history of previous head injury and the risk of developing AD.

Schimmel et al. showed that neuroinflammation following traumatic brain injury is a chronic response to an acute injury [114]. Simon et al. demonstrated that some individuals with traumatic brain injury develop chronic neuroinflammation, which can last for years after the injury, and is associated with activated microglia and the release of pro-inflammatory cytokines—a conclusion also supported by Xiong et al. and Zheng et al. [115–117].

2.19. Domestic Violence

Intimate partner violence (IPV; also termed spousal abuse or domestic violence) forms a sub-group of head trauma scenarios uniquely correlated with AD [118]. However, IPV is more than a focussed sub-type of head trauma. Unlike the head trauma typically seen during accidents or in professional athletes, IPV also comprehensively encompasses psychological, sexual and financial abuse and, not infrequently, is accompanied by alcohol or substance abuse; the nature of the physical violence in IPV is also different, frequently involving manual or ligature partial strangulation.

A 1990 case report by Roberts et al., describing a 76-year-old woman with dementia, connected IPV and AD [119]. A woman was found unconscious with head contusions; her relatives disclosed that her husband had been abusive for years. A post-mortem brain examination revealed morphological and immunological characteristics showing that the woman's IPV-associated brain trauma contributed significantly to the development and progression of her dementia. The consequences of traumatic brain injuries (TBIs) are significant, with evidence suggesting a single TBI may double one's likelihood of developing dementia. Traumatic brain injuries are highly prevalent amongst victims of IPV, arguably leaving hundreds of millions of women worldwide at increased risk for developing dementia.

The connection between IPV and AD is clear and involves multiple mechanisms including neuroinflammation. Newton et al. showed that IPV histories are associated with biologic mediators of inflammation, particularly elevated levels of IL-6 [120]. Similarly, Madison et al. showed that IPV is associated with augmented pro-inflammatory cytokine responses including IL-6 and TNFα [121].

2.20. Depression

Arguably, depression and dementia (AD) share a continuum as a single spectrum disorder: depression leads to dementia and dementia leads to depression. Depression is, thus, a risk factor for AD—an assertion supported by multiple studies. Moreover, emerging evidence is indicating that the time-point in life during which the depression occurs is crucial in determining the nature of this mutually triggering association between AD and depression. In particular, earlier-life depression is associated with a more-than-doubled increase in risk for AD and related dementias; in contrast, analyses of geriatric-onset depression are less definitive but, in general, they too support the notion of a depression–dementia co-dependency [122]. A variety of studies support these conclusions that depression is a risk factor for AD [123–128].

Multiple studies suggest that neuroinflammation is the key process linking depression to dementia [129]. In depression, chronic activation of innate immunity accelerates central inflammation, leading to higher levels of inflammatory cytokines, most consistently IL-1β, IL-6 and TNFα, which, in turn, correlates with greater depressive symptomatology [130]. Neuroinflammation is involved in the pathophysiology of depression through the actions of pro-inflammatory cytokines, which influence interneuronal cross-talk via serotonergic pathways as well as neurogenesis and neuroplasticity in mood-related cerebral regions; these cytokines also stimulate the hypothalamus–pituitary–adrenal axis, exerting influence on hormonal-mediated mood alterations [131].

2.21. Anxiety

Based on a comprehensive literature review, Becker et al. concluded that anxiety is a risk factor for AD (n = 26193, hazard ratio 1.53, 95% CI 1.16–2.01, p < 0.01) [132]. Similarly, based on a meta-analysis of prospective cohort studies, Santabárbara et al. evaluated nine prospective cohorts representing 29,608 participants and identified an overall relative risk of dementia of 1.24 (95% CI: 1.06–1.46) and a population fraction of dementia attributable to anxiety of 3.9%; they concluded that anxiety is extensively connected with an enhanced risk for AD [133].

The relationship between anxiety, neuroinflammation and AD is complex and bidirectional: anxiety causes neuroinflammation and neuroinflammation causes anxiety (analogous to the depression–dementia spectrum). Studies by Won and Kim suggest that anxiety disrupts the hypothalamic–pituitary–adrenal axis and affiliated autonomic nervous system activities; in turn, this mutually induces enhanced pro-inflammatory cytokine levels from activated microglia, particularly in prefrontal and limbic brain structures. The resulting enhanced neuroinflammatory conditions contribute to AD progression [134]. Conversely, based on animal and clinical studies, Zheng et al. and Guo et al. concluded that neuroinflammation induces anxiety by modulating neuronal plasticity in multiple brain regions but, particularly, the basolateral amygdala [135,136]. Thus, anxiety triggers neuroinflammation central to the pathogenesis of AD.

2.22. Insomnia

Sleep disorders, including insomnia, are a well-documented risk factor for AD [137]. In general, neurodegenerative diseases cause sleep disruption, also exemplified by clinical events such as "sundowning" and nocturnal wandering; conversely, chronic insomnia is itself a risk factor for neurodegenerative diseases, including AD. This is not surprising given that sleep has important roles in learning and memory consolidation. Also, sleep deprivation affects not only the symptoms but also the molecular pathogenesis of AD. Sleep contributes to the sequestration and removal of Aβ from neural tissue: Kang et al. showed in transgenic mice that chronic insomnia leads to Aβ accumulation and symptomatic disease progression [138]. Thus, multiple studies have now convincingly demonstrated that sleep deprivation is a risk factor for AD [139–144].

Neuroinflammation is the central cellular and molecular connection between insomnia and AD. Zhu et al. showed that disturbed sleep architecture increased pro-inflammatory IL-6 cytokine levels and induced the phenotypic activation of microglia in the mouse hippocampus, impairing learning and memory, which are hippocampus-dependent processes [145]. Zielinski and Gibbons described the neurotoxic pro-inflammatory role of the IL-1β and TNFα inflammatory cytokines and the NLRP3 inflammasome during periods of dysregulated sleep [146]. Chronic insomnia has also been associated with compromised structural integrity of the BBB, which permits increased entry of peripheral immune cells (macrophages) and inflammatory cytokines into the CNS, further contributing to the ongoing neuroinflammation implicated in AD pathogenesis [147]. Therefore, sleep impairment leads to neuroinflammation through increasing levels of pro-inflammatory cytokines (TNFα, IL-6 and IL-1β) and enzymes (COX), which catalyse inflammatory neurochemical processes.

2.23. Ethanol Abuse

Alcoholism is a substance abuse disorder clinically associated with multiple and varied cognitive problems, including acute intoxication, delirium, Wernicke's psychosis, alcoholic dementia and AD. Not surprisingly, chronic ethanol abuse has been identified as a risk factor for cognitive decline, AD and dementia [148]. Ethanol is a small lipophilic molecule capable of altering multiple neurochemical pathways, which subserve the cognition and memory processes essential to normal brain function; chronic ethanol toxicity, thus, shares and enhances negative effects on normal brain psychology with AD. In turn, this justifies the assertion that alcohol abuse increases the risk of developing AD [149–153].

Neuroinflammation is a major histochemical component of alcohol-induced neural damage [154]. Alcohol abuse triggers peripheral inflammation and central neuroinflammation; the receptor-mediated enabler of this diffuse inflammatory response is the up-regulation of the innate immunity TLR4 (Toll-Like Receptor 4) protein with subsequent microglial and inflammatory cytokine involvement. Based on mouse studies, Lowe et al.

established that chronic ethanol use and abuse promote the pathological entry of peripheral macrophages into the brain, with accompanying microglia activation mediated by stimulation of the CCR2/5 (C-C chemokine receptor types 2 and 5) immune receptor axis [155].

2.24. Social Isolation

Loneliness and social isolation are widespread and significant public health risks affecting many people and placing them at enhanced risk for AD. In an analysis of 502,506 British Biobank participants and 30,097 Canadian Longitudinal Study of Aging participants, Shafighi et al. evaluated risk factors for developing AD in the context of loneliness and aberrant social networking support; they identified strong links between social isolation and AD [156]. Similarly, in a study to establish Cox proportional hazard models with social isolation and loneliness as separate exposures, Shen et al. concluded that social isolation is a risk factor for AD that is independent of loneliness [157].

Neuroinflammation is a definite immunological concomitant of the psychosocial problems inflicted by social isolation. In a study on eight-week-old male C57BL/6 mice, Al Omran et al. showed that social isolation resulted in microglial activation and the release of pro-inflammatory cytokines [158]. Analogously, in a study with BALB/c mice, Ayilara and Owoyele demonstrated evidence of neuroinflammation manifesting as increased activated microglial count and elevated IL-1β and TNFα cytokine levels in a social isolation rearing model [159]. Also, Vu et al. showed that social isolation produces brain region-specific activation of the microglia state in C57Bl/6 mice [160].

2.25. Glaucoma

Glaucoma is the collective diagnostic term for a group of ocular diseases characterized by optic neuropathies linked to degeneration of the retinal ganglion cells; though glaucoma is conventionally conceptualized as a disorder of intraocular pressure, it is better regarded as primarily a disorder of neural tissue within the optic nerve, leading to visual impairment and blindness. Evidence of a link between AD and glaucoma has arisen from epidemiological analyses, revealing that people with AD have a significantly increased incidence of glaucoma [161]. Cesareo et al. studied 51 AD subjects and 67 sex-matched controls: subjects underwent measurements of intraocular pressure, visual field testing and retinal nerve fibre layer thickness assessment by slit-lamp biomicroscopy—patients with AD had a higher frequency of glaucoma-like alterations [162]. Crump et al. studied 324,730 persons diagnosed with glaucoma from 1995 to 2017 in Sweden and 3,247,300 age- and sex-matched population-based controls without prior dementia: in 16 million person-years of follow-up, 32,339 (10%) persons with glaucoma and 226,896 (7%) controls were diagnosed with dementia [163]. Persons with glaucoma had increased risks for AD (adjusted HR, 1.39; 95% CI, 1.35–1.43); among glaucoma subtypes, both primary open-angle and normal-tension glaucoma was associated with an increased risk for AD. Thus, people with glaucoma have an increased risk of developing AD [164,165].

Preclinical and clinical evidence supports the notion that glaucoma is a widespread neurodegenerative condition, whose shared pathogenic mechanism with AD is neuroinflammation. Williams et al. showed that the neuropathology of glaucoma extends beyond the visual pathways and involves pro-inflammatory neuroinflammation at both a cellular (microglia, astrocyte) and molecular (cytokine) level in other CNS locations [166]. Studies by Rolle et al., Rutigliani et al. and Soto and Howell reached similar conclusions [167–169].

2.26. Hearing Loss

Hearing loss at ages 45–65 is a significant risk factor for dementia, possibly accounting for 8 percent of all dementia cases; a 2020 *Lancet* report determined that hearing loss across a wide variety of types and aetiologies approximately doubles the risk of dementia, with even subclinical hearing loss enhancing AD risk [3]. Extensive studies by Lin et al. concluded that hearing loss is associated with increased cognitive decline and incident AD and other dementias in older adults [170]. Based on an analysis of a UK biobank cohort, Jiang et al. concluded that in people with hearing loss, restorative hearing aid use is associated with a reduced risk of dementia of a similar level to that of people without hearing loss, thereby highlighting the urgent need to take measures to address hearing loss as a remediable risk factor for AD [171].

Seicol et al. showed that age-related hearing loss is accompanied by chronic inflammation in neural structures, with elevated expression of pro-inflammatory cytokines and microglial activation [172]. Similarly, Frye et al. demonstrated that pro-inflammatory cytokines including TNFα and IL-1β, and chemokines including CCL2, are induced by hearing loss [173].

2.27. Noise Pollution

Despite their obvious interconnection, hearing loss and exposure to noise pollution are regarded as separate risk factors. Hearing loss caused by factors other than noise exposure is a risk factor for AD; chronic noise exposure of insufficient magnitude to cause obvious hearing loss is, likewise, a risk factor for AD. Epidemiological studies are increasingly identifying the association between external noise exposure (via noise pollution) and dementia [174]. Weuve et al., for example, showed that an increment of 10 A-weighted decibels (dBA) in noise corresponded to 36% and 29% higher odds of prevalent mild cognitive impairment (MCI; odds ratio (OR) = 1.36; 95% confidence interval (CI), 1.15 to 1.62) and AD (OR = 1.29, 95% CI, 1.08 to 1.55) [175]. Cantuaria et al. estimated that as many as 1216 out of the 8475 cases of dementia registered in Denmark in 2017 could be attributed to noise exposures, indicating a great potential for dementia prevention through reductions in ambient noise such as that arising from roadway traffic [176].

As with hearing loss, neuroinflammation is a central mechanistic player in the pathogenesis of noise-induced AD. Wang et al. showed that noise exposure is associated with elevated expression of pro-inflammatory cytokines and microglial activation in the primary auditory cortex; genetic knockout of TNFα or pharmacologically blocking TNFα expression prevented this neuroinflammation [177]. Similarly, Cui et al. showed that chronic noise exposure acts cumulatively to exacerbate neuroinflammation and AD pathology in the rat hippocampus [178].

2.28. Air Pollution

Based on a systematic literature review, Peters et al. concluded that greater exposure to PM2.5, NO2/NOx and CO was associated with an increased risk of dementia, where PM2.5 is airborne particulate matter $\leq$2.5 μ in size [179]. Subsequently, Peters and Li reaffirmed this observation, claiming that constituents of PM2.5, namely black carbon, organic matter, sulphates ($SO_4{}^{2-}$) and ammonium ($NH_4{}^+$), from traffic and fossil fuel combustion are significantly associated with the development of AD [180]. Also, a national cohort study (2000–2018) of long-term air pollution exposure and incident dementia in older adults in the United States showed that exposures to $PM_{2.5}$ and NO_2 are associated with an increased incidence of AD [181,182].

Campbell et al. showed that exposure to particulate matter in polluted air increases biomarkers of inflammation in the mouse brain, including activated microglia, and levels of pro-inflammatory cytokines such as IL-1β and TNFα [183]. Tin-Tin-Win-Shwe et al., likewise, showed changes in pro-inflammatory cytokine mRNA expressions in mice following nanoparticle air pollution exposure [184]. These data and others led Block and Calderón-Garcidueñas to conclude that the emerging evidence implicates air pollution as a chronic source of neuroinflammation, instigating AD with activation of microglia as key to this process [185].

2.29. Global Warming

In 2021, the World Health Organization (WHO) announced that climate change is the biggest global health threat to humanity's future. A 1.5 °C ambient temperature increase may seem trivial when one considers diurnal and seasonal variations, but it does induce subtle but tangible effects on neural pathways and mechanisms that underlie normal brain functioning; these pathways, including neuroinflammation, are implicated in neurodegeneration [186]. Thus, it is possible that global warming secondary to climate change will emerge as a risk factor for AD by facilitating a state of chronic neuroinflammation. In addition, climate warming puts people with AD at risk for symptom worsening and disease progression [187–189]. Gong et al. predicted a 4.5% increase in the risk of dementia hospital admission per 1 °C increase above 17 °C and a 300% increase in hospital admissions for AD by 2040 because of climate change [190]. Although risk factors such as diet and obesity are personally modifiable, risk factors such as climate change are problems which require societal solutions at an international level (Table 1).

Table 1. Risk factors for AD.

Constitutive Factors
1. Age (neuroinflammation, proteopathy, vascular)
2. Sex (neuroinflammation, proteopathy, vascular)
Personal Modifiable Factors
3. Arterial Hypertension (vascular, neuroinflammation, proteopathy)
4. Hypercholesterolemia (vascular, neuroinflammation, proteopathy)
5. Smoking (vascular, neuroinflammation, proteopathy)
6. Physical Inactivity (vascular, neuroinflammation, proteopathy)
7. Obesity (vascular, neuroinflammation, proteopathy)
8. Diet (vascular, neuroinflammation, proteopathy)
9. Diabetes Mellitus (vascular, neuroinflammation, proteopathy)
10. Cerebrovascular Disease (vascular, neuroinflammation, proteopathy)
11. Oral Hygiene (neuroinflammation, proteopathy)
12. Peptic Ulcer Disease (neuroinflammation, proteopathy)
13. Head Trauma (trauma, neuroinflammation, proteopathy)
14. Depression (neuroinflammation, proteopathy)
15. Anxiety (neuroinflammation, proteopathy)
16. Insomnia (neuroinflammation, proteopathy)
17. Ethanol Abuse (neuroinflammation, proteopathy)
18. Social Isolation (neuroinflammation)
19. Hearing Loss (neuroinflammation)
Societal Modifiable Factors
20. Domestic Violence (trauma, neuroinflammation, proteopathy)
21. Noise Pollution (neuroinflammation)
22. Air pollution (neuroinflammation)
23. Global Warming (neuroinflammation)
24. Educational Level (neuroinflammation, proteopathy)
Comorbidity or Concomitant Risk Factors
25. Systemic infection (neuroinflammation)
26. Systemic inflammation (neuroinflammation)
27. Chronic Pain (neuroinflammation)
28. Chronic Migraine (neuroinflammation, proteopathy)
29. Chronic Allergies (neuroinflammation)
30. Glaucoma (neuroinflammation, proteopathy)

As with the other risk factors, neuroinflammation is a key consideration in ascribing a mechanistic explanation for climate change as an AD risk factor. Given the relationship between ambient temperatures and inflammation, it is probable that neuroinflammation is part of the pathological spectrum response to global warming [191,192]. For example, in mice subjected to heat exposure, Lee et al. found: (1) an increased number of glial fibrillary acid protein (GFAP)- and macrophage-1 antigen (Mac-1)-positive cells, (2) up-regulated nuclear factor (NF)-κB, a master regulator of inflammation, and (3) marked increases in COX-2, inducible nitric oxide synthase (iNOS),and cytokine IL-1β and TNFα in the mouse hippocampus [193].

2.30. Educational Level

Lower education is associated with a greater risk for AD and related dementias [194]. The 2020 *Lancet* Commission that examined dementia risk factors found 7% of worldwide dementia cases could be prevented by increasing early-life education [3]. This analysis also concluded that higher childhood education levels and higher lifelong educational attainment could reduce AD and dementia risk. A focussed sub-type of educational attainment is the ability to speak multiple languages; multiple studies indicate that bilingualism or multilingualism offer a degree of protective delay against the development of AD [195–197].

The correlation of educational level with neuroinflammation is not as immediately apparent as for other risk factors, such as head trauma. Nonetheless, there are data clearly supporting a relationship between education and brain inflammatory markers. Steinvil et al. found a statistically significant inverse association between number of school years and high-sensitivity C-reactive protein (CRP), fibrinogen and erythrocyte sedimentation rate (ESR), concluding that level of education was inversely associated with inflammatory biomarkers, even within highly educated populations [198]. Similarly, Maurel et al. found a relationship between educational attainment and five inflammatory biomarkers (CRP, fibrinogen, IL-1β, IL-6 and TNFα), whereby a low educational attainment was associated with higher inflammation, even after adjusting for health behaviours and body mass index [199]. A 2015 study by Okonkwo and co-workers showed that older adults who completed at least 16 years of education had less evidence of AD biomarkers in their cerebrospinal fluid (CSF) than people with fewer years of education [200].

However, education is a complex societal phenomenon. Thus, it is also possible that education is associated with a higher socioeconomic status and quality of life (i.e., less obesity, better diet, better access to healthcare for hypertension, diabetes, depression, deafness) that helps keep people healthy and lowers AD risk.

3. Conclusions

The development of effective diagnostics and therapeutics for AD is one of humankind's pressing neuropharmacologic priorities. A hurdle in the successful attainment of these priorities is the immense cellular and molecular complexity of AD. This complexity is reflected by the equally complex diversity of risk factors associated with AD. However, more than merely mirroring disease complexity, risk factors also provide fundamental insights into the aetiology and pathogenesis of AD as a neurodegenerative disorder since they are central to disease initiation and subsequent propagation. Based on a systematic literature review, this analysis identified 30 risk factors for AD and then extended the analysis to further identify neuroinflammation as a unifying mechanism present in all 30 risk factors. Although other mechanisms (e.g., vasculopathy) were present in multiple risk factors, dysfunction of the neuroimmune–neuroinflammation axis was key to all 30 identified risk factors. Though the nature of the neuroinflammatory involvement varied, activation of microglia and the release of pro-inflammatory cytokines were common pathways shared by all risk factors. This observation provides further evidence for the importance of immunopathic mechanisms to aetiopathogenesis of AD.

Neuroinflammation is "bad for brain". The identification of these 30 risk factors for neuroinflammation (and, therefore, AD) is, thus, also a call to action. By 2050, more than

150 million people will be living with AD—the health and socioeconomic impacts of this statistic will be truly immense. Since humanity is struggling to devise curative therapeutics for AD, prophylactically addressing risk factors is and will continue to be an essential step in reducing the global burden of AD. This review identified 30 risk factors. Some are modifiable and can be addressed at the level of the individual (depression, diabetes, diet, educational level, excessive alcohol consumption, hearing impairment, hypertension, low social contact, obesity, oral hygiene, peptic ulcer disease, physical inactivity, smoking, traumatic brain injury); others need to be addressed at a societal or international level (air pollution, climate change, noise pollution, intimate partner violence). Meaningfully addressing these risk factors requires multi-level educational goals, targeting individuals, healthcare providers, school teachers, politicians and policy makers. Hopefully, we—individually and collectively—have the commitment to attain these goals, thereby reducing the neuroinflammation that mediates the transformation of lifestyle/societal circumstances into risk factors for a devastating disease.

Funding: This work was funded by grants from the Krembil Foundation, the Weston Foundation and a Harrington Scholar-Innovator Award.

Institutional Review Board Statement: Not applicable.

Informed Consent Statement: Not applicable.

Data Availability Statement: Not applicable.

Acknowledgments: D.F.W. acknowledges salary support from the Krembil Foundation.

Conflicts of Interest: The author declares no conflicts of interest.

References

1. Breijyeh, Z.; Karaman, R. Comprehensive Review on Alzheimer's disease: Causes and Treatment. *Molecules* **2020**, *25*, 5789. [CrossRef] [PubMed]
2. Knopman, D.S.; Amieva, H.; Petersen, R.C.; Chételat, G.; Holtzman, D.M.; Hyman, B.T.; Nixon, R.A.; Jones, D.T. Alzheimer disease. *Nat. Rev. Dis. Primers* **2021**, *7*, 33. [CrossRef] [PubMed]
3. Livingston, G.; Huntley, J.; Sommerlad, A.; Ames, D.; Ballard, C.; Banerjee, S.; Brayne, C.; Burns, A.; Cohen-Mansfield, J.; Cooper, C.; et al. Dementia prevention, intervention, and care: 2020 report of the Lancet Commission. *Lancet* **2020**, *396*, 413–446. [CrossRef]
4. Thakur, S.; Dhapola, R.; Sarma, P.; Medhi, B.; Reddy, D.H. Neuroinflammation in Alzheimer's disease: Current Progress in Molecular Signaling and Therapeutics. *Inflammation* **2023**, *46*, 1–17. [CrossRef] [PubMed]
5. Heneka, M.T.; Carson, M.J.; El Khoury, J.; Landreth, G.E.; Brosseron, F.; Feinstein, D.L.; Jacobs, A.H.; Wyss-Coray, T.; Vitorica, J.; Ransohoff, R.M.; et al. Neuroinflammation in Alzheimer's disease. *Lancet Neurol.* **2015**, *14*, 388–405. [CrossRef] [PubMed]
6. Katsel, P.; Haroutunian, V. Is Alzheimer disease a failure of mobilizing immune defense? Lessons from cognitively fit oldest-old. *Dialogues Clin. Neurosci.* **2019**, *21*, 7–19. [CrossRef]
7. Wu, K.M.; Zhang, Y.R.; Huang, Y.Y.; Dong, Q.; Tan, L.; Yu, J.T. The role of the immune system in Alzheimer's disease. *Ageing Res. Rev.* **2021**, *70*, 101409. [CrossRef]
8. Frost, G.R.; Jonas, L.A.; Li, Y.M. Friend, Foe or Both? Immune Activity in Alzheimer's disease. *Front. Aging Neurosci.* **2019**, *11*, 337. [CrossRef]
9. Jorfi, M.; Maaser-Hecker, A.; Tanzi, R.E. The neuroimmune axis of Alzheimer's disease. *Genome Med.* **2023**, *15*, 6. [CrossRef]
10. Princiotta Cariddi, L.; Mauri, M.; Cosentino, M.; Versino, M.; Marino, F. Alzheimer's disease: From Immune Homeostasis to Neuroinflammatory Condition. *Int. J. Mol. Sci.* **2022**, *23*, 13008. [CrossRef]
11. Gustavsson, A.; Norton, N.; Fast, T.; Frölich, L.; Georges, J.; Holzapfel, D.; Kirabali, T.; Krolak-Salmon, P.; Rossini, P.M.; Ferretti, M.T.; et al. Global estimates on the number of persons across the Alzheimer's disease continuum. *Alzheimer's Dement.* **2023**, *19*, 658–670. [CrossRef] [PubMed]
12. d'Avila, J.C.; Siqueira, L.D.; Mazeraud, A.; Azevedo, E.P.; Foguel, D.; Castro-Faria-Neto, H.C.; Sharshar, T.; Chrétien, F.; Bozza, F.A. Age-related cognitive impairment is associated with long-term neuroinflammation and oxidative stress in a mouse model of episodic systemic inflammation. *J. Neuroinflamm.* **2018**, *15*, 28. [CrossRef] [PubMed]
13. Andronie-Cioara, F.L.; Ardelean, A.I.; Nistor-Cseppento, C.D.; Jurcau, A.; Jurcau, M.C.; Pascalau, N.; Marcu, F. Molecular Mechanisms of Neuroinflammation in Aging and Alzheimer's disease Progression. *Int. J. Mol. Sci.* **2023**, *24*, 1869. [CrossRef] [PubMed]
14. Godbout, J.P.; Johnson, R.W. Age and neuroinflammation: A lifetime of psychoneuroimmune consequences. *Immunol. Allergy Clin. N. Am.* **2009**, *29*, 321–337. [CrossRef] [PubMed]

15. Beam, C.R.; Kaneshiro, C.; Jang, J.Y.; Reynolds, C.A.; Pedersen, N.L.; Gatz, M. Differences Between Women and Men in Incidence Rates of Dementia and Alzheimer's disease. *J. Alzheimer's Dis.* **2018**, *64*, 1077–1083. [CrossRef] [PubMed]
16. Liu, C.C.; Li, C.Y.; Sun, Y.; Hu, S.C. Gender and Age Differences and the Trend in the Incidence and Prevalence of Dementia and Alzheimer's disease in Taiwan: A 7-Year National Population-Based Study. *BioMed. Res. Int.* **2019**, *2019*, 5378540. [CrossRef]
17. Niu, H.; Álvarez-Álvarez, I.; Guillén-Grima, F.; Aguinaga-Ontoso, I. Prevalence and incidence of Alzheimer's disease in Europe: A meta-analysis. Prevalencia e incidencia de la enfermedad de Alzheimer en Europa: Metaanálisis. *Neurologia* **2017**, *32*, 523–532. [CrossRef]
18. Bekhbat, M.; Neigh, G.N. Sex differences in the neuro-immune consequences of stress: Focus on depression and anxiety. *Brain Behav. Immun.* **2018**, *67*, 1–12. [CrossRef]
19. Engler, H.; Benson, S.; Wegner, A.; Spreitzer, I.; Schedlowski, M.; Elsenbruch, S. Men and women differ in inflammatory and neuroendocrine responses to endotoxin but not in the severity of sickness symptoms. *Brain Behav. Immun.* **2016**, *52*, 18–26. [CrossRef]
20. Jacobson, D.L.; Gange, S.J.; Rose, N.R.; Graham, N.M. Epidemiology and estimated population burden of selected autoimmune diseases in the United States. *Clin. Immunol. Immunopathol.* **1997**, *84*, 223–243. [CrossRef]
21. Meier-Stephenson, F.S.; Meier-Stephenson, V.C.; Carter, M.D.; Meek, A.R.; Wang, Y.; Pan, L.; Chen, Q.; Jacobo, S.; Wu, F.; Lu, E.; et al. Alzheimer's disease as an autoimmune disorder of innate immunity endogenously modulated by tryptophan metabolites. *Alzheimer's Dement.* **2022**, *8*, e12283. [CrossRef] [PubMed]
22. Sierra, C. Hypertension and the Risk of Dementia. *Front. Cardiovasc. Med.* **2020**, *7*, 5. [CrossRef] [PubMed]
23. Gabin, J.M.; Tambs, K.; Saltvedt, I.; Sund, E.; Holmen, J. Association between blood pressure and Alzheimer disease measured up to 27 years prior to diagnosis: The HUNT Study. *Alzheimer's Res. Ther.* **2017**, *9*, 37. [CrossRef] [PubMed]
24. Bajwa, E.; Klegeris, A. Neuroinflammation as a mechanism linking hypertension with the increased risk of Alzheimer's disease. *Neural. Regen. Res.* **2022**, *17*, 2342–2346. [CrossRef] [PubMed]
25. Solé-Guardia, G.; Custers, E.; de Lange, A.; Clijncke, E.; Geenen, B.; Gutierrez, J.; Küsters, B.; Claassen, J.A.H.R.; de Leeuw, F.E.; Wiesmann, M.; et al. Association between hypertension and neurovascular inflammation in both normal-appearing white matter and white matter hyperintensities. *Acta Neuropathol. Commun.* **2023**, *11*, 2. [CrossRef] [PubMed]
26. Carnevale, D.; Mascio, G.; Ajmone-Cat, M.A.; D'andrea, I.; Cifelli, G.; Madonna, M.; Cocozza, G.; Frati, A.; Carullo, P.; Carnevale, L.; et al. Role of neuroinflammation in hypertension-induced brain amyloid pathology. *Neurobiol. Aging.* **2012**, *33*, 205.e19–205.e29. [CrossRef] [PubMed]
27. Mancini, G.; Dias, C.; Lourenço, C.F.; Laranjinha, J.; de Bem, A.; Ledo, A. A High Fat/Cholesterol Diet Recapitulates Some Alzheimer's disease-Like Features in Mice: Focus on Hippocampal Mitochondrial Dysfunction. *J. Alzheimer's Dis.* **2021**, *82*, 1619–1633. [CrossRef]
28. Ledreux, A.; Wang, X.; Schultzberg, M.; Granholm, A.C.; Freeman, L.R. Detrimental effects of a high fat/high cholesterol diet on memory and hippocampal markers in aged rats. *Behav. Brain Res.* **2016**, *312*, 294–304. [CrossRef]
29. Jin, P.; Pan, Y.; Pan, Z.; Xu, J.; Lin, M.; Sun, Z.; Chen, M.; Xu, M. Alzheimer-like brain metabolic and structural features in cholesterol-fed rabbit detected by magnetic resonance imaging. *Lipids Health Dis.* **2018**, *17*, 61, Erratum in *Lipids Health Dis.* **2018**, *17*, 204. [CrossRef]
30. Wu, M.; Zhai, Y.; Liang, X.; Chen, W.; Lin, R.; Ma, L.; Huang, Y.; Zhao, D.; Liang, Y.; Zhao, W.; et al. Connecting the Dots Between Hypercholesterolemia and Alzheimer's disease: A Potential Mechanism Based on 27-Hydroxycholesterol. *Front. Neurosci.* **2022**, *16*, 842814. [CrossRef]
31. Xu, C.; Apostolova, L.G.; Oblak, A.L.; Gao, S. Association of Hypercholesterolemia with Alzheimer's disease Pathology and Cerebral Amyloid Angiopathy. *J. Alzheimer's Dis.* **2020**, *73*, 1305–1311. [CrossRef] [PubMed]
32. Feringa, F.M.; van der Kant, R. Cholesterol and Alzheimer's disease; From Risk Genes to Pathological Effects. *Front. Aging Neurosci.* **2021**, *13*, 690372. [CrossRef]
33. Thirumangalakudi, L.; Prakasam, A.; Zhang, R.; Bimonte-Nelson, H.; Sambamurti, K.; Kindy, M.S.; Bhat, N.R. High cholesterol-induced neuroinflammation and amyloid precursor protein processing correlate with loss of working memory in mice. *J. Neurochem.* **2008**, *6*, 475–485. [CrossRef] [PubMed]
34. Chen, Y.; Yin, M.; Cao, X.; Hu, G.; Xiao, M. Pro- and Anti-inflammatory Effects of High Cholesterol Diet on Aged Brain. *Aging Dis.* **2018**, *9*, 374–390. [CrossRef] [PubMed]
35. Durazzo, T.C.; Mattsson, N.; Weiner, M.W. Smoking and increased Alzheimer's disease risk: A review of potential mechanisms. *Alzheimer's Dement.* **2014**, *10*, S122–S145. [CrossRef]
36. Jeong, S.; Park, J.; Han, K. Association of Changes in Smoking Intensity with Risk of Dementia in Korea. *JAMA Netw. Open.* **2023**, *6*, e2251506. [CrossRef]
37. Alrouji, M.; Manouchehrinia, A.; Gran, B.; Constantinescu, C.S. Effects of cigarette smoke on immunity, neuroinflammation and multiple sclerosis. *J. Neuroimmunol.* **2019**, *329*, 24–34. [CrossRef]
38. Liu, Y.; Li, H.; Wang, J. Association of Cigarette Smoking With Cerebrospinal Fluid Biomarkers of Neurodegeneration, Neuroinflammation, and Oxidation. *JAMA Netw. Open.* **2020**, *3*, e2018777. [CrossRef]
39. Meng, Q.; Lin, M.S.; Tzeng, I.S. Relationship Between Exercise and Alzheimer's disease: A Narrative Literature Review. *Front. Neurosci.* **2020**, *14*, 131. [CrossRef]

40. Chen, W.W.; Zhang, X.; Huang, W.J. Role of physical exercise in Alzheimer's disease. *Biomed. Rep.* **2016**, *4*, 403–407. [CrossRef]
41. Wang, M.; Zhang, H.; Liang, J.; Huang, J.; Chen, N. Exercise suppresses neuroinflammation for alleviating Alzheimer's disease. *J. Neuroinflamm.* **2023**, *20*, 76. [CrossRef] [PubMed]
42. Seo, D.Y.; Heo, J.W.; Ko, J.R.; Kwak, H.B. Exercise and Neuroinflammation in Health and Disease. *Int. Neurourol. J.* **2019**, *23* (Suppl. S2), S82–S92. [CrossRef] [PubMed]
43. Svensson, M.; Lexell, J.; Deierborg, T. Effects of Physical Exercise on Neuroinflammation, Neuroplasticity, Neurodegeneration, and Behavior: What We Can Learn From Animal Models in Clinical Settings. *Neurorehabil. Neural Repair.* **2015**, *29*, 577–589. [CrossRef] [PubMed]
44. Forny-Germano, L.; De Felice, F.G.; Vieira, M.N.D.N. The Role of Leptin and Adiponectin in Obesity-Associated Cognitive Decline and Alzheimer's disease. *Front. Neurosci.* **2019**, *12*, 1027. [CrossRef] [PubMed]
45. Flores-Cordero, J.A.; Pérez-Pérez, A.; Jiménez-Cortegana, C.; Alba, G.; Flores-Barragán, A.; Sánchez-Margalet, V. Obesity as a Risk Factor for Dementia and Alzheimer's disease: The Role of Leptin. *Int. J. Mol. Sci.* **2022**, *23*, 5202. [CrossRef] [PubMed]
46. Hayes, J.P.; Moody, J.N.; Roca, J.G.; Hayes, S.M. Body mass index is associated with smaller medial temporal lobe volume in those at risk for Alzheimer's disease. *NeuroImage Clin.* **2020**, *25*, 102156. [CrossRef] [PubMed]
47. Nordestgaard, L.T.; Tybjærg-Hansen, A.; Nordestgaard, B.G.; Frikke-Schmidt, R. Body Mass Index and Risk of Alzheimer's disease: A Mendelian Randomization Study of 399,536 Individuals. *J. Clin. Endo. Metab.* **2017**, *102*, 2310–2320. [CrossRef]
48. Emmerzaal, T.L.; Kiliaan, A.J.; Gustafson, D.R. 2003–2013: A decade of body mass index, Alzheimer's disease, and dementia. *J. Alzheimer's Dis.* **2015**, *43*, 739–755. [CrossRef]
49. García-Ptacek, S.; Faxén-Irving, G.; Cermáková, P.; Eriksdotter, M.; Religa, D. Body mass index in dementia. *Eur. J. Clin. Nutr.* **2014**, *68*, 1204–1209. [CrossRef]
50. Miller, A.A.; Spencer, S.J. Obesity and neuroinflammation: A pathway to cognitive impairment. *Brain Behav. Immun.* **2014**, *42*, 10–21. [CrossRef]
51. Henn, R.E.; Elzinga, S.E.; Glass, E.; Parent, R.; Guo, K.; Allouch, A.A.; Mendelson, F.E.; Hayes, J.; Webber-Davis, I.; Murphy, G.G.; et al. Obesity-induced neuroinflammation and cognitive impairment in young adult versus middle-aged mice. *Immun. Ageing* **2022**, *19*, 67. [CrossRef] [PubMed]
52. Xu Lou, I.; Ali, K.; Chen, Q. Effect of nutrition in Alzheimer's disease: A systematic review. *Front. Neurosci.* **2023**, *17*, 1147177. [CrossRef] [PubMed]
53. Tosatti, J.A.G.; Fontes, A.F.D.S.; Caramelli, P.; Gomes, K.B. Effects of Resveratrol Supplementation on the Cognitive Function of Patients with Alzheimer's disease: A Systematic Review of Randomized Controlled Trials. *Drugs Aging* **2022**, *39*, 285–295. [CrossRef]
54. Samadi, M.; Moradi, S.; Moradinazar, M.; Mostafai, R.; Pasdar, Y. Dietary pattern in relation to the risk of Alzheimer's disease: A systematic review. *Neurol. Sci.* **2019**, *40*, 2031–2043. [CrossRef] [PubMed]
55. Pistollato, F.; Iglesias, R.C.; Ruiz, R.; Aparicio, S.; Crespo, J.; Lopez, L.D.; Manna, P.P.; Giampieri, F.; Battino, M. Nutritional patterns associated with the maintenance of neurocognitive functions and the risk of dementia and Alzheimer's disease: A focus on human studies. *Pharmacol. Res.* **2018**, *131*, 32–43. [CrossRef] [PubMed]
56. Abduljawad, A.A.; Elawad, M.A.; Elkhalifa, M.E.M.; Ahmed, A.; Hamdoon, A.A.E.; Salim, L.H.M.; Ashraf, M.; Ayaz, M.; Hassan, S.S.U.; Bungau, S. Alzheimer's disease as a Major Public Health Concern: Role of Dietary Saponins in Mitigating Neurodegenerative Disorders and Their Underlying Mechanisms. *Molecules* **2022**, *27*, 6804. [CrossRef] [PubMed]
57. Kip, E.; Parr-Brownlie, L.C. Healthy lifestyles and wellbeing reduce neuroinflammation and prevent neurodegenerative and psychiatric disorders. *Front. Neurosci.* **2023**, *17*, 1092537. [CrossRef] [PubMed]
58. Bok, E.; Jo, M.; Lee, S.; Lee, B.R.; Kim, J.; Kim, H.J. Dietary Restriction and Neuroinflammation: A Potential Mechanistic Link. *Int. J. Mol. Sci.* **2019**, *20*, 464. [CrossRef]
59. Cavaliere, G.; Trinchese, G.; Penna, E.; Cimmino, F.; Pirozzi, C.; Lama, A.; Annunziata, C.; Catapano, A.; Mattace Raso, G.; Meli, R.; et al. High-Fat Diet Induces Neuroinflammation and Mitochondrial Impairment in Mice Cerebral Cortex and Synaptic Fraction. *Front. Cell. Neurosci.* **2019**, *13*, 509. [CrossRef]
60. Romanenko, M.; Kholin, V.; Koliada, A.; Vaiserman, A. Nutrition, Gut Microbiota, and Alzheimer's disease. *Front. Psych.* **2021**, *12*, 712673. [CrossRef]
61. Arvanitakis, Z.; Capuano, A.W.; Leurgans, S.E.; Bennett, D.A.; Schneider, J.A. Relation of cerebral vessel disease to Alzheimer's disease dementia and cognitive function in elderly people: A cross-sectional study. *Lancet Neurol.* **2016**, *15*, 934–943. [CrossRef] [PubMed]
62. Gottesman, R.F.; Schneider, A.L.; Zhou, Y.; Coresh, J.; Green, E.; Gupta, N.; Knopman, D.S.; Mintz, A.; Rahmim, A.; Sharrett, A.R.; et al. Association Between Midlife Vascular Risk Factors and Estimated Brain Amyloid Deposition. *JAMA* **2017**, *317*, 1443–1450. [CrossRef]
63. Honig, L.S.; Tang, M.X.; Albert, S.; Costa, R.; Luchsinger, J.; Manly, J.; Stern, Y.; Mayeux, R. Stroke and the risk of Alzheimer disease. *Arch. Neurol.* **2003**, *60*, 1707–1712. [CrossRef] [PubMed]
64. Wong, C.H.; Crack, P.J. Modulation of neuro-inflammation and vascular response by oxidative stress following cerebral ischemia-reperfusion injury. *Curr. Med. Chem.* **2008**, *15*, 1–14. [CrossRef] [PubMed]

65. Wu, L.; Xiong, X.; Wu, X.; Ye, Y.; Jian, Z.; Zhi, Z.; Gu, L. Targeting Oxidative Stress and Inflammation to Prevent Ischemia-Reperfusion Injury. *Front. Mol. Neurosci.* **2020**, *13*, 28. [CrossRef] [PubMed]

66. Drake, C.; Boutin, H.; Jones, M.S.; Denes, A.; McColl, B.W.; Selvarajah, J.R.; Hulme, S.; Georgiou, R.F.; Hinz, R.; Gerhard, A.; et al. Brain inflammation is induced by co-morbidities and risk factors for stroke. *Brain Behav. Immun.* **2011**, *25*, 1113–1122. [CrossRef]

67. Jurcau, A.; Simion, A. Neuroinflammation in Cerebral Ischemia and Ischemia/Reperfusion Injuries: From Pathophysiology to Therapeutic Strategies. *Int. J. Mol. Sci.* **2021**, *23*, 14. [CrossRef]

68. Jayaraj, R.L.; Azimullah, S.; Beiram, R. Diabetes as a risk factor for Alzheimer's disease in the Middle East and its shared pathological mediators. *Saudi J. Biol. Sci.* **2020**, *27*, 736–750. [CrossRef]

69. Barbagallo, M.; Dominguez, L.J. Type 2 diabetes mellitus and Alzheimer's disease. *World J. Diabetes* **2014**, *5*, 889–893. [CrossRef]

70. Whitmer, R.A. Type 2 diabetes and risk of cognitive impairment and dementia. *Curr. Neurol. Neurosci. Rep.* **2007**, *7*, 373–380. [CrossRef]

71. Van Dyken, P.; Lacoste, B. Impact of Metabolic Syndrome on Neuroinflammation and the Blood-Brain Barrier. *Front. Neurosci.* **2018**, *12*, 930. [CrossRef] [PubMed]

72. Vargas-Soria, M.; García-Alloza, M.; Corraliza-Gómez, M. Effects of diabetes on microglial physiology: A systematic review of in vitro, preclinical and clinical studies. *J. Neuroinflamm.* **2023**, *20*, 57. [CrossRef]

73. Kanagasingam, S.; von Ruhland, C.; Welbury, R.; Chukkapalli, S.S.; Singhrao, S.K. *Porphyromonas gingivalis* Conditioned Medium Induces Amyloidogenic Processing of the Amyloid-β Protein Precursor upon in vitro Infection of SH-SY5Y Cells. *J. Alzheimer's Dis. Rep.* **2022**, *6*, 577–587. [CrossRef] [PubMed]

74. Bouziane, A.; Lattaf, S.; Abdallaoui Maan, L. Effect of Periodontal Disease on Alzheimer's disease: A Systematic Review. *Cureus* **2023**, *15*, e46311. [CrossRef] [PubMed]

75. Gao, C.; Larvin, H.; Bishop, D.T.; Bunce, D.; Pavitt, S.; Wu, J.; Kang, J. Oral diseases are associated with cognitive function in adults over 60 years old. *Oral Dis.* **2023**. [CrossRef] [PubMed]

76. Bello-Corral, L.; Alves-Gomes, L.; Fernández-Fernández, J.A.; Fernández-García, D.; Casado-Verdejo, I.; Sánchez-Valdeón, L. Implications of gut and oral microbiota in neuroinflammatory responses in Alzheimer's disease. *Life Sci.* **2023**, *333*, 122132. [CrossRef] [PubMed]

77. Li, X.; Kiprowska, M.; Kansara, T.; Kansara, P.; Li, P. Neuroinflammation: A Distal Consequence of Periodontitis. *J. Dent. Res.* **2022**, *101*, 1441–1449. [CrossRef]

78. Luo, H.; Wu, B.; Kamer, A.R.; Adhikari, S.; Sloan, F.; Plassman, B.L.; Tan, C.; Qi, X.; Schwartz, M.D. Oral Health, Diabetes, and Inflammation: Effects of Oral Hygiene Behaviour. *Int. Dent. J.* **2022**, *72*, 484–490. [CrossRef]

79. Almarhoumi, R.; Alvarez, C.; Harris, T.; Tognoni, C.M.; Paster, B.J.; Carreras, I.; Dedeoglu, A.; Kantarci, A. Microglial cell response to experimental periodontal disease. *J. Neuroinflamm.* **2023**, *20*, 142. [CrossRef]

80. Hsu, C.C.; Hsu, Y.C.; Chang, K.H.; Lee, C.Y.; Chong, L.W.; Lin, C.L.; Kao, C.H. Association of Dementia and Peptic Ulcer Disease: A Nationwide Population-Based Study. *Am. J. Alzheimer's Dis. Dement.* **2016**, *31*, 389–394. [CrossRef]

81. Choi, H.G.; Soh, J.S.; Lim, J.S.; Sim, S.Y.; Jung, Y.J.; Lee, S.W. Peptic ulcer does not increase the risk of dementia: A nested case control study using a national sample cohort. *Medicine* **2020**, *99*, e21703. [CrossRef] [PubMed]

82. Huang, W.S.; Yang, T.Y.; Shen, W.C.; Lin, C.L.; Lin, M.C.; Kao, C.H. Association between Helicobacter pylori infection and dementia. *J. Clin. Neurosci.* **2014**, *21*, 1355–1358. [CrossRef] [PubMed]

83. Chang, Y.P.; Chiu, G.F.; Kuo, F.C.; Lai, C.L.; Yang, Y.H.; Hu, H.M.; Chang, P.Y.; Chen, C.Y.; Wu, D.C.; Yu, F.J. Eradication of Helicobacter pylori Is Associated with the Progression of Dementia: A Population-Based Study. *Gastroenterol. Res. Pract.* **2013**, *2013*, 175729. [CrossRef] [PubMed]

84. Noori, M.; Mahboobi, R.; Nabavi-Rad, A.; Jamshidizadeh, S.; Fakharian, F.; Yadegar, A.; Zali, M.R. *Helicobacter pylori* infection contributes to the expression of Alzheimer's disease-associated risk factors and neuroinflammation. *Heliyon* **2023**, *9*, e19607. [CrossRef] [PubMed]

85. Watanabe, T.; Higuchi, K.; Tanigawa, T. Mechanisms of peptic ulcer recurrence: Role of inflammation. *Inflammopharmacology* **2002**, *10*, 291–302. [CrossRef]

86. Rakic, S.; Hung, Y.M.A.; Smith, M.; So, D.; Tayler, H.M.; Varney, W.; Wild, J.; Harris, S.; Holmes, C.; Love, S.; et al. Systemic infection modifies the neuroinflammatory response in late stage Alzheimer's disease. *Acta Neuropathol. Commun.* **2018**, *6*, 88. [CrossRef]

87. Giridharan, V.V.; Catumbela, C.S.G.; Catalão, C.H.R.; Lee, J.; Ganesh, B.P.; Petronilho, F.; Dal-Pizzol, F.; Morales, R.; Barichello, T. Sepsis exacerbates Alzheimer's disease pathophysiology, modulates the gut microbiome, increases neuroinflammation and amyloid burden. *Mol. Psychiatry* **2023**. [CrossRef]

88. Lei, S.; Li, X.; Zhao, H.; Feng, Z.; Chun, L.; Xie, Y.; Li, J. Risk of Dementia or Cognitive Impairment in Sepsis Survivals: A Systematic Review and Meta-Analysis. *Front. Aging Neurosci.* **2022**, *14*, 839472. [CrossRef]

89. Holmes, C.; El-Okl, M.; Williams, A.L.; Cunningham, C.; Wilcockson, D.; Perry, V.H. Systemic infection, interleukin 1beta, and cognitive decline in Alzheimer's disease. *J. Neurol. Neurosurg. Psychiatry* **2003**, *74*, 788–789. [CrossRef]

90. Asby, D.; Boche, D.; Allan, S.; Love, S.; Miners, J.S. Systemic infection exacerbates cerebrovascular dysfunction in Alzheimer's disease. *Brain* **2021**, *144*, 1869–1883. [CrossRef]

91. Holmes, C.; Cunningham, C.; Zotova, E.; Woolford, J.; Dean, C.; Kerr, S.; Culliford, D.; Perry, V.H. Systemic inflammation and disease progression in Alzheimer disease. *Neurology* **2009**, *73*, 768–774. [CrossRef] [PubMed]

92. Walker, K.A.; Ficek, B.N.; Westbrook, R. Understanding the Role of Systemic Inflammation in Alzheimer's disease. *ACS Chem. Neurosci.* **2019**, *10*, 3340–3342. [CrossRef] [PubMed]
93. Xie, J.; Van Hoecke, L.; Vandenbroucke, R.E. The Impact of Systemic Inflammation on Alzheimer's disease Pathology. *Front. Immunol.* **2022**, *12*, 796867. [CrossRef] [PubMed]
94. Sangha, P.S.; Thakur, M.; Akhtar, Z.; Ramani, S.; Gyamfi, R.S. The Link Between Rheumatoid Arthritis and Dementia: A Review. *Cureus* **2020**, *12*, e7855. [CrossRef] [PubMed]
95. Trzeciak, P.; Herbet, M.; Dudka, J. Common Factors of Alzheimer's disease and Rheumatoid Arthritis-Pathomechanism and Treatment. *Molecules* **2021**, *26*, 6038. [CrossRef] [PubMed]
96. Joh, H.K.; Kwon, H.; Son, K.Y.; Yun, J.M.; Cho, S.H.; Han, K.; Park, J.H.; Cho, B. Allergic Diseases and Risk of Incident Dementia and Alzheimer's disease. *Ann. Neurol.* **2023**, *93*, 384–397. [CrossRef] [PubMed]
97. Bożek, A.; Bednarski, P.; Jarzab, J. Allergic rhinitis, bronchial asthma and other allergies in patients with Alzheimer's disease. *Postep. Dermatol. I Alergol.* **2016**, *33*, 353–358. [CrossRef]
98. Voisin, T.; Bouvier, A.; Chiu, I.M. Neuro-immune interactions in allergic diseases: Novel targets for therapeutics. *Int. Immunol.* **2017**, *29*, 247–261. [CrossRef]
99. Kabata, H.; Artis, D. Neuro-immune crosstalk and allergic inflammation. *J. Clin. Investig.* **2019**, *129*, 1475–1482. [CrossRef]
100. Mirotti, L.; Castro, J.; Costa-Pinto, F.A.; Russo, M. Neural pathways in allergic inflammation. *J. Allergy* **2010**, *2010*, 491928. [CrossRef]
101. Kim, J.; Ha, W.S.; Park, S.H.; Han, K.; Baek, M.S. Association between migraine and Alzheimer's disease: A nationwide cohort study. *Front. Aging Neurosci.* **2023**, *15*, 1196185. [CrossRef] [PubMed]
102. Hurh, K.; Jeong, S.H.; Kim, S.H.; Jang, S.Y.; Park, E.C.; Jang, S.I. Increased risk of all-cause, Alzheimer's, and vascular dementia in adults with migraine in Korea: A population-based cohort study. *J. Headache Pain* **2022**, *23*, 108. [CrossRef] [PubMed]
103. Morton, R.E.; St John, P.D.; Tyas, S.L. Migraine and the risk of all-cause dementia, Alzheimer's disease; and vascular dementia: A prospective cohort study in community-dwelling older adults. *Int. J. Geriatr. Psychiatry* **2019**, *34*, 1667–1676. [CrossRef] [PubMed]
104. Yang, F.C.; Lin, T.Y.; Chen, H.J.; Lee, J.T.; Lin, C.C.; Kao, C.H. Increased Risk of Dementia in Patients with Tension-Type Headache: A Nationwide Retrospective Population-Based Cohort Study. *PLoS ONE* **2016**, *11*, e0156097. [CrossRef] [PubMed]
105. Biscetti, L.; Cresta, E.; Cupini, L.M.; Calabresi, P.; Sarchielli, P. The putative role of neuroinflammation in the complex pathophysiology of migraine: From bench to bedside. *Neurobiol. Dis.* **2023**, *180*, 106072. [CrossRef] [PubMed]
106. Kursun, O.; Yemisci, M.; van den Maagdenberg, A.M.J.M.; Karatas, H. Migraine and neuroinflammation: The inflammasome perspective. *J. Headache Pain* **2021**, *22*, 55. [CrossRef] [PubMed]
107. Bornier, N.; Mulliez, A.; Chenaf, C.; Elyn, A.; Teixeira, S.; Authier, N.; Bertin, C.; Kerckhove, N. Chronic pain is a risk factor for incident Alzheimer's disease: A nationwide propensity-matched cohort using administrative data. *Front. Aging Neurosci.* **2023**, *15*, 1193108. [CrossRef]
108. Innes, K.E.; Sambamoorthi, U. The Potential Contribution of Chronic Pain and Common Chronic Pain Conditions to Subsequent Cognitive Decline, New Onset Cognitive Impairment, and Incident Dementia: A Systematic Review and Conceptual Model for Future Research. *J. Alzheimer's Dis.* **2020**, *78*, 1177–1195. [CrossRef]
109. Cao, S.; Fisher, D.W.; Yu, T.; Dong, H. The link between chronic pain and Alzheimer's disease. *J. Inflamm.* **2019**, *16*, 204. [CrossRef]
110. Vergne-Salle, P.; Bertin, P. Chronic pain and neuroinflammation. *Jt. Bone Spine* **2021**, *88*, 105222. [CrossRef]
111. Ji, R.R.; Xu, Z.Z.; Gao, Y.J. Emerging targets in neuroinflammation-driven chronic pain. *Nat. Rev. Drug. Disc.* **2014**, *13*, 533–548. [CrossRef] [PubMed]
112. Gottlieb, S. Head injury doubles the risk of Alzheimer's disease. *Br. Med. J.* **2000**, *321*, 1100.
113. Plassman, B.L.; Havlik, R.J.; Steffens, D.C.; Helms, M.J.; Newman, T.N.; Drosdick, D.; Phillips, C.; Gau, B.A.; Welsh-Bohmer, K.A.; Burke, J.R.; et al. Documented head injury in early adulthood and risk of Alzheimer's disease and other dementias. *Neurology* **2000**, *55*, 1158–1166. [CrossRef] [PubMed]
114. Schimmel, S.J.; Acosta, S.; Lozano, D. Neuroinflammation in traumatic brain injury: A chronic response to an acute injury. *Brain Circ.* **2017**, *3*, 135–142. [CrossRef] [PubMed]
115. Simon, D.W.; McGeachy, M.J.; Bayır, H.; Clark, R.S.; Loane, D.J.; Kochanek, P.M. The far-reaching scope of neuroinflammation after traumatic brain injury. *Nat. Rev. Neurol.* **2017**, *13*, 171–191. [CrossRef] [PubMed]
116. Xiong, Y.; Mahmood, A.; Chopp, M. Current understanding of neuroinflammation after traumatic brain injury and cell-based therapeutic opportunities. *Chin. J. Traumatol.* **2018**, *21*, 137–151. [CrossRef] [PubMed]
117. Zheng, R.Z.; Lee, K.Y.; Qi, Z.X.; Wang, Z.; Xu, Z.Y.; Wu, X.H.; Mao, Y. Neuroinflammation Following Traumatic Brain Injury: Take It Seriously or Not. *Front. Immunol.* **2022**, *13*, 855701. [CrossRef]
118. Mehr, J.B.; Bennett, E.R.; Price, J.L.; de Souza, N.L.; Buckman, J.F.; Wilde, E.A.; Tate, D.F.; Marshall, A.D.; Dams-O'Connor, K.; Esopenko, C. Intimate partner violence, substance use, and health comorbidities among women: A narrative review. *Front. Psychol.* **2023**, *13*, 1028375. [CrossRef]
119. Roberts, G.W.; Whitwell, H.L.; Acland, P.R.; Bruton, C.J. Dementia in a punch-drunk wife. *Lancet* **1990**, *335*, 918–919. [CrossRef]
120. Newton, T.L.; Fernandez-Botran, R.; Miller, J.J.; Lorenz, D.J.; Burns, V.E.; Fleming, K.N. Markers of inflammation in midlife women with intimate partner violence histories. *J. Women's Health* **2011**, *20*, 1871–1880. [CrossRef]

121. Madison, A.A.; Wilson, S.J.; Shrout, M.R.; Malarkey, W.B.; Kiecolt-Glaser, J.K. Intimate Partner Violence and Inflammaging: Conflict Tactics Predict Inflammation Among Middle-Aged and Older Adults. *Psychosom. Med.* **2023**. [CrossRef] [PubMed]
122. Byers, A.L.; Yaffe, K. Depression and risk of developing dementia. *Nat. Rev. Neurol.* **2011**, *7*, 323–331. [CrossRef] [PubMed]
123. Gatz, J.L.; Tyas, S.L.; St John, P.; Montgomery, P. Do depressive symptoms predict Alzheimer's disease and dementia? *J. Gerontol. Ser. A Biol. Sci. Med. Sci.* **2005**, *60*, 744–747. [CrossRef] [PubMed]
124. Chen, R.; Hu, Z.; Wei, L.; Qin, X.; McCracken, C.; Copeland, J.R. Severity of depression and risk for subsequent dementia: Cohort studies in China and the UK. *Br. J. Psychiatry* **2008**, *193*, 373–377. [CrossRef] [PubMed]
125. Byers, A.L.; Covinsky, K.E.; Barnes, D.E.; Yaffe, K. Dysthymia and depression increase risk of dementia and mortality among older veterans. *Am. J. Geriatr. Psych.* **2012**, *20*, 664–672. [CrossRef] [PubMed]
126. Wilson, R.S.; Barnes, L.L.; De Leon, C.M.; Aggarwal, N.T.; Schneider, J.S.; Bach, J.; Pilat, J.; Beckett, L.A.; Arnold, S.E.; Evans, D.A.; et al. Depressive symptoms, cognitive decline, and risk of Alzheimer's disease in older persons. *Neurology* **2002**, *59*, 364–370. [CrossRef] [PubMed]
127. Fuhrer, R.; Dufouil, C.; Dartigues, J.F. PAQUID Study. Exploring sex differences in the relationship between depressive symptoms and dementia incidence: Prospective results from the PAQUID Study. *J. Am. Geriat. Soc.* **2003**, *51*, 1055–1063. [CrossRef]
128. Geerlings, M.I.; Schmand, B.; Braam, A.W.; Jonker, C.; Bouter, L.M.; van Tilburg, W. Depressive symptoms and risk of Alzheimer's disease in more highly educated older people. *J. Am. Geriat. Soc.* **2000**, *48*, 1092–1097. [CrossRef]
129. Troubat, R.; Barone, P.; Leman, S.; Desmidt, T.; Cressant, A.; Atanasova, B.; Brizard, B.; El Hage, W.; Surget, A.; Belzung, C.; et al. Neuroinflammation and depression: A review. *Eur. J. Neurosci.* **2021**, *53*, 151–171. [CrossRef]
130. Hassamal, S. Chronic stress, neuroinflammation, and depression: An overview of pathophysiological mechanisms and emerging anti-inflammatories. *Front. Psychiatry* **2023**, *14*, 1130989. [CrossRef]
131. Jeon, S.W.; Kim, Y.K. The role of neuroinflammation and neurovascular dysfunction in major depressive disorder. *J. Inflamm. Res.* **2018**, *11*, 179–192. [CrossRef] [PubMed]
132. Becker, E.; Orellana Rios, C.L.; Lahmann, C.; Rücker, G.; Bauer, J.; Boeker, M. Anxiety as a risk factor of Alzheimer's disease and vascular dementia. *Br. J. Psychiatry* **2018**, *213*, 654–660. [CrossRef]
133. Santabárbara, J.; Lipnicki, D.M.; Olaya, B.; Villagrasa, B.; Bueno-Notivol, J.; Nuez, L.; López-Antón, R.; Gracia-García, P. Does Anxiety Increase the Risk of All-Cause Dementia? An Updated Meta-Analysis of Prospective Cohort Studies. *J. Clin. Med.* **2020**, *9*, 1791. [CrossRef] [PubMed]
134. Won, E.; Kim, Y.K. Neuroinflammation-Associated Alterations of the Brain as Potential Neural Biomarkers in Anxiety Disorders. *Int. J. Mol. Sci.* **2020**, *21*, 6546. [CrossRef] [PubMed]
135. Zheng, Z.H.; Tu, J.L.; Li, X.H.; Hua, Q.; Liu, W.Z.; Liu, Y.; Pan, B.X.; Hu, P.; Zhang, W.H. Neuroinflammation induces anxiety- and depressive-like behavior by modulating neuronal plasticity in the basolateral amygdala. *Brain Behav. Immun.* **2021**, *91*, 505–518. [CrossRef]
136. Guo, B.; Zhang, M.; Hao, W.; Wang, Y.; Zhang, T.; Liu, C. Neuroinflammation mechanisms of neuromodulation therapies for anxiety and depression. *Transl. Psychiatry* **2023**, *13*, 5. [CrossRef]
137. Minakawa, E.N.; Wada, K.; Nagai, Y. Sleep Disturbance as a Potential Modifiable Risk Factor for Alzheimer's disease. *Int. J. Mol. Sci.* **2019**, *20*, 803. [CrossRef]
138. Kang, J.E.; Lim, M.M.; Bateman, R.J.; Lee, J.J.; Smyth, L.P.; Cirrito, J.R.; Fujiki, N.; Nishino, S.; Holtzman, D.M. Amyloid-beta dynamics are regulated by orexin and the sleep-wake cycle. *Science* **2009**, *326*, 1005–1007. [CrossRef]
139. Bubu, O.M.; Brannick, M.; Mortimer, J.; Umasabor-Bubu, O.; Sebastião, Y.V.; Wen, Y.; Schwartz, S.; Borenstein, A.R.; Wu, Y.; Morgan, D.; et al. Sleep, Cognitive impairment, and Alzheimer's disease: A Systematic Review and Meta-Analysis. *Sleep* **2017**, *40*, zsw032. [CrossRef]
140. Sadeghmousavi, S.; Eskian, M.; Rahmani, F.; Rezaei, N. The effect of insomnia on development of Alzheimer's disease. *J. Neuroinflamm.* **2020**, *17*, 289. [CrossRef]
141. Shamim, S.A.; Warriach, Z.I.; Tariq, M.A.; Rana, K.F.; Malik, B.H. Insomnia: Risk Factor for Neurodegenerative Diseases. *Cureus* **2019**, *11*, e6004. [CrossRef] [PubMed]
142. Johar, H.; Kawan, R.; Emeny, R.T.; Ladwig, K.H. Impaired Sleep Predicts Cognitive Decline in Old People: Findings from the Prospective KORA Age Study. *Sleep* **2016**, *39*, 217–226. [CrossRef] [PubMed]
143. Benito-León, J.; Bermejo-Pareja, F.; Vega, S.; Louis, E.D. Total daily sleep duration and the risk of dementia: A prospective population-based study. *Eur. J. Neurol.* **2009**, *16*, 990–997. [CrossRef] [PubMed]
144. Cricco, M.; Simonsick, E.M.; Foley, D.J. The impact of insomnia on cognitive functioning in older adults. *J. Am. Geriat. Soc.* **2001**, *49*, 1185–1189. [CrossRef] [PubMed]
145. Zhu, B.; Dong, Y.; Xu, Z.; Gompf, H.S.; Ward, S.A.; Xue, Z.; Miao, C.; Zhang, Y.; Chamberlin, N.L.; Xie, Z. Sleep disturbance induces neuroinflammation and impairment of learning and memory. *Neurobiol. Dis.* **2012**, *48*, 348–355. [CrossRef] [PubMed]
146. Zielinski, M.R.; Gibbons, A.J. Neuroinflammation, Sleep, and Circadian Rhythms. *Front. Cell. Infect. Microbiol.* **2022**, *12*, 853096. [CrossRef]
147. Herrero Babiloni, A.; Baril, A.-A.; Charlebois-Plante, C.; Jodoin, M.; Sanchez, E.; De Baets, L.; Arbour, C.; Lavigne, G.J.; Gosselin, N.; De Beaumont, L. The Putative Role of Neuroinflammation in the Interaction between Traumatic Brain Injuries, Sleep, Pain and Other Neuropsychiatric Outcomes: A State-of-the-Art Review. *J. Clin. Med.* **2023**, *12*, 1793. [CrossRef]

148. Rehm, J.; Hasan, O.S.M.; Black, S.E.; Shield, K.D.; Schwarzinger, M. Alcohol use and dementia: A systematic scoping review. *Alzheimer's Res. Ther.* **2019**, *11*, 1. [CrossRef]

149. Evert, D.L.; Oscar-Berman, M. Alcohol-Related Cognitive Impairments: An Overview of How Alcoholism May Affect the Workings of the Brain. *Alcohol Health Res. World* **1995**, *19*, 89–96.

150. Smith, D.M.; Atkinson, R.M. Alcoholism and dementia. *Int. J. Addict.* **1995**, *30*, 1843–1869. [CrossRef]

151. Tyas, S.L. Are tobacco and alcohol use related to Alzheimer's disease? A critical assessment of the evidence and its implications. *Addict. Biol.* **1996**, *1*, 237–254. [CrossRef] [PubMed]

152. Tyas, S.L. Alcohol use and the risk of developing Alzheimer's disease. *Alcohol Res. Health J. Natl. Inst. Alcohol Abus. Alcoholism.* **2001**, *25*, 299–306.

153. Jeon, K.H.; Han, K.; Jeong, S.M.; Park, J.; Yoo, J.E.; Yoo, J.; Lee, J.; Kim, S.; Shin, D.W. Changes in Alcohol Consumption and Risk of Dementia in a Nationwide Cohort in South Korea. *JAMA Netw. Open* **2023**, *6*, e2254771. [CrossRef] [PubMed]

154. Orio, L.; Alen, F.; Pavón, F.J.; Serrano, A.; García-Bueno, B. Oleoylethanolamide, Neuroinflammation, and Alcohol Abuse. *Front. Mol. Neurosci.* **2019**, *11*, 490. [CrossRef] [PubMed]

155. Lowe, P.P.; Morel, C.; Ambade, A.; Iracheta-Vellve, A.; Kwiatkowski, E.; Satishchandran, A.; Furi, I.; Cho, Y.; Gyongyosi, B.; Catalano, D.; et al. Chronic alcohol-induced neuroinflammation involves CCR2/5-dependent peripheral macrophage infiltration and microglia alterations. *J. Neuroinflamm.* **2020**, *17*, 296. [CrossRef] [PubMed]

156. Shafighi, K.; Villeneuve, S.; Rosa Neto, P.; Badhwar, A.; Poirier, J.; Sharma, V.; Medina, Y.I.; Silveira, P.P.; Dube, L.; Glahn, D.; et al. Social isolation is linked to classical risk factors of Alzheimer's disease-related dementias. *PLoS ONE* **2023**, *18*, e0280471. [CrossRef] [PubMed]

157. Shen, C.; Rolls, E.; Cheng, W.; Kang, J.; Dong, G.; Xie, C.; Zhao, X.M.; Sahakian, B.; Feng, J. Associations of Social Isolation and Loneliness with Later Dementia. *Neurology* **2022**, *99*, e164–e175. [CrossRef] [PubMed]

158. Al Omran, A.J.; Shao, A.S.; Watanabe, S.; Zhang, Z.; Zhang, J.; Xue, C.; Watanabe, J.; Davies, D.L.; Shao, X.M.; Liang, J. Social Isolation Induces Neuroinflammation and Microglia Overactivation, While Dihydromyricetin Prevents and Improves Them. *Res. Sq.* **2021**, rs.3.rs-923871. [CrossRef]

159. Ayilara, G.O.; Owoyele, B.V. Neuroinflammation and microglial expression in brains of social-isolation rearing model of schizophrenia. *IBRO Neurosci. Rep.* **2023**, *15*, 31–41. [CrossRef]

160. Vu, A.P.; Lam, D.; Denney, C.; Lee, K.V.; Plemel, J.R.; Jackson, J. Social isolation produces a sex- and brain region-specific alteration of microglia state. *Eur. J. Neurosci.* **2023**, *57*, 1481–1497. [CrossRef]

161. Wostyn, P.; Audenaert, K.; De Deyn, P.P. Alzheimer's disease and glaucoma: Is there a causal relationship? *Br. J. Ophthalmol.* **2009**, *93*, 1557–1559. [CrossRef] [PubMed]

162. Cesareo, M.; Martucci, A.; Ciuffoletti, E.; Mancino, R.; Cerulli, A.; Sorge, R.P.; Martorana, A.; Sancesario, G.; Nucci, C. Association Between Alzheimer's disease and Glaucoma: A Study Based on Heidelberg Retinal Tomography and Frequency Doubling Technology Perimetry. *Front. Neurosci.* **2015**, *9*, 479. [CrossRef] [PubMed]

163. Crump, C.; Sundquist, J.; Sieh, W.; Sundquist, K. Risk of Alzheimer's disease and Related Dementias in Persons With Glaucoma: A National Cohort Study. *Ophthalmology* **2023**. [CrossRef] [PubMed]

164. Sugiyama, T. Glaucoma and Alzheimer's disease: Their clinical similarity and future therapeutic strategies for glaucoma. *World J. Ophthalmol.* **2014**, *4*, 47–51. [CrossRef]

165. Mancino, R.; Martucci, A.; Cesareo, M.; Giannini, C.; Corasaniti, M.T.; Bagetta, G.; Nucci, C. Glaucoma and Alzheimer Disease: One Age-Related Neurodegenerative Disease of the Brain. *Curr. Neuropharmacol.* **2018**, *16*, 971–977. [CrossRef] [PubMed]

166. Williams, P.A.; Marsh-Armstrong, N.; Howell, G.R. Lasker/IRRF Initiative on Astrocytes and Glaucomatous Neurodegeneration Participants. Neuroinflammation in glaucoma: A new opportunity. *Exp. Eye Res.* **2017**, *157*, 20–27. [CrossRef] [PubMed]

167. Rolle, T.; Ponzetto, A.; Malinverni, L. The Role of Neuroinflammation in Glaucoma: An Update on Molecular Mechanisms and New Therapeutic Options. *Front. Neurol.* **2021**, *11*, 612422. [CrossRef]

168. Soto, I.; Howell, G.R. The complex role of neuroinflammation in glaucoma. *Cold Spring Harb. Perspect. Med.* **2014**, *4*, a017269. [CrossRef]

169. Rutigliani, C.; Tribble, J.R.; Hagström, A.; Lardner, E.; Jóhannesson, G.; Stålhammar, G.; Williams, P.A. Widespread retina and optic nerve neuroinflammation in enucleated eyes from glaucoma patients. *Acta Neuropathol. Commun.* **2022**, *10*, 118. [CrossRef]

170. Lin, F.R.; Pike, J.R.; Albert, M.S.; Arnold, M.; Burgard, S.; Chisolm, T.; Couper, D.; Deal, J.A.; Goman, A.M.; Glynn, N.W.; et al. Hearing intervention versus health education control to reduce cognitive decline in older adults with hearing loss in the USA (ACHIEVE): A multicentre, randomised controlled trial. *Lancet* **2023**, *402*, 786–797. [CrossRef]

171. Jiang, F.; Mishra, S.R.; Shrestha, N.; Ozaki, A.; Virani, S.S.; Bright, T.; Kuper, H.; Zhou, C.; Zhu, D. Association between hearing aid use and all-cause and cause-specific dementia: An analysis of the UK Biobank cohort. *Lancet Public Health* **2023**, *8*, e329–e338. [CrossRef] [PubMed]

172. Seicol, B.J.; Lin, S.; Xie, R. Age-Related Hearing Loss Is Accompanied by Chronic Inflammation in the Cochlea and the Cochlear Nucleus. *Front. Aging Neurosci.* **2022**, *14*, 846804. [CrossRef] [PubMed]

173. Frye, M.D.; Ryan, A.F.; Kurabi, A. Inflammation associated with noise-induced hearing loss. *J. Acoust. Soc. Am.* **2019**, *146*, 4020. [CrossRef] [PubMed]

174. Huang, L.; Zhang, Y.; Wang, Y.; Lan, Y. Relationship Between Chronic Noise Exposure, Cognitive Impairment, and Degenerative Dementia: Update on the Experimental and Epidemiological Evidence and Prospects for Further Research. *J. Alzheimer's Dis.* **2021**, *79*, 1409–1427. [CrossRef] [PubMed]
175. Weuve, J.; D'Souza, J.; Beck, T.; Evans, D.A.; Kaufman, J.D.; Rajan, K.B.; de Leon, C.F.M.; Adar, S.D. Long-term community noise exposure in relation to dementia, cognition, and cognitive decline in older adults. *Alzheimer's Dement.* **2021**, *17*, 525–533. [CrossRef] [PubMed]
176. Cantuaria, M.L.; Waldorff, F.B.; Wermuth, L.; Pedersen, E.R.; Poulsen, A.H.; Thacher, J.D.; Raaschou-Nielsen, O.; Ketze, M.; Khan, J.; Valencia, V.H.; et al. Residential exposure to transportation noise in Denmark and incidence of dementia: National cohort study. *BMJ* **2021**, *374*, n1954. [CrossRef] [PubMed]
177. Wang, W.; Zhang, L.S.; Zinsmaier, A.K.; Patterson, G.; Leptich, E.J.; Shoemaker, S.L.; Yatskievych, T.A.; Gibboni, R.; Pace, E.; Luo, H.; et al. Neuroinflammation mediates noise-induced synaptic imbalance and tinnitus in rodent models. *PLoS Biol.* **2019**, *17*, e3000307. [CrossRef]
178. Cui, B.; Li, K.; Gai, Z.; She, X.; Zhang, N.; Xu, C.; Chen, X.; An, G.; Ma, Q.; Wang, R. Chronic Noise Exposure Acts Cumulatively to Exacerbate Alzheimer's disease-Like Amyloid-β Pathology and Neuroinflammation in the Rat Hippocampus. *Sci. Rep.* **2015**, *5*, 12943. [CrossRef]
179. Peters, R.; Ee, N.; Peters, J.; Booth, A.; Mudway, I.; Anstey, K.J. Air Pollution and Dementia: A Systematic Review. *J. Alzheimer's Dis.* **2019**, *70*, S145–S163. [CrossRef]
180. Peters, A. Ambient air pollution and Alzheimer's disease: The role of the composition of fine particles. *Proc. Natl. Acad. Sci. USA* **2023**, *120*, e2220028120. [CrossRef]
181. Shi, L. Incident dementia and long-term exposure to constituents of fine particle air pollution: A national cohort study in the United States. *Proc. Natl. Acad. Sci. USA* **2022**, *120*, e2211282119. [CrossRef] [PubMed]
182. Shi, L.; Steenland, K.; Li, H.; Liu, P.; Zhang, Y.; Lyles, R.H.; Requia, W.J.; Ilango, S.D.; Chang, H.H.; Wingo, T.; et al. A national cohort study (2000–2018) of long-term air pollution exposure and incident dementia in older adults in the United States. *Nat. Comm.* **2021**, *12*, 6754. [CrossRef] [PubMed]
183. Campbell, A.; Oldham, M.; Becaria, A.; Bondy, S.C.; Meacher, D.; Sioutas, C.; Misra, C.; Mendez, L.B.; Kleinman, M. Particulate matter in polluted air may increase biomarkers of inflammation in mouse brain. *Neurotoxicology* **2005**, *26*, 133–140. [CrossRef] [PubMed]
184. Mitsushima, D.; Yamamoto, S.; Fukushima, A.; Funabashi, T.; Kobayashi, T.; Fujimaki, H. Changes in neurotransmitter levels and proinflammatory cytokine mRNA expressions in the mice olfactory bulb following nanoparticle exposure. *Toxicol. Appl. Pharmacol.* **2008**, *226*, 192–198. [CrossRef]
185. Block, M.L.; Calderón-Garcidueñas, L. Air pollution: Mechanisms of neuroinflammation and CNS disease. *Trends Neurosci.* **2009**, *32*, 506–516. [CrossRef] [PubMed]
186. Bongioanni, P.; Del Carratore, R.; Corbianco, S.; Diana, A.; Cavallini, G.; Masciandaro, S.M.; Dini, M.; Buizza, R. Climate change and neurodegenerative diseases. *Environ. Res.* **2021**, *201*, 111511. [CrossRef] [PubMed]
187. Zuelsdorff, M.; Limaye, V.S. A Framework for Assessing the Effects of Climate Change on Dementia Risk and Burden. *Gerontologist* **2023**, gnad082. [CrossRef]
188. Stella, A.B.; Galmonte, A.; Deodato, M.; Ozturk, S.; Reis, J.; Manganotti, P. Climate Change and Global Warming: Are Individuals with Dementia—Including Alzheimer's disease—At a Higher Risk? *Curr. Alzheimer Res.* **2023**, *20*, 209–212. [CrossRef]
189. Ruszkiewicz, J.A.; Tinkov, A.A.; Skalny, A.V.; Siokas, V.; Dardiotis, E.; Tsatsakis, A.; Bowman, A.B.; da Rocha, J.B.T.; Aschner, M. Brain diseases in changing climate. *Environ. Res.* **2019**, *177*, 108637. [CrossRef]
190. Gong, J.; Part, C.; Hajat, S. Current and future burdens of heat-related dementia hospital admissions in England. *Environ. Int.* **2022**, *159*, 107027. [CrossRef]
191. O'Donnell, S. The neurobiology of climate change. *Sci. Nat.* **2018**, *105*, 11. [CrossRef] [PubMed]
192. Habibi, L.; Perry, G.; Mahmoudi, M. Global warming and neurodegenerative disorders: Speculations on their linkage. *BioImpacts* **2014**, *4*, 167–170. [CrossRef] [PubMed]
193. Lee, W.; Moon, M.; Kim, H.G.; Lee, T.H.; Oh, M.S. Heat stress-induced memory impairment is associated with neuroinflammation in mice. *J. Neuroinflamm.* **2015**, *12*, 102. [CrossRef] [PubMed]
194. Sharp, E.S.; Gatz, M. Relationship between education and dementia: An updated systematic review. *Alzheimer Dis. Assoc. Disord.* **2011**, *25*, 289–304. [CrossRef] [PubMed]
195. Klimova, B.; Valis, M.; Kuca, K. Bilingualism as a strategy to delay the onset of Alzheimer's disease. *Clin. Interv. Aging* **2017**, *12*, 1731–1737. [CrossRef] [PubMed]
196. Craik, F.I.; Bialystok, E.; Freedman, M. Delaying the onset of Alzheimer disease: Bilingualism as a form of cognitive reserve. *Neurology* **2010**, *75*, 1726–1729. [CrossRef] [PubMed]
197. Liu, H.; Wu, L. Lifelong Bilingualism Functions as an Alternative Intervention for Cognitive Reserve against Alzheimer's disease. *Front. Psychiatry* **2021**, *12*, 696015. [CrossRef]
198. Steinvil, A.; Shirom, A.; Melamed, S.; Toker, S.; Justo, D.; Saar, N.; Shapira, I.; Berliner, S.; Rogowski, O. Relation of educational level to inflammation-sensitive biomarker level. *Am. J. Cardiol.* **2008**, *102*, 1034–1039. [CrossRef]

199. Maurel, M.; Castagné, R.; Berger, E.; Bochud, M.; Chadeau-Hyam, M.; Fraga, S.; Gandini, M.; Hutri-Kähönen, N.; Jalkanen, S.; Kivimäki, M.; et al. Patterning of educational attainment across inflammatory markers: Findings from a multi-cohort study. *Brain Behav. Immun.* **2020**, *90*, 303–310. [CrossRef]
200. Almeida, R.P.; Schultz, S.A.; Austin, B.P.; Boots, E.A.; Dowling, N.; Gleason, C.E.; Bendlin, B.B.; Sager, M.A.; Hermann, B.P.; Zetterberg, H.; et al. Effect of Cognitive Reserve on Age-Related Changes in Cerebrospinal Fluid Biomarkers of Alzheimer Disease. *JAMA Neurol.* **2015**, *72*, 699–706. [CrossRef]

Disclaimer/Publisher's Note: The statements, opinions and data contained in all publications are solely those of the individual author(s) and contributor(s) and not of MDPI and/or the editor(s). MDPI and/or the editor(s) disclaim responsibility for any injury to people or property resulting from any ideas, methods, instructions or products referred to in the content.

Review

The Diverse Roles of Reactive Astrocytes in the Pathogenesis of Amyotrophic Lateral Sclerosis

Kangqin Yang [1], Yang Liu [1] and Min Zhang [1,2,*

[1] Department of Neurology and Psychiatry, Tongji Hospital, Tongji Medical College, Huazhong University of Science and Technology, Wuhan 430030, China; kangkangy1999@163.com (K.Y.); liu_yang2014@hust.edu.cn (Y.L.)

[2] Hubei Key Laboratory of Neural Injury and Functional Reconstruction, Huazhong University of Science and Technology, Wuhan 430030, China

* Correspondence: zhang_min_3464@126.com

Abstract: Astrocytes displaying reactive phenotypes are characterized by their ability to remodel morphologically, molecularly, and functionally in response to pathological stimuli. This process results in the loss of their typical astrocyte functions and the acquisition of neurotoxic or neuroprotective roles. A growing body of research indicates that these reactive astrocytes play a pivotal role in the pathogenesis of amyotrophic lateral sclerosis (ALS), involving calcium homeostasis imbalance, mitochondrial dysfunction, abnormal lipid and lactate metabolism, glutamate excitotoxicity, etc. This review summarizes the characteristics of reactive astrocytes, their role in the pathogenesis of ALS, and recent advancements in astrocyte-targeting strategies.

Keywords: reactive astrocytes; amyotrophic lateral sclerosis; pathogenesis; astrocyte-targeting strategies

Citation: Yang, K.; Liu, Y.; Zhang, M. The Diverse Roles of Reactive Astrocytes in the Pathogenesis of Amyotrophic Lateral Sclerosis. *Brain Sci.* **2024**, *14*, 158. https://doi.org/10.3390/brainsci14020158

Academic Editors: Andrew Clarkson, Junhui Wang, Hongxing Wang and Jing Sun

Received: 21 December 2023
Revised: 17 January 2024
Accepted: 29 January 2024
Published: 4 February 2024

Copyright: © 2024 by the authors. Licensee MDPI, Basel, Switzerland. This article is an open access article distributed under the terms and conditions of the Creative Commons Attribution (CC BY) license (https://creativecommons.org/licenses/by/4.0/).

1. Introduction

Astrocytes, the most numerous and giant glial cells in the central nervous system (CNS), possess the unique ability to divide and proliferate throughout life. The cytosol of astrocytes exhibits a distinctive star-shaped morphology, housing a critical structural component known as the glial filament. Comprised of glial fibrillary acidic protein (GFAP), this intermediate filament is essential to the cytoskeleton and serves as a standard marker for astrocytes. Importantly, it is not entirely exclusive to astrocytes but also labels neural stem cells [1–3]. Apart from organizing the blood–brain barrier (BBB) and supporting, sequestering, and isolating neurons [1], these cells perform a multitude of vital biological functions. These include the metabolism, synthesis, and secretion of neurotrophic factors, regulation of neurotransmitters and calcium homeostasis, maintenance of mitochondrial function, participation in nervous system and circuit development, and regulation of the immune status of the CNS [1,2,4–7]. However, these functions are partially or wholly lost in reactive astrocytes.

Amyotrophic lateral sclerosis (ALS) is a fatal illness characterized by the degeneration of upper and lower motor neurons (MNs), with an average survival of 3–5 years [8,9]. Current treatments, such as riluzole, edaravone, AMX0035, and tofersen, can only temporarily extend survival [10–15]. The exact etiology and pathogenesis of the disease remain unknown, with proposed causes including neuroinflammation, oxidative stress, mitochondrial dysfunction, glutamate excitotoxicity, calcium homeostasis imbalance, metabolic abnormality, etc. [8,16,17]. The involvement of non-cell-autonomous processes, particularly reactive astrocytes, in the pathogenesis of ALS has been recognized [18,19]. In this review, we aim to summarize the contribution of reactive astrocytes in the pathogenesis of ALS and identify potential therapeutic targets.

2. Astrocytes in Pathological Conditions

2.1. Definition of Reactive Astrocytes

In the past, the response of astrocytes to abnormal events such as trauma, ischemia, infection, and tumor, epilepsy, and neurodegenerative and demyelinating diseases has been described using various terms, including astrocytosis, astrogliosis, reactive gliosis, astrocyte activation, astrocyte reactivity, astrocyte re-activation, and astrocyte reaction [20]. These terms can be replaced by the standard term "reactive astrocytes", proposed to define the response process to pathological conditions, that is, under pathological conditions of the CNS, such as infection, trauma, or neurodegenerative disease, astrocytes undergo morphological, biochemical, transcriptional regulatory, molecular, and functional remodeling, ultimately losing most of their normal astrocytic functions and acquiring new neurotoxic or neuroprotective functions [20–23]. Compared to normal astrocytes, reactive astrocytes exhibit distinct morphological changes, including hypertrophy, elongation, process extension towards the injury site, and overlap of some three-dimensional structural domains [20,24]. It should be noted that the plasticity of healthy astrocytes, which are constantly activated by physiological signals from the CNS, should not be confused with changes in astrocyte responsiveness to pathological stimuli [20]. In this paper, the term "reactive astrocytes" and the above definition of reactive astrocytes will be used (Table 1).

Table 1. Comparison between healthy astrocytes and reactive astrocytes.

Characteristic	Healthy Astrocytes	Reactive Astrocytes
Morphology	Star-shaped morphology Multiple branches with numerous fine processes	Hypertrophy Process elongation Overlap of some structures in three-dimensional space
Molecular Aspect	Low GFAP expression	Increased GFAP expression
Biochemistry	Release of neurotrophic factors	Increased release of pro-inflammatory factors Increased production of ROS Activated complement cascades
Transcriptional Regulation	Steady-state regulation of gene expression	Upregulation of genes associated with neuroinflammation
Function	Neuron trophic support	No or decreased trophic support or active neurotoxicity
	Neurotransmitter uptake and recycling	Decreased neurotransmitter uptake and/or recycling
	Synapse formation, maturation, and function	Decreased synapse formation and altered neuronal activity
	Regulation of blood and glymphatic flow	Increased immune cell infiltration and blood–brain barrier maintenance and/or repair
	Interaction and coordination with immune cells	Proliferate and form scars or borders
	Stable and rhythmic calcium transients	Corral peripheral immune cells and/or amplify inflammatory responses Irregular calcium transients, decreased gap junction coupling Abnormal cellular metabolism Newly acquired neurotoxic or neuroprotective functions, depending on context

2.2. Subsets/Heterogeneity of Reactive Astrocytes

In earlier studies, reactive astrocytes have been divided into neurotoxic and neuroprotective phenotypes, also known as A1 and A2 cell subpopulations [22] (Table 2). However, caution has been advised against using the oversimplified terms "neurotoxic" or "neuroprotective" when characterizing astrocyte phenotypes [20]. This is due to the limitations of these binary classifications in capturing the heterogeneity and diverse functions of reactive astrocytes, especially with advancements in technologies like transcriptomics and high-throughput sequencing [20,21,25]. Transcriptomic studies have revealed that reactive astrocyte phenotypes exhibit significant variability across different regions of the CNS and in response to various pathological stimuli [20,25]. In addition, a recent review [26] introduced an alternative classification scheme for astrocyte reactivity phenotypes. It categorizes them broadly into non-proliferative astrogliosis and proliferative astrogliosis. The former subtype typically occurs in neural tissue responding to pathology while maintaining its fundamental tissue architecture without overt damage [26]. This can be observed in tissue regions distant from focal lesions resulting from stroke, trauma, autoimmune attack, neurodegenerative changes, or diffuse neuroinflammation induced by peripheral exposure to microbial antigens such as lipopolysaccharide (LPS) [27–30]. On the other hand, the latter subtype exhibits anisomorphic features characterized by loss of domain, significant structural reorganization, and the potential diffuse alterations or development of new compact "limitans" borders surrounding evident fibrotic tissue lesions [31]. These changes can occur due to stroke, extensive trauma, infections, foreign bodies (including medical implants), autoimmune inflammation, neoplasms, or profound neurodegenerative processes [31–33]. This classification also possesses certain limitations similar to the previous categorization [26]. Therefore, a broader range of molecules is required to characterize these cells accurately [20]. When identifying astrocyte subpopulations, consideration must be given to multidimensional factors such as location, morphology, gene expression levels, specific cellular functions, and their demonstrated impact on pathological hallmarks to fully appreciate their heterogeneity [34].

Table 2. Comparison between A1 astrocytes and A2 astrocytes.

Characteristic	A1 Astrocytes	A2 Astrocytes
Morphology	Hypertrophy, long dendrites	Hypertrophy, few dendrites
Marker	C3, GBP2, Serping1	PTX3, S100a10, SphK1, tm4sf1, S1Pr3, Tweak
Signaling pathway	Activated NF-κB, JAK/STAT3	Activated PK2/PKR1, JAK/STAT3, FGF2/FGFR1, CXCR7/PI3K/Akt
Cellular functions	Neurotoxic effect: upregulate pro-inflammatory factors; associated with neurodegeneration and chronic neuropathic pain	Neuroprotective effect: upregulate neurotrophic factors and pro-synaptic thrombospondins; promote neuronal growth and support synaptic repair

2.3. Marker of Reactive Astrocytes

With the progression of sequencing technology, numerous astrocyte markers have been identified. However, the specific functions of these markers and their practical implications still need to be clarified, necessitating further exploration through in vivo and in vitro experiments. Several established markers commonly used to label reactive astrocytes and emerging markers with slightly defined functions are described below.

2.3.1. GFAP

As previously delineated, GFAP is an essential protein component of astrocyte intermediate filaments, contributing to the cytoskeletal organization, and serves as the most extensively employed marker of reactive astrocytes [2,7,20]. A prevalent attribute of numerous reactive astrocytes, albeit not all, present in various CNS disorders is the elevation of GFAP due to the upregulation of GFAP mRNA and protein rather than local recruitment or proliferation of astrocytes [34,35]. GFAP is also a sensitive indicator of an early injury response, detectable even without apparent neuronal death [36]. Furthermore, the severity of the injury is correlated with the quantity of GFAP expression in reactive astrocytes [20,37]. Notably, elevated GFAP levels are necessary but insufficient for reactive astrocyte classification, suggesting that increased GFAP levels occur due to pathological stimuli and regional differences in astrocytes, initial GFAP levels, and physiological stimuli [20]. For instance, in a healthy mouse brain, astrocytes in the hippocampus exhibit higher levels of GFAP than those in the cortex, thalamus, or striatum. However, this does not imply that astrocytes in the hippocampus are inherently more reactive [38–40]. Additionally, GFAP expression is influenced by physiological stimuli such as physical activity, exposure to enriched environments, glucocorticoids, and fluctuations in the circadian rhythm in the suprachiasmatic nucleus of the optic cross [41,42]. Moreover, GFAP is not exclusively derived from astrocytes but can also be produced by progenitor cells, depending on the stage of development [3]. Hence, changes in GFAP levels reflect a response to pathological stimuli and adaptation to physiological stimuli and regional differences.

2.3.2. Complement C3

In the central nervous system, complement component 3 (C3) is primarily synthesized by astrocytes [43] and has been used as a marker for type A1 or neurotoxic astrocytes [22]. In conjunction with GFAP, it labels reactive astrocytes and exerts neurotoxicity in neurodegenerative diseases, such as Alzheimer's disease (AD), ALS, multiple sclerosis (MS), and Parkinson's disease (PD), as well as in infectious diseases and spinal cord injury [44–47]. Under physiologic conditions, C3 secreted by astrocytes is involved in the complement cascade and mediates synaptic elimination during the development of the CNS [48]. However, when the CNS is subjected to infections, injuries, and neurodegenerative diseases, the secretion of C3 by reactive astrocytes significantly increases, activating the complement cascades abnormally to eliminate normal synapses and resulting in the loss of neurons and damage to cognitive function [47]. Additionally, C3 also enhances superoxide production and mediates oxidative stress injury in the nervous system [49]. The oxidative stress induced by LPS in chronic neuroinflammation is significantly reduced without C3 expression [49].

2.3.3. Other Markers

TNF-related apoptosis-inducing ligand (TRAIL), also known as tumor necrosis factor superfamily member 10 (TNFSF10), is a member of the TNF superfamily, contributing to apoptosis through both extrinsic and intrinsic signaling pathways. Initially, it was believed to specifically target tumor cells and was absent in the healthy CNS. However, recent studies have demonstrated that astrocytes, microglia, and neurons can express TRAIL within the CNS under pathological conditions [50]. Notably, the function of TRAIL$^+$ astrocytes is versatile. For instance, in experimental autoimmune encephalomyelitis (EAE), which is a mouse model used for studying MS, TRAIL$^+$ astrocytes, acting as anti-inflammation astrocytes, induce CD4$^+$ T cell apoptosis to alleviate neuroinflammation [51]. Conversely, TRAIL$^+$ astrocytes exhibit toxicity towards neurons in the AD mouse model, directly inducing the apoptosis of neurons [52,53].

Reactive astrocytes also express several nonspecific markers [20], including those associated with the cytoskeleton (nestin, synemin, vimentin), cellular metabolism (ALDOC, BLBP/FABP7, MAO-B, TSPO), membrane channels (EAAT1 and 2, KIR4. 1), secreted proteins (CHI3L1/YKL40, Lcn2, Serpina3n/ACT, MT, THBS-1), signal transduction and transcription (NFAT, NTRK2/TrkB IL17R, S100B, SOX9, STAT3), molecular chaperones (CRYAB, HSPB1/HSP27), etc. It should be highlighted that the markers mentioned above are not exhaustive, and they have yet to be employed as a unique marker of reactive astrocytes due to their inability to distinguish between specific types of reactive astrocytes. Consequently, numerous additional markers need to be identified.

2.4. Functions of Reactive Astrocytes

Reactive astrocytes exhibit a significant loss of functional capabilities compared to their regular counterparts, characterized by reduced trophic support, neurotransmitter uptake, synapse formation, gap junction coupling, and altered neuronal activity [20,54]. These deficits are accompanied by increased immune cell infiltration and irregular calcium transients [54]. Reactive astrocytes can acquire novel neurotoxic or neuroprotective functions depending on the specific pathological condition [21,23,54]. For instance, in a transgenic mouse model targeting reactive astrocyte ablation, CNS tissue experiences a significantly more severe disruption, demyelination, neuronal and oligodendrocyte death, and pronounced motor deficits, along with an inability to repair the blood–brain barrier, in comparison to the non-transgenic mouse model after mild or moderate stab or crush spinal cord injury [55]. Notably, the functions of reactive astrocytes are not always constant. After cerebral ischemia, these cells exhibit a protective role in the early stage by secreting neurotrophic substances and antioxidants and forming glial scars to limit the spread of the immune response. However, in the later stages, the formation of a glial scar impedes neurological recovery [56]. In summary, when the CNS experiences secondary degeneration caused by trauma or ischemia, reactive astrocytes may provide protection in the early stage by repairing the blood–brain barrier, restricting neuroinflammation, and preserving motor functions. Nonetheless, astrocytic scar formation is detrimental to axonal regeneration in the later stages [57]. Conversely, in neurodegenerative diseases (NDs) such as AD, PD, MS, Huntington's disease (HD), and ALS, abnormal protein accumulation, excessive production of inflammatory factors and reactive oxygen species (ROS), and disruption of ion homeostasis and metabolism in reactive astrocytes contribute to a persistent inflammatory environment and neuron death [34,54]. The extent of the protective influence of reactive astrocytes in NDs during various disease stages has yet to be thoroughly investigated.

2.5. Link between Reactive Astrocytes and Environmental Elements

The development of neuropathological conditions, such as ALS, is influenced by environmental factors, such as the presence of heavy metals and pesticides. In today's rapidly industrializing and modernizing world, heavy metal pollution has emerged as a well-recognized public health concern that impacts daily life, including food, water sources, air quality, and occupational exposure [58,59]. Excessive intake of heavy metals can lead to neurotoxicity and subsequent neurological disorders [60–62]. As previously mentioned, astrocytes play a pivotal role in maintaining homeostasis in the central nervous system and safeguarding neurons against various types of harm caused by heavy metal accumulation [7,61]. However, this protective function also makes astrocytes susceptible to the neurotoxicity of heavy metals. This vulnerability manifests through distributions in blood–brain barrier integrity, elevated levels of ROS, pro-inflammatory factors, impairment of mitochondrial respiration, and abnormalities in glutamate and lipid metabolism. These effects have been demonstrated through numerous experiments conducted both in vivo and in vitro [60–67]. For instance, Shi Fan et al.'s study [63] involved exposing rats to drinking water containing lead acetate (PbAc) for nine continuous weeks, which impaired learning memory and exploratory abilities. Additionally, expression levels of GFAP, along with other genes associated with reactive astrocytes affected by neurotoxicity, were significantly elevated compared to the control group. Subsequent experiments involved administration of PbAc to MA-c cells, an astrocyte cell line, confirmed these findings while revealing that NF-κB transcription factor regulates astrocyte activation following lead exposure [63]. Despite substantial evidence from experimental studies supporting these conclusions, however, little information is available regarding the cumulative effects of heavy metals on human astrocytes [61]. Given the indispensable role of pesticides in agricultural development, humans are exposed to these chemicals through various means such as occupational activities, agricultural practices, domestic use, and air, water, soil, and food contamination. Similar to heavy metals, this exposure leads to disruption of the BBB and activation of astrocytes, based on numerous studies conducted in vivo and in vitro [68–74]. Furthermore, research has indicated that Parkinson's disease protein 7 (PARK7/DJ-1), found in astrocytes, plays a crucial role in regulating the neurotoxic consequences caused by the pesticide rotenone [75–77]. Additionally, maintaining astrocyte homeostasis is closely associated with other environmental factors such as gut microbiota, composition intake of dietary components, and air pollution [78–81].

3. Reactive Astrocytes Are Toxic to MNs in ALS

Reactive astrocytes in ALS display morphological modifications, such as hypertrophy, process elongation, and partial overlap of specific features of these cells (protrusions, branches, or other morphological characteristics) in three-dimensional space [20]. Moreover, these cells undergo chemical remodeling characterized by increased transcription of pro-inflammatory factors and oxidation particles [82]. They also exhibit molecular remodeling, with markedly elevated expression of GFAP and C3 compared to naive astrocytes [83]. Numerous in vivo and in vitro studies have corroborated the significant elevation of C3 mRNA and protein levels in mouse models of ALS, as well as in patients with familial and sporadic ALS (SALS and FALS) [22,83–85]. Furthermore, inhibition of C3 release from reactive astrocytes has been shown to reduce neuronal damage [86]. One previous study has demonstrated that astrocytes derived from both sporadic and familial ALS patients exhibit an equivalent level of toxicity toward motor neurons [82]. In this investigation, SALS and FALS astrocytes were obtained from postmortem spinal cord neural progenitor cells (NPCs), which were then supplemented with 10% fetal bovine serum to induce differentiation into astrocytic. This innovative model system was employed to elucidate the underlying molecular mechanisms and evaluate potential therapeutic approaches for SALS [82]. The researchers co-cultured these differentiated astrocytes with mouse embryonic stem-cell-derived motor neurons, observing an accelerated demise of motor neurons in this co-culture setup while noting a significant upregulation of twenty-two inflammatory genes in both FALS and SALS astrocytes [82]. Additionally, when these were co-cultured with GABAergic neurons, they did not exert any influence on them. These findings suggest that astrocytes target and impair motor neurons [19,82,87].

Frontotemporal dementia (FTD) encompasses a spectrum of disorders characterized by the progressive degeneration of the frontal and temporal lobes in the brain, resulting in alterations in personality, behavior, and language. Coexistence or shared clinical, genetic (*SOD1*, *TARDBP*, and *C9ORF72* et al.), and pathological features (TDP-43 inclusions in astrocytes and neurons) have been observed between ALS and FTD [88]. Previous studies have demonstrated that astrocytes from postmortem patients with ALS/FTD also exert detrimental effects on motor neurons [89–91]. Astrocyte dysfunction specific to ALS/FTD is comparable to that seen in ALS without FTD; however, there are some distinctions [92]. For example, individuals with both ALS and FTD exhibit more pronounced increased blood–brain barrier permeability associated with poor prognosis compared to those solely affected by ALS [93]. In a clinical trial comparing GFAP levels within serum samples taken from participants experiencing cognitive and/or behavioral impairment or FTD versus those diagnosed only with ALS, the results showed significant variation [88]; however, no difference was noted among various types of clinical presentations for individuals diagnosed with only ALS regarding GFAP levels within their serum samples [88].

Recent investigations have indicated distinct roles for astrocytes derived from both the motor cortex (MC) and the spinal cord (SC) of newborn $SOD1^{G93A}$ mice during disease progression stages [94–96]. Spectrophotometric and cytofluorimetric analyses revealed elevated redox stress, reduced antioxidant capacity, and relative mitochondrial respiratory uncoupling in MC $SOD1^{G93A}$ astrocytes. In contrast, SC mutated cells exhibited enhanced resistance against oxidative damage, attributed to augmented antioxidant defense [94]. However, the most extensively studied astrocytes throughout this manuscript are predominantly derived from the spinal cord.

In general, the detrimental effects of reactive astrocytes are not only mediated by mitochondrial dysfunction, calcium homeostasis dysregulation, and endoplasmic reticulum stress but also exacerbated by metabolic dysfunction. This exacerbation triggers neuroinflammatory responses and releases various toxic factors, including polyphosphate, glutamate, lactate, and lipids, which directly act on motor neurons.

3.1. Mitochondrial Dysfunction

Prior research has observed a disruption of the mitochondrial respiratory chain in astrocytes derived from $SOD1^{G93A}$ rats, characterized by diminished oxygen utilization, absence of ADP-dependent respiratory control, and reduced membrane potential. This leads to elevated oxygen radicals and nitric oxide levels, contributing to motor neuron demise [97–99]. There is a notable reduction in motor neuron survival when employing mitochondrial respiration inhibitors in non-transgenic astrocyte cultures, particularly azide-dependent inhibition of cytochrome c oxidase and fluorocitrate-dependent inhibition of aconitase [97]. This finding suggests that the survival of motor neurons depends on the mitochondrial function in astrocytes. Furthermore, the neurotoxic phenotype can be mitigated by restoring mitochondrial respiration in astrocytes by using antioxidants Mito-Q and Mito-CP [97]. In other in vivo studies [100,101], it was found that the administration of dichloroacetate (DCA), a drug enhancing mitochondrial function by stimulating the activity of the pyruvate dehydrogenase (PDH) complex, improves survival and motor performance while reducing MN degeneration and gliosis in the $SOD1^{G93A}$ rat model. Furthermore, when DCA was used in astrocyte–motor neuron co-cultures (consisting of astrocytes derived from the spinal cord of an $SOD1^{G93A}$ rat and motor neurons derived from embryonic day 15 rats), phosphorylation of PDH decreased, leading to enhanced mitochondrial coupling and increased motor neuron survival [101].

3.2. Disturbance of Ca^{2+} Homeostasis

The resting state of astrocytes is contingent on Ca^{2+} signals, such as local Ca^{2+} fluctuations or Ca^{2+} waves, to execute their pathological or physiological functions, including the secretion of neurotrophic or neurotoxic factors [102]. The endoplasmic reticulum (ER) is widely regarded as the most crucial and metabolically relevant reservoir and buffering system for intracellular Ca^{2+}. It has been recognized that perturbations in astrocyte Ca^{2+} homeostasis can exert toxic effects on ALS motor neurons, but the exact mechanism remains complex and elusive [103]. Recent findings suggest that alterations in astrocyte store-operated Ca^{2+} entry (SOCE) might underlie abnormal gliotransmitter secretion and astrocyte-mediated neurotoxicity in ALS [104,105]. In this study, SOCE in $SOD1^{G93A}$ mouse spinal cord primary astrocytes was compared with that in wild-type (WT) controls, revealing an increase in SOCE in $SOD1^{G93A}$ astrocytes concurrent with a decline in ER Ca^{2+}-ATPase and ER Ca^{2+} concentrations, resulting in abnormally high intracellular Ca^{2+} variations that potentially harm MNs [105]. Another study [6] found that astrocytes with TDP-43 inclusions exhibit reduced monocarboxylate transporter one and noradrenergic cAMP and Ca^{2+} signaling. These changes play a pivotal role in modulating cellular metabolism, contributing to excessive accumulation of lipid droplets and increased glycolysis and lactate. These findings indicate that astrocytes with TDP-43 inclusions are unable to support neurons. Other astrocyte ion imbalances, such as K^+ and Na^+, leading to motor neuron death, are also observed in ALS [106–109].

3.3. PolyP

Polyphosphate (polyP), an inorganic neuroactive compound that potentiates the activity of Nav and Kv channels, is synthesized by astrocytes, functioning as a glial messenger to facilitate communications between astrocytes and neurons [110]. Elevated polyP levels were observed in induced pluripotent stem cell (iPSC)-derived astrocytes from mice and postmortem patients with various ALS/FTD-associated mutations (*SOD1*, *TARDBP*, and *C9ORF72*) [89,90]. Similarly, spinal cord sections from patients with familial and sporadic ALS displayed abundant polyP staining signals [89]. Furthermore, polyP levels were increased in the astrocyte-conditioned media (ACM) from ALS/FTD. In contrast, motor neuron death was significantly decreased by degradation or neutralization of polyP within ALS/FTD astrocytes or ACM, suggesting that excessive astrocytic polyP could be a critical factor for non-cell autonomous MN degeneration and a potential therapeutic target for ALS/FTD [89,90]. Elevated polyP levels were also detected in the cerebrospinal fluid (CSF) of ALS patients, indicating that polyP might serve as a novel biomarker for ALS/FTD [89]. However, it is noteworthy that in an in vivo experiment [89], injecting viral vectors into the intracerebroventricular compartments of $SOD1^{G93A}$ suckling mice to disrupt polyP production did not delay disease onset or extend the survival of $SOD1^{G93A}$ mice, despite a significant reduction in polyP deposition in astrocytes and neurons, suggesting that astrocyte-derived polyP might be involved in ALS pathogenesis in conjunction with other factors. Consequently, further exploration and research into the role of polyP in ALS pathogenesis are warranted.

3.4. Glutamate

Glutamate (Glu), a major excitatory neurotransmitter in the CNS [111], exhibits overexcitation and excitotoxic effects when its inter-synaptic concentration is abnormally high. This occurs through the activation of glutamate receptors in the postsynaptic membrane, potentially leading to irreversible neuronal damage. Astrocytes play a role in the metabolism of glutamate released into the synaptic cleft [112]. As there is a lack of extracellular glutamate metabolizing enzyme, glutamate can only be taken up by astrocytes with the assistance of glutamate transporters to maintain normal glutamate levels in the extracellular fluid. Five types of excitatory amino acid transporters (EAATs) have been identified, with EAAT1 and EAAT2 primarily found in astrocytes and the other three mainly in neurons. EAAT2, also known as glutamate transporter 1 (GLT-1), accounts for over 90% of glutamate uptake into the cell [113,114]. Under physiological conditions, astrocytes convert glutamate to glutamine, providing energy for motor neurons, serving as a neurotransmitter precursor, and contributing to the amino acid balance in the central nervous system [115]. In many animal models (ALS $SOD1^{G93A}$ mouse and rat models, as well as TDP43 ALS mouse models) and patients with ALS, a significant reduction in the expression of the *EAAT2* gene in reactive astrocytes has been observed, leading to an accumulation of excess glutamate in the extracellular synaptic cleft. This results in excitatory cytotoxicity of ALS spinal cord motor neurons and impaired motor neuron survival [112,115]. Riluzole, the first drug approved by the FDA for the treatment of ALS, functions by reducing neuronal excitability by blocking glutamatergic neurotransmission in the CNS and activating postsynaptic glutamate receptors to promote glutamate uptake [13].

Previously, membralin (Tmem259 or C19orf6), an innovative component of the ER-associated degradation (ERAD) machinery, was identified, significantly reducing Aβ production by limiting the excessive activation of the γ-secretase complex. Recently, a study has demonstrated a significant reduction in the expression of membralin in astrocytes from the spinal cord of ALS postmortem patients and $SOD1^{G93A}$ mice [112]. Furthermore, the absence of membralin has been found to significantly impact *EAAT2* expression through the TNF-α/TNFR1/NFκB pathway, dramatically increasing extracellular glutamate and glutamatergic motor neuron toxicity [112]. Conversely, the elevation of membralin expression through transduction of adeno-associated virus (AAV)-membralin in $SOD1^{G93A}$ mice has demonstrated that increased membralin expression can reverse the neurotoxic effect, prolong mouse survival, reduce glial cell proliferation, and enhance *EAAT2* expression [112]. Collectively, these findings underscored the crucial role of membralin in astrocyte-regulated glutamate homeostasis and EAAT2-mediated glutamate excitotoxicity in ALS.

Previous research has demonstrated that the treatment of astrocytes expressing the $SOD1^{G93A}$ mutation, when co-cultured with motor neurons in the presence of glutamate, leads to decreased levels of lactate, creatine, creatinine, deoxycarnitine, L-acetylcarnitine, and nicotinamide adenine dinucleotide (NAD) and elevated glucose levels [116]. As NAD is essential for both glycolysis and lactate dehydrogenase activity, the observed reduction in lactate and increased glucose levels in ALS can be attributed to impaired glycolysis resulting from reduced NAD levels [116,117].

3.5. Fatty Acids

In the CNS, surplus fatty acids are primarily stored in astrocytes, predominantly as lipid droplets. This storage increases under conditions of hypoxia, cellular stress, and exposure to high levels of exogenous free fatty acids [118]. When released from the ApoE-positive lipid granules of overactive neurons, toxic lipids are taken up by neighboring astrocytes via endocytosis. These transferred fatty acids are utilized as metabolic intermediates to enhance mitochondrial oxidation and detoxification in astrocytes, thereby preventing the accumulation of toxic fatty acids in neurons.

It is well established that the brain is an energy-intensive organ primarily fueled by glucose. However, recent studies have indicated that approximately 20% of the brain's total energy requirement is derived from fatty acid β-oxidation in astrocytes [118]. Notably, fatty acid oxidation for energy generation is a double-edged sword, as it produces more energy but consumes more oxygen, potentially exposing cells to oxidative stress and exacerbating the production of reactive oxygen species if fatty acid β-oxidation persists [119]. Previous studies have demonstrated that lipids served as the primary energy in ALS due to the high energy demands of neurons and impaired glucose metabolism. However, as mentioned above, this metabolic switch may generate additional oxidative stress products, leading to the demise of motor neurons [120,121]. One study found that astrocytes expressing mutant TDP43 exhibit a more significant accumulation of lipid droplets [6].

Fatty-acid-binding proteins (FABPs) are vital regulators of lipid metabolism, energy homeostasis, and inflammation by controlling nuclear receptor uptake, transport, and ligand availability [122]. FABP7, the predominant brain FABP isoform [123], is primarily expressed in astrocytes, safeguarding these cells against ROS toxicity through the formation of lipid droplets [124] and playing an essential role in reactive astrocyte proliferation associated with CNS injury [125]. Moreover, FABP7 regulates astrocyte responses to external stimuli by controlling lipid raft function [126]. These findings suggest that FABP7 exhibits neuroprotective effects in reactive astrocytes [127]. However, in the spinal cords of $SOD1^{G93A}$ and $SOD1^{H46R/H48Q}$ mice, FABP7 expression is upregulated in grey matter astrocytes. It ultimately harms motor neuron survival by promoting NF-κB-driven pro-inflammatory responses in astrocytes [122].

In the CNS, polyunsaturated fatty acids, particularly arachidonic acid, are primarily produced and secreted by astrocytes, which further synthesize prostaglandin E2 (PGE2), an inflammatory molecule contributing to neuroinflammation and motor neuron death. The levels of PGE2 in the CSF of most ALS patients are elevated 10-fold [128–132]. Cyclooxygenase 2 (COX2) catalyzes the conversion of arachidonic acid to PGE2; thus, blocking COX2 specifically with celecoxib or rofecoxib may slow down the development and progression of ALS [133,134].

Astrocytes represent the predominant cell type for cholesterol production in the CNS, with their biosynthesis being governed by the transcription factor sterol regulatory element-binding protein-2 (SREBP2), which is notably elevated in ALS [133]. Overexpression of SREBP2 in the CNS results in an accumulation of cholesterol and neutral lipids and the emergence of ALS-like symptoms, including progressive hind limb paralysis, spasticity, and shortened lifespan in mice [120,133].

Recent findings also indicated that astrocyte-mediated cell death is triggered by astrocytes' secretion of saturated lipids. Furthermore, in vitro and in vivo models of acute axonal injury induced by astrocytes can be mitigated by explicitly silencing the expression of the saturated lipase *ELOVL1* in astrocytes, thereby preventing the formation of long-chain saturated lipids [135]. Increased expression of astrocyte *ELOVL1* or an elevation in the production of long-chain saturated free fatty acids has been reported in ALS, warranting further investigation [133].

4. Elements Leading to Astrocyte Activation in ALS

In general, the activation of inflammatory factors, upregulation of peroxiredoxin 6 (*PRDX6*), and gene mutation in astrocytes are the main contributors to the reactive transformation of astrocytes in ALS (Figure 1). This transformation is mediated by pro-inflammatory factors, such as IL-1α, TNFα, and C1q, secreted by reactive microglia. These factors trigger the conversion of quiescent astrocytes into reactive astrocytes, characterized by a notable upregulation of C3 expression, leading to the death of neurons and oligodendrocytes [22]. In the $IL\text{-}1\alpha{-}/{-}\ TNF\alpha{-}/{-}\ C1q{-}/{-}\ SOD1^{G93A}$ mouse model, the knockout of *IL-1α*, *TNFα*, and *C1q* significantly reduced the proportion of $C3^+$ reactive astrocytes, improved motor function, and extended survival in $SOD1^{G93A}$ mice [83]. The treatment of low doses of IL-1α, TNFα, and C1q to $SOD1^{G93A}$ -expressing astrocytes and WT astrocytes elicited a notable increase in immune activation and astrocyte reactivity-associated genes upregulated in $SOD1^{G93A}$ astrocytes [83]. This finding suggests that $SOD1^{G93A}$ astrocytes can generate a significant response to minor damage. Concurrently, it underscores the crucial role of mutant *SOD1* in transforming astrocytes into reactive astrocytes and demonstrates the cellular autonomy of astrocytes [83]. Furthermore, a recent study has shown that the expression of *PRDX6* in the spinal cord of $SOD1^{G93A}$ mice is also involved in the induction of A1-type astrocytes and the excessive production of inflammatory cytokines through a calcium-dependent phospholipase A-dependent mechanism [84].

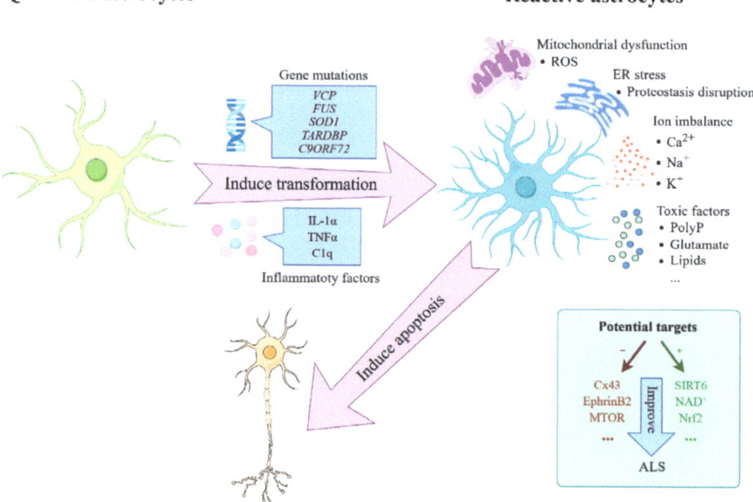

Figure 1. Astrocytes transforming from quiescence to reactivity are toxic to MNs in ALS. Under conditions of gene mutations (*VCP, FUS, SOD1, TARDBP, C9ORF72*) or inflammatory factors (IL-1α, TNFα, C1q) in in vivo or in vitro models, astrocytes transform from quiescence to reactivity in ALS. Reactive astrocytes, ultimately resulting in the apoptosis of motor neurons (MNs), are characterized by mitochondrial dysfunction, ER stress, ion imbalance, secretion of toxic factors, etc. Additionally, inhibiting Cx43, EphrinB2, and MTOR or enhancing SIRT6, NAD^+, and Nrf2 can mitigate motor neuron loss.

Apart from ALS-linked pathogenic variants in *SOD1* [136,137], astrocytes undergo a reactive transformation in response to various pathogenic mutations, such as *VCP*, *FUS*, *TARDBP*, and *C9ORF72* [138,139]. Despite their molecular and functional heterogeneity at early stages, treatment with inflammatory factors ultimately transforms all astrocytes into C3-related reactive astrocytes. In contrast, when WT astrocytes were treated with a conditioned medium obtained from *SOD1* or *VCP* mutant hiPSC-derived astrocytes, no increase in C3 expression or reactive astrocyte transformation was observed [19]. This suggests that the transformation is not caused by the secretion of mutant astrocytes but rather by the autonomy of these cells, further emphasizing the significance of the mutation in the pathogenesis of the disease [19,139]. Moreover, inhibiting the expression of both mutant *SOD1* and wild-type *SOD1* in FALS/SALS astrocytes using a lentivirus encoding a short hairpin (sh) RNA leads to a remarkable reduction in astrocyte-mediated motor neuron toxicity [82,87]. This finding implies that suppressing *SOD1* expression in hiPSC-derived astrocytes could be a potential therapeutic target for both FALS patients with *SOD1* mutations and SALS patients. The newly approved drug tofersen, an antisense oligonucleotide that reduces SOD1 protein synthesis, holds promise [14].

Although the role of reactive astrocytes in the pathogenesis of ALS has been highlighted, it is important to note that they are not a trigger for MN death but crucial contributors. Therefore, the occurrence of neurodegeneration depends on the pathology of neurons.

5. Potential Targets on Astrocytes for the Treatment of ALS

At present, the available treatment options and their therapeutic effects for ALS are minimal. As previously mentioned, astrocytes play an essential role in the pathogenesis and disease progression of ALS. To date, riluzole remains the sole drug that targets astrocytes [13]. Therefore, there is an urgent imperative to identify novel potential therapeutic targets for astrocytes.

5.1. GDNF

The neurotrophic factor glial-cell-line-derived neurotrophic factor (GDNF), secreted by astrocytes, plays a crucial role in neuronal survival and synaptic promotion. However, the function of reactive astrocytes is impaired in ALS models and patients, resulting in motor neuron death. Consequently, delivering CNS GDNF or transplanting healthy astrocytes may potentially improve motor function in ALS patients [140,141].

A combination of stem cell and gene therapy was employed in a phase I/IIa clinical trial led by Dr Clive Svendsen's team [140]. Neural progenitor cells, which were genetically engineered to express GDNF protein, were transplanted into the dorsal and ventral horns of the lumbar segment of the spinal cord in ALS patients. These cells were then transformed into supportive glial cells. The neural precursor cells can give rise to new supporting glial cells, releasing the protective protein GDNF, collectively aiding in preserving motor neurons [140]. This "double whammy" approach concurrently employs the generated new glial cells and GDNF protein to support the survival of dying motor neurons in the face of the disease.

The limited half-life of GDNF in plasma, its inability to directly cross the blood–brain barrier during subcutaneous administration, and its poor penetration into the brain and spinal cord during intrathecal injection trials render it challenging to achieve a therapeutic effect using these approaches. Consequently, in this trial [140], the stem cell product CNS10-NPC-GDNF was safely delivered into the dorsal and ventral horns of the lumbar segment of the spinal cord in ALS patients using a novel in-house-developed injection device. After a single transplant via this innovative method, neural progenitor cells survived up to 42 months and continued to generate new glial cells and GDNF proteins. The results indicated that the rate of leg strength decline was slower on the treated side than on the untreated side, although this difference was not statistically significant. Furthermore, this cell transplantation did not cause substantial adverse effects on muscle strength in the treated leg compared to the untreated side. However, in some patients, many of these cells reached sensory areas in the spinal cord, potentially leading to pain [140]. Overall, this clinical trial demonstrated the safety of this approach, but further assessments of efficacy are required. The team is also currently utilizing these GDNF-secreting stem cells in another ALS clinical trial (https://clinicaltrials.gov/ct2/show/NCT05306457, accessed on 3 November 2023) by transplanting them into the "hand-knob" area of the motor cortex of patients with ALS. Ongoing progress in the efficacy and safety of stem cell combined gene therapy for ALS patients should be expected.

5.2. AstroRx®

AstroRx®, an allogeneic cell-based product derived from human embryonic stem cells, is generated under cGMP conditions in Kadimastem's GMP facility via standard procedures and assessed according to stringent criteria by external qualified certified GLP laboratory (Hylabs laboratories, Jerusalem, Israel) [141]. It exhibits functional, healthy astrocyte effects, such as clearing excessive glutamate, reducing oxidative stress, secreting various neuroprotective factors, and acting as an immunomodulator. In a phase I/IIa clinical trial involving intrathecally injected human astrocytes (AstroRx®), the rate of ALSFRS-R worsening within the first three months post-treatment was significantly reduced, accompanied by fewer adverse events, regardless of whether subjects received high or low doses of healthy astrocytes. These positive results warrant further exploration of repeated intrathecal administration of AstroRx®, such as every three months [141].

Regarding the two emerging clinical trials mentioned above, each has its merits and drawbacks. After a single treatment with stem cells combined with gene therapy, neural progenitor cells can survive for extended periods, differentiate into new glial cells, continuously produce GDNF proteins, and directly act on a specific group of neurons. In contrast, direct intrathecal injection of astrocytes has a shorter duration of action and may necessitate injections every three months. Most importantly, the ability of healthy astrocytes to perform their normal function in the CNS within an inflammatory environment remains unclear. Meanwhile, there are existing challenges associated with stem cell transplantation, including immune rejection, abnormal hyperplasia, ethical concerns, etc. In terms of efficacy, intrathecal injection of astrocytes has been preliminarily validated in phase I/IIa clinical trials. However, the effect of stem cells combined with gene therapy remains uncertain in current clinical trials. Overall, these research advancements are promising and deserve further investigation.

5.3. Cx43

In recent years, foundational research has unearthed additional potential astrocyte targets. Cx43, an essential astrocyte connectivity protein, together with its hemichannels, facilitates communication between astrocytes within the central nervous system [142]. Increased expression of Cx43 has been observed in animal models of ALS, the cerebrospinal fluid of ALS patients, and postmortem samples, indicating its toxicity towards neurons [143]. In vitro experiments such as co-culturing and blocking Cx43 and its hemichannels corroborate this. In vivo experiments have revealed that the removal of Cx43, specifically from astrocytes in $SOD1^{G93A}$ mice, resulted in a spatial (in the cervical and lumbar spinal cords) and temporal (at the pre-symptomatic, symptomatic, and end stages) deceleration of disease progression, as well as protection for motor neurons, and an increase in survival rate [143]. Tonabersat, a drug candidate capable of blocking Cx43 hemichannels and crossing the blood–brain barrier [144], has been shown to provide neuroprotection by reducing neuronal death when co-cultured with human induced pluripotent stem-cell-derived astrocytes (hiPSC-A) derived from both familial and sporadic ALS patients using control motor neurons (hiPSC-MNs) [144]. Administration of tonabersat intraperitoneally at 10 mg/kg once daily to $SOD1^{G93A}$ mice exhibited potential for enhancing motor function [143]. Notably, the expression of Cx43 in astrocytes remains unaltered by tonabersat, whereas the expression of GFAP and Iba-1 significantly decreased [143]. The drug has also been investigated in the context of migraine and epilepsy [144]. In conclusion, the targeted blockade of astrocyte Cx43 and the integration of tonabersat into ALS clinical trials are worth considering.

5.4. EphrinB2

Expression of ephrinB2, a transmembrane signaling molecule, is significantly elevated in astrocytes within the spinal cord of $SOD1^{G93A}$ mice and ALS patients [145,146]. Delivery of viral-mediated shRNA to astrocytes in the cervical segment of the spinal cord selectively represses ephrinB2 expression, thereby mitigating motor neuron loss and preserving respiratory function by sustaining motor neuron innervation of the diaphragm [146]. This study suggests that the upregulation of ephrinB2 is both a transcellular signaling mechanism for astrocyte pathogenicity in ALS and a promising therapeutic target.

5.5. NAD$^+$, Nrf2, and SIRT6

Preliminary findings suggest that NAD$^+$, Nrf2, and SIRT6 synthesized by astrocytes confer neuroprotection against ALS, with SIRT6 playing a pivotal role. An increase in NAD$^+$ availability is known to enhance resistance to oxidative stress and reduce mitochondrial ROS production in various cell types and disease models [147]. Nrf2 activation is critical for regulating antioxidant defenses and protecting neighboring neurons in co-culture and in vivo settings [148,149]. Furthermore, elevating total NAD$^+$ levels in astrocytes activates Nrf2 and SIRT6 in these cells, while SIRT6 overexpression further activates Nrf2. Decreased expression of NAD$^+$, Nrf2, and SIRT6 has been observed in the spinal cords of ALS patients. In animal models of ALS, NAD$^+$ depletion does not affect survival, but administering biologically active NAD$^+$ precursors significantly improves motor performance and extends survival [150]. In addition, upregulating Nrf2 in astrocytes has been demonstrated to promote neuronal survival in in vitro co-culture studies [151] and in an ALS mouse model [152]. However, silencing SIRT6 expression in an in vitro cell culture model did not prevent astrocyte neurotoxicity towards motor neurons, even with pre-supplementation of NAD$^+$ precursors [153]. Thus, SIRT6 plays a crucial role in this neuroprotective effect. Overall, enhancing SIRT6 and Nrf2 activity and administering NAD$^+$ precursors that abolish the neurotoxic phenotype of astrocytes expressing the ALS-associated mutation $SOD1$ are potential therapy approaches [154–156].

5.6. MTOR

The mechanistic target of rapamycin kinase (MTOR) is a regulator of numerous extracellular and intracellular signals that participate in cellular metabolism, growth, proliferation, survival, and macro-autophagy/autophagy [157]. Activation of the MTOR pathway has been demonstrated to be elevated in $SOD1^{G93A}$ mutant hiPSC-derived astrocytes, resulting in the suppression of macro-autophagy/autophagy, aberrant cell proliferation, and an increased reactivity of the astrocytes [158]. Concurrently, MTOR pathway activation is correlated with post-transcriptional upregulation of the insulin-like growth factor 1 receptor (IGF1R). Therefore, inhibition of the IGF1R-MTOR pathway decreases cell proliferation and the reactivity of mutant $SOD1^{G93A}$ astrocytes, thereby mitigating their toxicity towards motor neurons. These findings suggest that modulation of the IGF1R-MTOR pathway in astrocytes may represent a plausible therapeutic target for ALS [158].

6. Conclusions and Future Prospects

The pathogenesis of ALS is complex and involves multiple pathophysiological mechanisms, among which reactive astrocytes play a crucial role. Therefore, therapeutic strategies targeting astrocytes, such as inhibiting their reactive transformation at each stage of ALS, obstructing the pathways through which transformed reactive astrocytes exert toxic effects on neurons, and replacing reactive astrocytes with normal astrocytes, could potentially offer ALS patients the prospect of prolonged survival and improved motor and respiratory functions.

Despite extensive research confirming the non-cell-autonomous functions of reactive astrocytes in ALS, numerous queries necessitate future investigation. These encompass: (1) the practical applications of these findings for ALS patients; (2) the development of more precise and scientifically grounded classification methods to identify reactive astrocytes in ALS; (3) investigation into potential variations in morphology, molecular composition, functionality, and gene expression of reactive astrocytes among different clinical subtypes of ALS. Understanding these distinctions could offer insights into disease onset, affected regions, progression, motor function impairment, respiratory function decline, prognosis, and other aspects. (4) By utilizing advanced techniques such as positron emission tomography (PET), specific markers and tracers with defined functions can be employed to dynamically monitor the changes and migration patterns of astrocytes within various regions of the central nervous system during pre-symptomatic stages as well as symptomatic and end-stage phases in both animal and human patients with ALS. This approach aims to visualize the involvement of reactive astrocytes directly in living organisms throughout the progression of ALS. It is anticipated that further comprehensive research will address these inquiries while enhancing our understanding of the role of reactive astrocytes in ALS.

Author Contributions: K.Y. prepared the manuscript and created the figures and tables, and Y.L. and M.Z. reviewed the manuscript. All authors have read and agreed to the published version of the manuscript.

Funding: This research was funded by the National Natural Science Foundation of China (grant number 82271478).

Institutional Review Board Statement: Not applicable.

Informed Consent Statement: Not applicable.

Data Availability Statement: Not applicable.

Acknowledgments: Thanks for all the references. The figures were created using Figdraw (www.figdraw.com, accessed on 30 October 2023).

Conflicts of Interest: The authors declare no conflicts of interest.

Abbreviations

BBB: blood–brain barrier; ALS: amyotrophic lateral sclerosis; SALS: sporadic amyotrophic lateral sclerosis; FALS: familial amyotrophic lateral sclerosis; FTD: frontotemporal dementia; CNS: central nervous system; GFAP: glial fibrillary acidic protein; MNs: motor neurons; AD: Alzheimer's disease; MS: multiple sclerosis; PD: Parkinson's disease; LPS: lipopolysaccharide; TRAIL: TNF-related apoptosis-inducing ligand; NDs: neurodegenerative diseases; HD: Huntington's disease; ROS: reactive oxygen species; SOD1: superoxide dismutase-1; ER: endoplasmic reticulum; SOCE: store-operated Ca^{2+} entry; polyP: polyphosphate; ACM: astrocyte conditioned media; iPSC: induced pluripotent stem cell; FTD: frontotemporal dementia; TARDBP: transactive response DNA-binding protein; C9ORF72: chromosome 9 open reading frame 72; CSF: cerebrospinal fluid; EAATs: excitatory amino acid transporters; GLT-1: glutamate transporter 1; ERAD: ER-associated degradation; NAD: nicotinamide adenine dinucleotide; FABPs: fatty-acid-binding proteins; PGE2: prostaglandin E2; COX2: cyclooxygenase 2; SREBP2: sterol regulatory element-binding protein-2; PRDX6: peroxiredoxin 6; MTOR: mechanistic target of rapamycin kinase; IGF1R: insulin-like growth factor 1 receptor; SIRT6: sirtuin 6; Nrf2: nuclear factor erythroid 2-related factor 2; NPCs: neural progenitor cells; GDNF: glial-cell-derived neurotrophic factor; Cx43: connexin 43; VCP: valosin-containing protein; TDP-43: TAR DNA binding protein 43; FUS: fused in sarcoma; iPSC: induced pluripotent stem cell; LPS: lipopolysaccharide; C3: complement component 3; TNFSF 10: tumor necrosis factor superfamily member 10; EAE: experimental autoimmune encephalomyelitis; PbAc: lead acetate; PARK7/DJ-1: Parkinson's disease protein 7; MC: motor cortex; SC: spinal cortex; DCA: dichloroacetate; PDH: pyruvate dehydrogenase; WT: wild type; Glu: glutamate.

References

1. Allen, N.J.; Lyons, D.A. Glia as architects of central nervous system formation and function. *Science* **2018**, *362*, 181–185. [CrossRef]
2. Hasel, P.; Liddelow, S.A. Astrocytes. *Curr. Biol.* **2021**, *31*, R326–R327. [CrossRef] [PubMed]
3. Malatesta, P.; Hack, M.A.; Hartfuss, E.; Kettenmann, H.; Klinkert, W.; Kirchhoff, F.; Götz, M. Neuronal or glial progeny: Regional differences in radial glia fate. *Neuron* **2003**, *37*, 751–764. [CrossRef] [PubMed]
4. Hart, C.G.; Karimi-Abdolrezaee, S. Recent insights on astrocyte mechanisms in CNS homeostasis, pathology, and repair. *J. Neurosci. Res.* **2021**, *99*, 2427–2462. [CrossRef] [PubMed]
5. Sofroniew, M.V.; Vinters, H.V. Astrocytes: Biology and pathology. *Acta Neuropathol.* **2010**, *119*, 7–35. [CrossRef]
6. Velebit, J.; Horvat, A.; Smolič, T.; Prpar Mihevc, S.; Rogelj, B.; Zorec, R.; Vardjan, N. Astrocytes with TDP-43 inclusions exhibit reduced noradrenergic cAMP and Ca^{2+} signaling and dysregulated cell metabolism. *Sci. Rep.* **2020**, *10*, 6003. [CrossRef]
7. Verkhratsky, A.; Nedergaard, M. Physiology of Astroglia. *Physiol. Rev.* **2018**, *98*, 239–389. [CrossRef]
8. Kiernan, M.C.; Vucic, S.; Cheah, B.C.; Turner, M.R.; Eisen, A.; Hardiman, O.; Burrell, J.R.; Zoing, M.C. Amyotrophic lateral sclerosis. *Lancet* **2011**, *377*, 942–955. [CrossRef]
9. Xu, L.; Liu, T.; Liu, L.; Yao, X.; Chen, L.; Fan, D.; Zhan, S.; Wang, S. Global variation in prevalence and incidence of amyotrophic lateral sclerosis: A systematic review and meta-analysis. *J. Neurol.* **2020**, *267*, 944–953. [CrossRef]
10. Aschenbrenner, D.S. New Drug Approved For ALS. *Am. J. Nurs.* **2023**, *123*, 22–23. [CrossRef]
11. Brooks, B.R.; Berry, J.D.; Ciepielewska, M.; Liu, Y.; Zambrano, G.S.; Zhang, J.; Hagan, M. Intravenous edaravone treatment in ALS and survival: An exploratory, retrospective, administrative claims analysis. *EClinicalMedicine* **2022**, *52*, 101590. [CrossRef] [PubMed]
12. Fels, J.A.; Dash, J.; Leslie, K.; Manfredi, G.; Kawamata, H. Effects of PB-TURSO on the transcriptional and metabolic landscape of sporadic ALS fibroblasts. *Ann. Clin. Transl. Neurol.* **2022**, *9*, 1551–1564. [CrossRef]
13. Jaiswal, M.K. Riluzole and edaravone: A tale of two amyotrophic lateral sclerosis drugs. *Med. Res. Rev.* **2019**, *39*, 733–748. [CrossRef]
14. Miller, T.; Cudkowicz, M.; Shaw, P.J.; Andersen, P.M.; Atassi, N.; Bucelli, R.C.; Genge, A.; Glass, J.; Ladha, S.; Ludolph, A.L.; et al. Phase 1–2 Trial of Antisense Oligonucleotide Tofersen for SOD1 ALS. *N. Engl. J. Med.* **2020**, *383*, 109–119. [CrossRef]
15. Paganoni, S.; Macklin, E.A.; Hendrix, S.; Berry, J.D.; Elliott, M.A.; Maiser, S.; Karam, C.; Caress, J.B.; Owegi, M.A.; Quick, A.; et al. Trial of Sodium Phenylbutyrate-Taurursodiol for Amyotrophic Lateral Sclerosis. *N. Engl. J. Med.* **2020**, *383*, 919–930. [CrossRef]
16. Mejzini, R.; Flynn, L.L.; Pitout, I.L.; Fletcher, S.; Wilton, S.D.; Akkari, P.A. ALS Genetics, Mechanisms, and Therapeutics: Where Are We Now? *Front. Neurosci.* **2019**, *13*, 1310. [CrossRef]
17. Taylor, J.P.; Brown, R.H.; Cleveland, D.W., Jr. Decoding ALS: From genes to mechanism. *Nature* **2016**, *539*, 197–206. [CrossRef]
18. Van Harten, A.C.M.; Phatnani, H.; Przedborski, S. Non-cell-autonomous pathogenic mechanisms in amyotrophic lateral sclerosis. *Trends Neurosci.* **2021**, *44*, 658–668. [CrossRef]
19. Taha, D.M.; Clarke, B.E.; Hall, C.E.; Tyzack, G.E.; Ziff, O.J.; Greensmith, L.; Kalmar, B.; Ahmed, M.; Alam, A.; Thelin, E.P.; et al. Astrocytes display cell autonomous and diverse early reactive states in familial amyotrophic lateral sclerosis. *Brain* **2022**, *145*, 481–489. [CrossRef]
20. Escartin, C.; Galea, E.; Lakatos, A.; O'Callaghan, J.P.; Petzold, G.C.; Serrano-Pozo, A.; Steinhäuser, C.; Volterra, A.; Carmignoto, G.; Agarwal, A.; et al. Reactive astrocyte nomenclature, definitions, and future directions. *Nat. Neurosci.* **2021**, *24*, 312–325. [CrossRef]
21. Escartin, C.; Guillemaud, O.; Carrillo-de Sauvage, M.A. Questions and (some) answers on reactive astrocytes. *Glia* **2019**, *67*, 2221–2247. [CrossRef]
22. Liddelow, S.A.; Guttenplan, K.A.; Clarke, L.E.; Bennett, F.C.; Bohlen, C.J.; Schirmer, L.; Bennett, M.L.; Münch, A.E.; Chung, W.S.; Peterson, T.C.; et al. Neurotoxic reactive astrocytes are induced by activated microglia. *Nature* **2017**, *541*, 481–487. [CrossRef] [PubMed]
23. Pekny, M.; Wilhelmsson, U.; Pekna, M. The dual role of astrocyte activation and reactive gliosis. *Neurosci. Lett.* **2014**, *565*, 30–38. [CrossRef] [PubMed]
24. Schiweck, J.; Eickholt, B.J.; Murk, K. Important Shapeshifter: Mechanisms Allowing Astrocytes to Respond to the Changing Nervous System During Development, Injury and Disease. *Front. Cell Neurosci.* **2018**, *12*, 261. [CrossRef] [PubMed]
25. Spurgat, M.S.; Tang, S.J. Single-Cell RNA-Sequencing: Astrocyte and Microglial Heterogeneity in Health and Disease. *Cells* **2022**, *11*, 2021. [CrossRef]
26. Verkhratsky, A.; Butt, A.; Li, B.; Illes, P.; Zorec, R.; Semyanov, A.; Tang, Y.; Sofroniew, M.V. Astrocytes in human central nervous system diseases: A frontier for new therapies. *Signal Transduct. Target Ther.* **2023**, *8*, 396. [CrossRef]
27. Diaz-Castro, B.; Bernstein, A.M.; Coppola, G.; Sofroniew, M.V.; Khakh, B.S. Molecular and functional properties of cortical astrocytes during peripherally induced neuroinflammation. *Cell Rep.* **2021**, *36*, 109508. [CrossRef]
28. Hasel, P.; Rose, I.V.L.; Sadick, J.S.; Kim, R.D.; Liddelow, S.A. Neuroinflammatory astrocyte subtypes in the mouse brain. *Nat. Neurosci.* **2021**, *24*, 1475–1487. [CrossRef]

29. Nam, M.H.; Cho, J.; Kwon, D.H.; Park, J.Y.; Woo, J.; Lee, J.M.; Lee, S.; Ko, H.Y.; Won, W.; Kim, R.G.; et al. Excessive Astrocytic GABA Causes Cortical Hypometabolism and Impedes Functional Recovery after Subcortical Stroke. *Cell Rep.* **2020**, *32*, 107861. [CrossRef]

30. Wilhelmsson, U.; Bushong, E.A.; Price, D.L.; Smarr, B.L.; Phung, V.; Terada, M.; Ellisman, M.H.; Pekny, M. Redefining the concept of reactive astrocytes as cells that remain within their unique domains upon reaction to injury. *Proc. Natl. Acad. Sci. USA* **2006**, *103*, 17513–17518. [CrossRef]

31. Wanner, I.B.; Anderson, M.A.; Song, B.; Levine, J.; Fernandez, A.; Gray-Thompson, Z.; Ao, Y.; Sofroniew, M.V. Glial scar borders are formed by newly proliferated, elongated astrocytes that interact to corral inflammatory and fibrotic cells via STAT3-dependent mechanisms after spinal cord injury. *J. Neurosci.* **2013**, *33*, 12870–12886. [CrossRef]

32. Krawczyk, M.C.; Haney, J.R.; Pan, L.; Caneda, C.; Khankan, R.R.; Reyes, S.D.; Chang, J.W.; Morselli, M.; Vinters, H.V.; Wang, A.C.; et al. Human Astrocytes Exhibit Tumor Microenvironment-, Age-, and Sex-Related Transcriptomic Signatures. *J. Neurosci.* **2022**, *42*, 1587–1603. [CrossRef]

33. O'Shea, T.M.; Wollenberg, A.L.; Kim, J.H.; Ao, Y.; Deming, T.J.; Sofroniew, M.V. Foreign body responses in mouse central nervous system mimic natural wound responses and alter biomaterial functions. *Nat. Commun.* **2020**, *11*, 6203. [CrossRef]

34. Qian, K.; Jiang, X.; Liu, Z.Q.; Zhang, J.; Fu, P.; Su, Y.; Brazhe, N.A.; Liu, D.; Zhu, L.Q. Revisiting the critical roles of reactive astrocytes in neurodegeneration. *Mol. Psychiatry* **2023**, *28*, 2697–2706. [CrossRef]

35. Serrano-Pozo, A.; Gómez-Isla, T.; Growdon, J.H.; Frosch, M.P.; Hyman, B.T. A phenotypic change but not proliferation underlies glial responses in Alzheimer disease. *Am. J. Pathol.* **2013**, *182*, 2332–2344. [CrossRef] [PubMed]

36. Brenner, M.; Kisseberth, W.C.; Su, Y.; Besnard, F.; Messing, A. GFAP promoter directs astrocyte-specific expression in transgenic mice. *J. Neurosci.* **1994**, *14*, 1030–1037. [CrossRef] [PubMed]

37. Ben Haim, L.; Rowitch, D.H. Functional diversity of astrocytes in neural circuit regulation. *Nat. Rev. Neurosci.* **2017**, *18*, 31–41. [CrossRef]

38. Brenner, M.; Messing, A. Regulation of GFAP Expression. *ASN Neuro* **2021**, *13*, 1759091420981206. [CrossRef] [PubMed]

39. Hou, J.; Bi, H.; Ge, Q.; Teng, H.; Wan, G.; Yu, B.; Jiang, Q.; Gu, X. Heterogeneity analysis of astrocytes following spinal cord injury at single-cell resolution. *FASEB J.* **2022**, *36*, e22442. [CrossRef] [PubMed]

40. Middeldorp, J.; Hol, E.M. GFAP in health and disease. *Prog. Neurobiol.* **2011**, *93*, 421–443. [CrossRef] [PubMed]

41. Rodríguez, J.J.; Terzieva, S.; Olabarria, M.; Lanza, R.G.; Verkhratsky, A. Enriched environment and physical activity reverse astrogliodegeneration in the hippocampus of AD transgenic mice. *Cell Death Dis.* **2013**, *4*, e678. [CrossRef]

42. O'Callaghan, J.P.; Brinton, R.E.; McEwen, B.S. Glucocorticoids regulate the synthesis of glial fibrillary acidic protein in intact and adrenalectomized rats but do not affect its expression following brain injury. *J. Neurochem.* **1991**, *57*, 860–869. [CrossRef]

43. Lévi-Strauss, M.; Mallat, M. Primary cultures of murine astrocytes produce C3 and factor B, two components of the alternative pathway of complement activation. *J. Immunol.* **1987**, *139*, 2361–2366. [CrossRef] [PubMed]

44. Shah, A.; Kishore, U.; Shastri, A. Complement System in Alzheimer's Disease. *Int. J. Mol. Sci.* **2021**, *22*, 13647. [CrossRef] [PubMed]

45. Li, Y.; Fang, S.C.; Zhou, L.; Mo, X.M.; Guo, H.D.; Deng, Y.B.; Yu, H.H.; Gong, W.Y. Complement Receptor 3 Pathway and NMDA Receptor 2B Subunit Involve Neuropathic Pain Associated with Spinal Cord Injury. *J. Pain Res.* **2022**, *15*, 1813–1823. [CrossRef] [PubMed]

46. Nitkiewicz, J.; Borjabad, A.; Morgello, S.; Murray, J.; Chao, W.; Emdad, L.; Fisher, P.B.; Potash, M.J.; Volsky, D.J. HIV induces expression of complement component C3 in astrocytes by NF-κB-dependent activation of interleukin-6 synthesis. *J. Neuroinflamm.* **2017**, *14*, 23. [CrossRef]

47. Gharagozloo, M.; Smith, M.D.; Jin, J.; Garton, T.; Taylor, M.; Chao, A.; Meyers, K.; Kornberg, M.D.; Zack, D.J.; Ohayon, J.; et al. Complement component 3 from astrocytes mediates retinal ganglion cell loss during neuroinflammation. *Acta Neuropathol.* **2021**, *142*, 899–915. [CrossRef]

48. Stevens, B.; Allen, N.J.; Vazquez, L.E.; Howell, G.R.; Christopherson, K.S.; Nouri, N.; Micheva, K.D.; Mehalow, A.K.; Huberman, A.D.; Stafford, B.; et al. The classical complement cascade mediates CNS synapse elimination. *Cell* **2007**, *131*, 1164–1178. [CrossRef]

49. Zhou, R.; Chen, S.H.; Zhao, Z.; Tu, D.; Song, S.; Wang, Y.; Wang, Q.; Feng, J.; Hong, J.S. Complement C3 Enhances LPS-Elicited Neuroinflammation and Neurodegeneration via the Mac1/NOX2 Pathway. *Mol. Neurobiol.* **2023**, *60*, 5167–5183. [CrossRef]

50. Huang, Y.; Erdmann, N.; Peng, H.; Zhao, Y.; Zheng, J. The role of TNF related apoptosis-inducing ligand in neurodegenerative diseases. *Cell Mol. Immunol.* **2005**, *2*, 113–122.

51. Sanmarco, L.M.; Wheeler, M.A.; Gutiérrez-Vázquez, C.; Polonio, C.M.; Linnerbauer, M.; Pinho-Ribeiro, F.A.; Li, Z.; Giovannoni, F.; Batterman, K.V.; Scalisi, G.; et al. Gut-licensed IFNγ⁺ NK cells drive LAMP1⁺TRAIL⁺ anti-inflammatory astrocytes. *Nature* **2021**, *590*, 473–479. [CrossRef] [PubMed]

52. Burgaletto, C.; Platania, C.B.M.; Di Benedetto, G.; Munafò, A.; Giurdanella, G.; Federico, C.; Caltabiano, R.; Saccone, S.; Conti, F.; Bernardini, R.; et al. Targeting the miRNA-155/TNFSF10 network restrains inflammatory response in the retina in a mouse model of Alzheimer's disease. *Cell Death Dis.* **2021**, *12*, 905. [CrossRef] [PubMed]

53. Cantarella, G.; Di Benedetto, G.; Puzzo, D.; Privitera, L.; Loreto, C.; Saccone, S.; Giunta, S.; Palmeri, A.; Bernardini, R. Neutralization of TNFSF10 ameliorates functional outcome in a murine model of Alzheimer's disease. *Brain* **2015**, *138*, 203–216. [CrossRef] [PubMed]

54. Patani, R.; Hardingham, G.E.; Liddelow, S.A. Functional roles of reactive astrocytes in neuroinflammation and neurodegeneration. *Nat. Rev. Neurol.* **2023**, *19*, 395–409. [CrossRef] [PubMed]

55. Faulkner, J.R.; Herrmann, J.E.; Woo, M.J.; Tansey, K.E.; Doan, N.B.; Sofroniew, M.V. Reactive astrocytes protect tissue and preserve function after spinal cord injury. *J. Neurosci.* **2004**, *24*, 2143–2155. [CrossRef] [PubMed]

56. Dong, R.; Chen, M.; Liu, J.; Kang, J.; Zhu, S. Temporospatial effects of acyl-ghrelin on activation of astrocytes after ischaemic brain injury. *J. Neuroendocrinol.* **2019**, *31*, e12767. [CrossRef] [PubMed]

57. Hara, M.; Kobayakawa, K.; Ohkawa, Y.; Kumamaru, H.; Yokota, K.; Saito, T.; Kijima, K.; Yoshizaki, S.; Harimaya, K.; Nakashima, Y.; et al. Interaction of reactive astrocytes with type I collagen induces astrocytic scar formation through the integrin-N-cadherin pathway after spinal cord injury. *Nat. Med.* **2017**, *23*, 818–828. [CrossRef]

58. Chowdhury, S.; Mazumder, M.A.J.; Al-Attas, O.; Husain, T. Heavy metals in drinking water: Occurrences, implications, and future needs in developing countries. *Sci. Total Environ.* **2016**, *569–570*, 476–488. [CrossRef]

59. Uddin, M.M.; Zakeel, M.C.M.; Zavahir, J.S.; Marikar, F.; Jahan, I. Heavy Metal Accumulation in Rice and Aquatic Plants Used as Human Food: A General Review. *Toxics* **2021**, *9*, 360. [CrossRef]

60. Calabrese, G.; Molzahn, C.; Mayor, T. Protein interaction networks in neurodegenerative diseases: From physiological function to aggregation. *J. Biol. Chem.* **2022**, *298*, 102062. [CrossRef]

61. Li, B.; Xia, M.; Zorec, R.; Parpura, V.; Verkhratsky, A. Astrocytes in heavy metal neurotoxicity and neurodegeneration. *Brain Res.* **2021**, *1752*, 147234. [CrossRef]

62. Murumulla, L.; Bandaru, L.J.M.; Challa, S. Heavy Metal Mediated Progressive Degeneration and Its Noxious Effects on Brain Microenvironment. *Biol. Trace Elem. Res.* **2023**. [CrossRef]

63. Fan, S.; Weixuan, W.; Han, H.; Liansheng, Z.; Gang, L.; Jierui, W.; Yanshu, Z. Role of NF-κB in lead exposure-induced activation of astrocytes based on bioinformatics analysis of hippocampal proteomics. *Chem. Biol. Interact.* **2023**, *370*, 110310. [CrossRef]

64. Fathabadi, B.; Dehghanifiroozabadi, M.; Aaseth, J.; Sharifzadeh, G.; Nakhaee, S.; Rajabpour-Sanati, A.; Amirabadizadeh, A.; Mehrpour, O. Comparison of Blood Lead Levels in Patients with Alzheimer's Disease and Healthy People. *Am. J. Alzheimer's Dis. Other Demen.* **2018**, *33*, 541–547. [CrossRef]

65. Gerhardsson, L.; Blennow, K.; Lundh, T.; Londos, E.; Minthon, L. Concentrations of metals, beta-amyloid and tau-markers in cerebrospinal fluid in patients with Alzheimer's disease. *Dement Geriatr. Cogn. Disord.* **2009**, *28*, 88–94. [CrossRef]

66. Szabo, S.T.; Harry, G.J.; Hayden, K.M.; Szabo, D.T.; Birnbaum, L. Comparison of Metal Levels between Postmortem Brain and Ventricular Fluid in Alzheimer's Disease and Nondemented Elderly Controls. *Toxicol. Sci.* **2016**, *150*, 292–300. [CrossRef] [PubMed]

67. Wu, L.; Li, S.; Li, C.; He, B.; Lv, L.; Wang, J.; Wang, J.; Wang, W.; Zhang, Y. The role of regulatory T cells on the activation of astrocytes in the brain of high-fat diet mice following lead exposure. *Chem. Biol. Interact.* **2022**, *351*, 109740. [CrossRef] [PubMed]

68. Cresto, N.; Forner-Piquer, I.; Baig, A.; Chatterjee, M.; Perroy, J.; Goracci, J.; Marchi, N. Pesticides at brain borders: Impact on the blood-brain barrier, neuroinflammation, and neurological risk trajectories. *Chemosphere* **2023**, *324*, 138251. [CrossRef] [PubMed]

69. Hsu, S.S.; Jan, C.R.; Liang, W.Z. Uncovering malathion (an organophosphate insecticide) action on Ca^{2+} signal transduction and investigating the effects of BAPTA-AM (a cell-permeant Ca^{2+} chelator) on protective responses in glial cells. *Pestic. Biochem. Physiol.* **2019**, *157*, 152–160. [CrossRef] [PubMed]

70. Klement, W.; Oliviero, F.; Gangarossa, G.; Zub, E.; De Bock, F.; Forner-Piquer, I.; Blaquiere, M.; Lasserre, F.; Pascussi, J.M.; Maurice, T.; et al. Life-long Dietary Pesticide Cocktail Induces Astrogliosis along with Behavioral Adaptations and Activates p450 Metabolic Pathways. *Neuroscience* **2020**, *446*, 225–237. [CrossRef]

71. Liu, L.; Koo, Y.; Russell, T.; Gay, E.; Li, Y.; Yun, Y. Three-dimensional brain-on-chip model using human iPSC-derived GABAergic neurons and astrocytes: Butyrylcholinesterase post-treatment for acute malathion exposure. *PLoS ONE* **2020**, *15*, e0230335. [CrossRef]

72. Ravid, O.; Elhaik Goldman, S.; Macheto, D.; Bresler, Y.; De Oliveira, R.I.; Liraz-Zaltsman, S.; Gosselet, F.; Dehouck, L.; Beeri, M.S.; Cooper, I. Blood-Brain Barrier Cellular Responses Toward Organophosphates: Natural Compensatory Processes and Exogenous Interventions to Rescue Barrier Properties. *Front. Cell Neurosci.* **2018**, *12*, 359. [CrossRef]

73. Voorhees, J.R.; Remy, M.T.; Erickson, C.M.; Dutca, L.M.; Brat, D.J.; Pieper, A.A. Occupational-like organophosphate exposure disrupts microglia and accelerates deficits in a rat model of Alzheimer's disease. *NPJ Aging Mech. Dis.* **2019**, *5*, 3. [CrossRef]

74. Bharatiya, R.; Bratzu, J.; Lobina, C.; Corda, G.; Cocco, C.; De Deurwaerdere, P.; Argiolas, A.; Melis, M.R.; Sanna, F. The pesticide fipronil injected into the substantia nigra of male rats decreases striatal dopamine content: A neurochemical, immunohistochemical and behavioral study. *Behav. Brain Res.* **2020**, *384*, 112562. [CrossRef]

75. Mullett, S.J.; Di Maio, R.; Greenamyre, J.T.; Hinkle, D.A. DJ-1 expression modulates astrocyte-mediated protection against neuronal oxidative stress. *J. Mol. Neurosci.* **2013**, *49*, 507–511. [CrossRef]

76. Mullett, S.J.; Hinkle, D.A. DJ-1 knock-down in astrocytes impairs astrocyte-mediated neuroprotection against rotenone. *Neurobiol. Dis.* **2009**, *33*, 28–36. [CrossRef]

77. Mullett, S.J.; Hinkle, D.A. DJ-1 deficiency in astrocytes selectively enhances mitochondrial Complex I inhibitor-induced neurotoxicity. *J. Neurochem.* **2011**, *117*, 375–387. [CrossRef] [PubMed]

78. Li, Z.H.; Jiang, Y.Y.; Long, C.Y.; Peng, Q.; Yue, R.S. The gut microbiota-astrocyte axis: Implications for type 2 diabetic cognitive dysfunction. *CNS Neurosci. Ther.* **2023**, *29* (Suppl. S1), 59–73. [CrossRef] [PubMed]

79. Zhao, Y.F.; Wei, D.N.; Tang, Y. Gut Microbiota Regulate Astrocytic Functions in the Brain: Possible Therapeutic Consequences. *Curr. Neuropharmacol.* **2021**, *19*, 1354–1366. [CrossRef] [PubMed]

80. Gómez-Budia, M.; Konttinen, H.; Saveleva, L.; Korhonen, P.; Jalava, P.I.; Kanninen, K.M.; Malm, T. Glial smog: Interplay between air pollution and astrocyte-microglia interactions. *Neurochem. Int.* **2020**, *136*, 104715. [CrossRef] [PubMed]

81. Navarro, A.; García, M.; Rodrigues, A.S.; Garcia, P.V.; Camarinho, R.; Segovia, Y. Reactive astrogliosis in the dentate gyrus of mice exposed to active volcanic environments. *J. Toxicol. Environ. Health A* **2021**, *84*, 213–226. [CrossRef] [PubMed]

82. Haidet-Phillips, A.M.; Hester, M.E.; Miranda, C.J.; Meyer, K.; Braun, L.; Frakes, A.; Song, S.; Likhite, S.; Murtha, M.J.; Foust, K.D.; et al. Astrocytes from familial and sporadic ALS patients are toxic to motor neurons. *Nat. Biotechnol.* **2011**, *29*, 824–828. [CrossRef]

83. Guttenplan, K.A.; Weigel, M.K.; Adler, D.I.; Couthouis, J.; Liddelow, S.A.; Gitler, A.D.; Barres, B.A. Knockout of reactive astrocyte activating factors slows disease progression in an ALS mouse model. *Nat. Commun.* **2020**, *11*, 3753. [CrossRef] [PubMed]

84. Yamamuro-Tanabe, A.; Mukai, Y.; Kojima, W.; Zheng, S.; Matsumoto, N.; Takada, S.; Mizuhara, M.; Kosuge, Y.; Ishimaru, Y.; Yoshioka, Y. An Increase in Peroxiredoxin 6 Expression Induces Neurotoxic A1 Astrocytes in the Lumbar Spinal Cord of Amyotrophic Lateral Sclerosis Mice Model. *Neurochem. Res.* **2023**, *48*, 3571–3584. [CrossRef] [PubMed]

85. Ganesalingam, J.; An, J.; Shaw, C.E.; Shaw, G.; Lacomis, D.; Bowser, R. Combination of neurofilament heavy chain and complement C3 as CSF biomarkers for ALS. *J. Neurochem.* **2011**, *117*, 528–537. [CrossRef]

86. Shi, Q.; Chowdhury, S.; Ma, R.; Le, K.X.; Hong, S.; Caldarone, B.J.; Stevens, B.; Lemere, C.A. Complement C3 deficiency protects against neurodegeneration in aged plaque-rich APP/PS1 mice. *Sci. Transl. Med.* **2017**, *9*, eaaf6295. [CrossRef] [PubMed]

87. Marchetto, M.C.; Muotri, A.R.; Mu, Y.; Smith, A.M.; Cezar, G.G.; Gage, F.H. Non-cell-autonomous effect of human SOD1 G37R astrocytes on motor neurons derived from human embryonic stem cells. *Cell Stem Cell* **2008**, *3*, 649–657. [CrossRef]

88. Falzone, Y.M.; Domi, T.; Mandelli, A.; Pozzi, L.; Schito, P.; Russo, T.; Barbieri, A.; Fazio, R.; Volontè, M.A.; Magnani, G.; et al. Integrated evaluation of a panel of neurochemical biomarkers to optimize diagnosis and prognosis in amyotrophic lateral sclerosis. *Eur. J. Neurol.* **2022**, *29*, 1930–1939. [CrossRef]

89. Arredondo, C.; Cefaliello, C.; Dyrda, A.; Jury, N.; Martinez, P.; Díaz, I.; Amaro, A.; Tran, H.; Morales, D.; Pertusa, M.; et al. Excessive release of inorganic polyphosphate by ALS/FTD astrocytes causes non-cell-autonomous toxicity to motoneurons. *Neuron* **2022**, *110*, 1656–1670.e1612. [CrossRef]

90. Rojas, F.; Aguilar, R.; Almeida, S.; Fritz, E.; Corvalán, D.; Ampuero, E.; Abarzúa, S.; Garcés, P.; Amaro, A.; Diaz, I.; et al. Mature iPSC-derived astrocytes of an ALS/FTD patient carrying the TDP43(A90V) mutation display a mild reactive state and release polyP toxic to motoneurons. *Front. Cell Dev. Biol.* **2023**, *11*, 1226604. [CrossRef]

91. Szebényi, K.; Wenger, L.M.D.; Sun, Y.; Dunn, A.W.E.; Limegrover, C.A.; Gibbons, G.M.; Conci, E.; Paulsen, O.; Mierau, S.B.; Balmus, G.; et al. Human ALS/FTD brain organoid slice cultures display distinct early astrocyte and targetable neuronal pathology. *Nat. Neurosci.* **2021**, *24*, 1542–1554. [CrossRef]

92. Valori, C.F.; Sulmona, C.; Brambilla, L.; Rossi, D. Astrocytes: Dissecting Their Diverse Roles in Amyotrophic Lateral Sclerosis and Frontotemporal Dementia. *Cells* **2023**, *12*, 1450. [CrossRef]

93. Li, J.Y.; Cai, Z.Y.; Sun, X.H.; Shen, D.C.; Yang, X.Z.; Liu, M.S.; Cui, L.Y. Blood-brain barrier dysfunction and myelin basic protein in survival of amyotrophic lateral sclerosis with or without frontotemporal dementia. *Neurol. Sci.* **2022**, *43*, 3201–3210. [CrossRef]

94. Marini, C.; Cossu, V.; Kumar, M.; Milanese, M.; Cortese, K.; Bruno, S.; Bellese, G.; Carta, S.; Zerbo, R.A.; Torazza, C.; et al. The Role of Endoplasmic Reticulum in the Differential Endurance against Redox Stress in Cortical and Spinal Astrocytes from the Newborn SOD1(G93A) Mouse Model of Amyotrophic Lateral Sclerosis. *Antioxidants* **2021**, *10*, 1392. [CrossRef]

95. Cistaro, A.; Cuccurullo, V.; Quartuccio, N.; Pagani, M.; Valentini, M.C.; Mansi, L. Role of PET and SPECT in the study of amyotrophic lateral sclerosis. *Biomed. Res. Int.* **2014**, *2014*, 237437. [CrossRef]

96. Liu, Y.; Jiang, H.; Qin, X.; Tian, M.; Zhang, H. PET imaging of reactive astrocytes in neurological disorders. *Eur. J. Nucl. Med. Mol. Imaging* **2022**, *49*, 1275–1287. [CrossRef]

97. Cassina, P.; Cassina, A.; Pehar, M.; Castellanos, R.; Gandelman, M.; de León, A.; Robinson, K.M.; Mason, R.P.; Beckman, J.S.; Barbeito, L.; et al. Mitochondrial dysfunction in SOD1G93A-bearing astrocytes promotes motor neuron degeneration: Prevention by mitochondrial-targeted antioxidants. *J. Neurosci.* **2008**, *28*, 4115–4122. [CrossRef]

98. Markovinovic, A.; Greig, J.; Martín-Guerrero, S.M.; Salam, S.; Paillusson, S. Endoplasmic reticulum-mitochondria signaling in neurons and neurodegenerative diseases. *J. Cell Sci.* **2022**, *135*, jcs248534. [CrossRef]

99. Smith, E.F.; Shaw, P.J.; De Vos, K.J. The role of mitochondria in amyotrophic lateral sclerosis. *Neurosci. Lett.* **2019**, *710*, 132933. [CrossRef]

100. Miquel, E.; Cassina, A.; Martínez-Palma, L.; Bolatto, C.; Trías, E.; Gandelman, M.; Radi, R.; Barbeito, L.; Cassina, P. Modulation of astrocytic mitochondrial function by dichloroacetate improves survival and motor performance in inherited amyotrophic lateral sclerosis. *PLoS ONE* **2012**, *7*, e34776. [CrossRef]

101. Martínez-Palma, L.; Miquel, E.; Lagos-Rodríguez, V.; Barbeito, L.; Cassina, A.; Cassina, P. Mitochondrial Modulation by Dichloroacetate Reduces Toxicity of Aberrant Glial Cells and Gliosis in the SOD1G93A Rat Model of Amyotrophic Lateral Sclerosis. *Neurotherapeutics* **2019**, *16*, 203–215. [CrossRef] [PubMed]

102. Verkhratsky, A. Astroglial Calcium Signaling in Aging and Alzheimer's Disease. *Cold Spring Harb. Perspect. Biol.* **2019**, *11*, a035188. [CrossRef] [PubMed]

103. Tedeschi, V.; Petrozziello, T.; Secondo, A. Ca^{2+} dysregulation in the pathogenesis of amyotrophic lateral sclerosis. *Int. Rev. Cell. Mol. Biol.* **2021**, *363*, 21–47. [PubMed]

104. Kawamata, H.; Ng, S.K.; Diaz, N.; Burstein, S.; Morel, L.; Osgood, A.; Sider, B.; Higashimori, H.; Haydon, P.G.; Manfredi, G.; et al. Abnormal intracellular calcium signaling and SNARE-dependent exocytosis contributes to SOD1G93A astrocyte-mediated toxicity in amyotrophic lateral sclerosis. *J. Neurosci.* **2014**, *34*, 2331–2348. [CrossRef]

105. Norante, R.P.; Peggion, C.; Rossi, D.; Martorana, F.; De Mario, A.; Lia, A.; Massimino, M.L.; Bertoli, A. ALS-Associated SOD1(G93A) Decreases SERCA Pump Levels and Increases Store-Operated Ca^{2+} Entry in Primary Spinal Cord Astrocytes from a Transgenic Mouse Model. *Int. J. Mol. Sci.* **2019**, *20*, 5151. [CrossRef]

106. Nakagawa, T.; Otsubo, Y.; Yatani, Y.; Shirakawa, H.; Kaneko, S. Mechanisms of substrate transport-induced clustering of a glial glutamate transporter GLT-1 in astroglial-neuronal cultures. *Eur. J. Neurosci.* **2008**, *28*, 1719–1730. [CrossRef]

107. Rojas, F.; Cortes, N.; Abarzua, S.; Dyrda, A.; van Zundert, B. Astrocytes expressing mutant SOD1 and TDP43 trigger motoneuron death that is mediated via sodium channels and nitroxidative stress. *Front. Cell Neurosci.* **2014**, *8*, 24. [CrossRef]

108. Zhao, C.; Devlin, A.C.; Chouhan, A.K.; Selvaraj, B.T.; Stavrou, M.; Burr, K.; Brivio, V.; He, X.; Mehta, A.R.; Story, D.; et al. Mutant C9orf72 human iPSC-derived astrocytes cause non-cell autonomous motor neuron pathophysiology. *Glia* **2020**, *68*, 1046–1064. [CrossRef]

109. Stevenson, R.; Samokhina, E.; Mangat, A.; Rossetti, I.; Purushotham, S.S.; Malladi, C.S.; Morley, J.W.; Buskila, Y. Astrocytic K^{+} clearance during disease progression in amyotrophic lateral sclerosis. *Glia* **2023**, *71*, 2456–2472. [CrossRef] [PubMed]

110. Kornberg, A.; Rao, N.N.; Ault-Riché, D. Inorganic polyphosphate: A molecule of many functions. *Annu. Rev. Biochem.* **1999**, *68*, 89–125. [CrossRef] [PubMed]

111. Perry, T.L.; Hansen, S.; Berry, K.; Mok, C.; Lesk, D. Free amino acids and related compounds in biopsies of human brain. *J. Neurochem.* **1971**, *18*, 521–528. [CrossRef] [PubMed]

112. Jiang, L.L.; Zhu, B.; Zhao, Y.; Li, X.; Liu, T.; Pina-Crespo, J.; Zhou, L.; Xu, W.; Rodriguez, M.J.; Yu, H.; et al. Membralin deficiency dysregulates astrocytic glutamate homeostasis leading to ALS-like impairment. *J. Clin. Investig.* **2019**, *129*, 3103–3120. [CrossRef] [PubMed]

113. Karki, P.; Smith, K.; Johnson, J.; Jr Aschner, M.; Lee, E.Y. Genetic dys-regulation of astrocytic glutamate transporter EAAT2 and its implications in neurological disorders and manganese toxicity. *Neurochem. Res.* **2015**, *40*, 380–388. [CrossRef] [PubMed]

114. Lee, J.; Hyeon, S.J.; Im, H.; Ryu, H.; Kim, Y.; Ryu, H. Astrocytes and Microglia as Non-cell Autonomous Players in the Pathogenesis of ALS. *Exp. Neurobiol.* **2016**, *25*, 233–240. [CrossRef] [PubMed]

115. Pardo, A.C.; Wong, V.; Benson, L.M.; Dykes, M.; Tanaka, K.; Rothstein, J.D.; Maragakis, N.J. Loss of the astrocyte glutamate transporter GLT1 modifies disease in SOD1(G93A) mice. *Exp. Neurol.* **2006**, *201*, 120–130. [CrossRef]

116. Madji Hounoum, B.; Mavel, S.; Coque, E.; Patin, F.; Vourc'h, P.; Marouillat, S.; Nadal-Desbarats, L.; Emond, P.; Corcia, P.; Andres, C.R.; et al. Wildtype motoneurons, ALS-Linked SOD1 mutation and glutamate profoundly modify astrocyte metabolism and lactate shuttling. *Glia* **2017**, *65*, 592–605. [CrossRef]

117. Kane, D.A. Lactate oxidation at the mitochondria: A lactate-malate-aspartate shuttle at work. *Front. Neurosci.* **2014**, *8*, 366. [CrossRef] [PubMed]

118. Ebert, D.; Haller, R.G.; Walton, M.E. Energy contribution of octanoate to intact rat brain metabolism measured by 13C nuclear magnetic resonance spectroscopy. *J. Neurosci.* **2003**, *23*, 5928–5935. [CrossRef]

119. Liu, L.; MacKenzie, K.R.; Putluri, N.; Maletić-Savatić, M.; Bellen, H.J. The Glia-Neuron Lactate Shuttle and Elevated ROS Promote Lipid Synthesis in Neurons and Lipid Droplet Accumulation in Glia via APOE/D. *Cell Metab.* **2017**, *26*, 719–737.e716. [CrossRef]

120. Agrawal, I.; Lim, Y.S.; Ng, S.Y.; Ling, S.C. Deciphering lipid dysregulation in ALS: From mechanisms to translational medicine. *Transl. Neurodegener.* **2022**, *11*, 48. [CrossRef]

121. Palamiuc, L.; Schlagowski, A.; Ngo, S.T.; Vernay, A.; Dirrig-Grosch, S.; Henriques, A.; Boutillier, A.L.; Zoll, J.; Echaniz-Laguna, A.; Loeffler, J.P.; et al. A metabolic switch toward lipid use in glycolytic muscle is an early pathologic event in a mouse model of amyotrophic lateral sclerosis. *EMBO Mol. Med.* **2015**, *7*, 526–546. [CrossRef]

122. Killoy, K.M.; Harlan, B.A.; Pehar, M.; Vargas, M.R. FABP7 upregulation induces a neurotoxic phenotype in astrocytes. *Glia* **2020**, *68*, 2693–2704. [CrossRef]

123. Boneva, N.B.; Mori, Y.; Kaplamadzhiev, D.B.; Kikuchi, H.; Zhu, H.; Kikuchi, M.; Tonchev, A.B.; Yamashima, T. Differential expression of FABP 3, 5, 7 in infantile and adult monkey cerebellum. *Neurosci. Res.* **2010**, *68*, 94–102. [CrossRef]

124. Islam, A.; Kagawa, Y.; Miyazaki, H.; Shil, S.K.; Umaru, B.A.; Yasumoto, Y.; Yamamoto, Y.; Owada, Y. FABP7 Protects Astrocytes Against ROS Toxicity via Lipid Droplet Formation. *Mol. Neurobiol.* **2019**, *56*, 5763–5779. [CrossRef]

125. Hara, T.; Abdulaziz Umaru, B.; Sharifi, K.; Yoshikawa, T.; Owada, Y.; Kagawa, Y. Fatty Acid Binding Protein 7 is Involved in the Proliferation of Reactive Astrocytes, but not in Cell Migration and Polarity. *Acta Histochem. Cytochem.* **2020**, *53*, 73–81. [CrossRef]

126. Kagawa, Y.; Yasumoto, Y.; Sharifi, K.; Ebrahimi, M.; Islam, A.; Miyazaki, H.; Yamamoto, Y.; Sawada, T.; Kishi, H.; Kobayashi, S.; et al. Fatty acid-binding protein 7 regulates function of caveolae in astrocytes through expression of caveolin-1. *Glia* **2015**, *63*, 780–794. [CrossRef]

127. Xiong, X.Y.; Tang, Y.; Yang, Q.W. Metabolic changes favor the activity and heterogeneity of reactive astrocytes. *Trends Endocrinol. Metab.* **2022**, *33*, 390–400. [CrossRef] [PubMed]

128. Almer, G.; Teismann, P.; Stevic, Z.; Halaschek-Wiener, J.; Deecke, L.; Kostic, V.; Przedborski, S. Increased levels of the pro-inflammatory prostaglandin PGE2 in CSF from ALS patients. *Neurology* **2002**, *58*, 1277–1279. [CrossRef] [PubMed]

129. Chaves-Filho, A.B.; Pinto, I.F.D.; Dantas, L.S.; Xavier, A.M.; Inague, A.; Faria, R.L.; Medeiros, M.H.G.; Glezer, I.; Yoshinaga, M.Y.; Miyamoto, S. Alterations in lipid metabolism of spinal cord linked to amyotrophic lateral sclerosis. *Sci. Rep.* **2019**, *9*, 11642. [CrossRef]

130. Iłzecka, J. Prostaglandin E2 is increased in amyotrophic lateral sclerosis patients. *Acta Neurol. Scand.* **2003**, *108*, 125–129. [CrossRef] [PubMed]

131. Moore, S.A.; Yoder, E.; Murphy, S.; Dutton, G.R.; Spector, A.A. Astrocytes, not neurons, produce docosahexaenoic acid (22:6 omega-3) and arachidonic acid (20:4 omega-6). *J. Neurochem.* **1991**, *56*, 518–524. [CrossRef] [PubMed]

132. Stella, N.; Tencé, M.; Glowinski, J.; Prémont, J. Glutamate-evoked release of arachidonic acid from mouse brain astrocytes. *J. Neurosci.* **1994**, *14*, 568–575. [CrossRef]

133. Ng, W.; Ng, S.Y. Remodeling of astrocyte secretome in amyotrophic lateral sclerosis: Uncovering novel targets to combat astrocyte-mediated toxicity. *Transl. Neurodegener.* **2022**, *11*, 54. [CrossRef] [PubMed]

134. Zou, Y.H.; Guan, P.P.; Zhang, S.Q.; Guo, Y.S.; Wang, P. Rofecoxib Attenuates the Pathogenesis of Amyotrophic Lateral Sclerosis by Alleviating Cyclooxygenase-2-Mediated Mechanisms. *Front. Neurosci.* **2020**, *14*, 817. [CrossRef]

135. Guttenplan, K.A.; Weigel, M.K.; Prakash, P.; Wijewardhane, P.R.; Hasel, P.; Rufen-Blanchette, U.; Münch, A.E.; Blum, J.A.; Fine, J.; Neal, M.C.; et al. Neurotoxic reactive astrocytes induce cell death via saturated lipids. *Nature* **2021**, *599*, 102–107. [CrossRef] [PubMed]

136. Nagai, M.; Re, D.B.; Nagata, T.; Chalazonitis, A.; Jessell, T.M.; Wichterle, H.; Przedborski, S. Astrocytes expressing ALS-linked mutated SOD1 release factors selectively toxic to motor neurons. *Nat. Neurosci.* **2007**, *10*, 615–622. [CrossRef]

137. Di Giorgio, F.P.; Boulting, G.L.; Bobrowicz, S.; Eggan, K.C. Human embryonic stem cell-derived motor neurons are sensitive to the toxic effect of glial cells carrying an ALS-causing mutation. *Cell Stem Cell* **2008**, *3*, 637–648. [CrossRef]

138. Barton, S.K.; Lau, C.L.; Chiam, M.D.F.; Tomas, D.; Muyderman, H.; Beart, P.M.; Turner, B.J. Mutant TDP-43 Expression Triggers TDP-43 Pathology and Cell Autonomous Effects on Primary Astrocytes: Implications for Non-cell Autonomous Pathology in ALS. *Neurochem. Res.* **2020**, *45*, 1451–1459. [CrossRef]

139. Birger, A.; Ben-Dor, I.; Ottolenghi, M.; Turetsky, T.; Gil, Y.; Sweetat, S.; Perez, L.; Belzer, V.; Casden, N.; Steiner, D.; et al. Human iPSC-derived astrocytes from ALS patients with mutated C9ORF72 show increased oxidative stress and neurotoxicity. *EBioMedicine* **2019**, *50*, 274–289. [CrossRef]

140. Baloh, R.H.; Johnson, J.P.; Avalos, P.; Allred, P.; Svendsen, S.; Gowing, G.; Roxas, K.; Wu, A.; Donahue, B.; Osborne, S.; et al. Transplantation of human neural progenitor cells secreting GDNF into the spinal cord of patients with ALS: A phase 1/2a trial. *Nat. Med.* **2022**, *28*, 1813–1822. [CrossRef]

141. Gotkine, M.; Caraco, Y.; Lerner, Y.; Blotnick, S.; Wanounou, M.; Slutsky, S.G.; Chebath, J.; Kuperstein, G.; Estrin, E.; Ben-Hur, T.; et al. Safety and efficacy of first-in-man intrathecal injection of human astrocytes (AstroRx®) in ALS patients: Phase I/IIa clinical trial results. *J. Transl. Med.* **2023**, *21*, 122. [CrossRef]

142. Xing, L.; Yang, T.; Cui, S.; Chen, G. Connexin Hemichannels in Astrocytes: Role in CNS Disorders. *Front. Mol. Neurosci.* **2019**, *12*, 23. [CrossRef]

143. Almad, A.A.; Taga, A.; Joseph, J.; Gross, S.K.; Welsh, C.; Patankar, A.; Richard, J.P.; Rust, K.; Pokharel, A.; Plott, C.; et al. Cx43 hemichannels contribute to astrocyte-mediated toxicity in sporadic and familial ALS. *Proc. Natl. Acad. Sci. USA* **2022**, *119*, e2107391119. [CrossRef]

144. Damodaram, S.; Thalakoti, S.; Freeman, S.E.; Garrett, F.G.; Durham, P.L. Tonabersat inhibits trigeminal ganglion neuronal-satellite glial cell signaling. *Headache* **2009**, *49*, 5–20. [CrossRef]

145. Pasquale, E.B. Eph-ephrin bidirectional signaling in physiology and disease. *Cell* **2008**, *133*, 38–52. [CrossRef]

146. Urban, M.W.; Charsar, B.A.; Heinsinger, N.M.; Markandaiah, S.S.; Sprimont, L.; Zhou, W.; Brown, E.V.; Henderson, N.T.; Thomas, S.J.; Ghosh, B.; et al. EphrinB2 knockdown in cervical spinal cord preserves diaphragm innervation in a mutant SOD1 mouse model of ALS. *eLife* **2024**, *12*, RP89298. [CrossRef]

147. Harlan, B.A.; Pehar, M.; Sharma, D.R.; Beeson, G.; Beeson, C.C.; Vargas, M.R. Enhancing NAD+ Salvage Pathway Reverts the Toxicity of Primary Astrocytes Expressing Amyotrophic Lateral Sclerosis-linked Mutant Superoxide Dismutase 1 (SOD1). *J. Biol. Chem.* **2016**, *291*, 10836–10846. [CrossRef]

148. Baxter, P.S.; Hardingham, G.E. Adaptive regulation of the brain's antioxidant defences by neurons and astrocytes. *Free Radic. Biol. Med.* **2016**, *100*, 147–152. [CrossRef]

149. Vargas, M.R.; Johnson, J.A. The Nrf2-ARE cytoprotective pathway in astrocytes. *Expert Rev. Mol. Med.* **2009**, *11*, e17. [CrossRef]

150. Harlan, B.A.; Killoy, K.M.; Pehar, M.; Liu, L.; Auwerx, J.; Vargas, M.R. Evaluation of the NAD⁺ biosynthetic pathway in ALS patients and effect of modulating NAD⁺ levels in hSOD1-linked ALS mouse models. *Exp. Neurol.* **2020**, *327*, 113219. [CrossRef]

151. Vargas, M.R.; Pehar, M.; Cassina, P.; Martínez-Palma, L.; Thompson, J.A.; Beckman, J.S.; Barbeito, L. Fibroblast growth factor-1 induces heme oxygenase-1 via nuclear factor erythroid 2-related factor 2 (Nrf2) in spinal cord astrocytes: Consequences for motor neuron survival. *J. Biol. Chem.* **2005**, *280*, 25571–25579. [CrossRef] [PubMed]

152. Vargas, M.R.; Johnson, D.A.; Sirkis, D.W.; Messing, A.; Johnson, J.A. Nrf2 activation in astrocytes protects against neurodegeneration in mouse models of familial amyotrophic lateral sclerosis. *J. Neurosci.* **2008**, *28*, 13574–13581. [CrossRef] [PubMed]

153. Harlan, B.A.; Pehar, M.; Killoy, K.M.; Vargas, M.R. Enhanced SIRT6 activity abrogates the neurotoxic phenotype of astrocytes expressing ALS-linked mutant SOD1. *FASEB J.* **2019**, *33*, 7084–7091. [CrossRef]

154. De la Rubia, J.E.; Drehmer, E.; Platero, J.L.; Benlloch, M.; Caplliure-Llopis, J.; Villaron-Casales, C.; de Bernardo, N.; AlarcÓn, J.; Fuente, C.; Carrera, S.; et al. Efficacy and tolerability of EH301 for amyotrophic lateral sclerosis: A randomized, double-blind, placebo-controlled human pilot study. *Amyotroph Lateral Scler. Frontotemporal. Degener.* **2019**, *20*, 115–122. [CrossRef] [PubMed]
155. Fels, J.A.; Casalena, G.; Konrad, C.; Holmes, H.E.; Dellinger, R.W.; Manfredi, G. Gene expression profiles in sporadic ALS fibroblasts define disease subtypes and the metabolic effects of the investigational drug EH301. *Hum. Mol. Genet.* **2022**, *31*, 3458–3477. [CrossRef]
156. Obrador, E.; Salvador, R.; Marchio, P.; López-Blanch, R.; Jihad-Jebbar, A.; Rivera, P.; Vallés, S.L.; Banacloche, S.; Alcácer, J.; Colomer, N.; et al. Nicotinamide Riboside and Pterostilbene Cooperatively Delay Motor Neuron Failure in ALS SOD1(G93A) Mice. *Mol. Neurobiol.* **2021**, *58*, 1345–1371. [CrossRef]
157. Laplante, M.; Sabatini, D.M. mTOR signaling at a glance. *J. Cell Sci.* **2009**, *122*, 3589–3594. [CrossRef]
158. Granatiero, V.; Sayles, N.M.; Savino, A.M.; Konrad, C.; Kharas, M.G.; Kawamata, H.; Manfredi, G. Modulation of the IGF1R-MTOR pathway attenuates motor neuron toxicity of human ALS SOD1(G93A) astrocytes. *Autophagy* **2021**, *17*, 4029–4042. [CrossRef]

Disclaimer/Publisher's Note: The statements, opinions and data contained in all publications are solely those of the individual author(s) and contributor(s) and not of MDPI and/or the editor(s). MDPI and/or the editor(s) disclaim responsibility for any injury to people or property resulting from any ideas, methods, instructions or products referred to in the content.

Review

Neuroinflammation in Glioblastoma: Progress and Perspectives

Xin Li, Wenting Gou and Xiaoqin Zhang *

Department of Pathology, School of Medicine, South China University of Technology, Guangzhou 510006, China
* Correspondence: mczhxq@scut.edu.cn

Abstract: Glioblastoma is the most common and malignant primary brain tumor, with high morbidity and mortality. Despite an aggressive, multimodal treatment regimen, including surgical resection followed by chemotherapy and radiotherapy, the prognosis of glioblastoma patients remains poor. One formidable challenge to advancing glioblastoma therapy is the complexity of the tumor microenvironment. The tumor microenvironment of glioblastoma is a highly dynamic and heterogeneous system that consists of not only cancerous cells but also various resident or infiltrating inflammatory cells. These inflammatory cells not only provide a unique tumor environment for glioblastoma cells to develop and grow but also play important roles in regulating tumor aggressiveness and treatment resistance. Targeting the tumor microenvironment, especially neuroinflammation, has increasingly been recognized as a novel therapeutic approach in glioblastoma. In this review, we discuss the components of the tumor microenvironment in glioblastoma, focusing on neuroinflammation. We discuss the interactions between different tumor microenvironment components as well as their functions in regulating glioblastoma pathogenesis and progression. We will also discuss the anti-tumor microenvironment interventions that can be employed as potential therapeutic targets.

Keywords: glioblastoma; neuroinflammation; tumor microenvironment (TME); therapy

Citation: Li, X.; Gou, W.; Zhang, X. Neuroinflammation in Glioblastoma: Progress and Perspectives. *Brain Sci.* **2024**, *14*, 687. https://doi.org/ 10.3390/brainsci14070687

Academic Editor: Álmos Klekner

Received: 10 May 2024
Revised: 25 June 2024
Accepted: 7 July 2024
Published: 9 July 2024

Copyright: © 2024 by the authors. Licensee MDPI, Basel, Switzerland. This article is an open access article distributed under the terms and conditions of the Creative Commons Attribution (CC BY) license (https:// creativecommons.org/licenses/by/ 4.0/).

1. Introduction

Glioblastoma, also known as glioblastoma multiforme (GBM), is the most common and lethal primary brain tumor in adults, with an aggressive nature and poor treatment response [1,2]. Despite a multimodal treatment regimen including surgical resection, radiation, and chemotherapy, the prognosis of glioblastoma patients remains poor, with a median survival of only 12–15 months [3,4]. One formidable challenge in advancing glioblastoma therapy is the complexity of the tumor microenvironment (TME) [3–6]. The TME of glioblastoma is a highly dynamic and heterogeneous system that not only consists of cancerous cells but also various types of non-cancerous cells, the predominant part of which are resident or infiltrating inflammatory cells [7–10]. Over the past decades, the heterogeneous nature of glioblastoma has been extensively studied and regarded as a key factor in the poor treatment efficacy of the disease. However, most of these studies are cancer cell-centric, which may underestimate the role of the tumor microenvironment, especially neuroinflammation, in glioblastoma pathogenesis and progression.

Neuroinflammation is the inflammatory response of the brain characterized by the infiltration of various immune cells and the release of inflammation-related cytokines, chemokines, and growth factors. In glioblastoma, neuroinflammation also has the typical features of enhanced vascularization, hypoxic tumor microenvironment, and immune-suppressive milieu. All these characteristics, together with the presence of the blood–brain barrier, make the neuroinflammatory microenvironment in glioblastoma a unique pathologic process. Accumulating evidence has suggested that the neuroinflammatory microenvironment plays an important role in glioblastoma progression, invasion, and treatment response [11–13]. It can affect the biological behavior of tumor cells directly through inflammation–tumor cell interactions or indirectly via the release of related cytokines, chemokines, and growth factors. Neuroinflammation has increasingly been recognized as a key player and potential therapeutic target in glioblastoma [11,14,15]. A deeper understanding of the inflammatory microenvironment in glioblastoma and its interactions with the cancer cells could provide the basis for more efficient therapies.

In this review, we discuss the known and emerging concepts related to the role of the tumor microenvironment in glioblastoma carcinogenesis and progression, focusing on neuroinflammation. We discuss the components of the tumor microenvironment in glioblastoma, especially neuroinflammation-related components, and their impacts on tumor invasion and progression. We also review the anti-tumor microenvironment interventions that can potentially be employed as therapeutic targets in glioblastoma.

2. The Blood–Brain Barrier and Vasculature in Glioblastoma

Anatomically, the human brain is protected by a natural membrane called the blood–brain barrier (BBB), which is composed of endothelial cells connected by tight junctions and surrounded by pericytes, astrocytes, and the basement membrane [16,17]. Under normal conditions, the protective BBB has a major role in maintaining normal brain function by preventing toxins and pathogens from entering the brain through circulation [16,17]. The traditional belief even holds that the human brain is immunologically privileged since the BBB is impermeable to immune cells. This perceived dogma has recently been changed with the breakthrough findings that the human brain actually possesses a conventional and functional lymphatic system like other organs [18,19]. Moreover, it has been increasingly accepted that the integrity of BBB is actually compromised under pathological conditions, e.g., tumor-associated inflammation in glioblastoma, which leads to increased permeability of BBB and the infiltration of inflammatory cells into the brain [20,21].

These changes, together with other factors, e.g., the rapid growth of glioblastoma cells, contribute greatly to the proliferation of microvasculature, one of the most characterized pathologic features of glioblastoma [22]. A variety of angiogenic factors and chemokines have been described as being involved in the formation of vasculature in glioblastoma; the elevated level of vascular endothelial growth factor (VEGF) is regarded as the predominant one [22]. Additionally, hypoxia-inducible factor (HIF), which can actually enhance the expression of VEGF, is also reported to be a key player in vessel formation [22]. However, it is important to note that the vasculature in glioblastoma is poorly organized, hyper-dilated, and has leaky vessels. This abnormal vasculature leads to the leaking of blood components into tumor tissues and also attracts inflammatory cells, which release proangiogenic factors, thereby further enhancing the endothelial cell proliferation in glioblastoma TME.

3. Composition of Tumor Microenvironment in Glioblastoma

It is now widely accepted that the tumor microenvironment (TME) of glioblastoma is a complex and dynamic system consisting of both non-cellular and cellular components [5,7–11,23] (Figure 1). The non-cellular components include an extracellular matrix (ECM) in which cells are embedded and various soluble factors (e.g., growth factors, cytokines, and chemokines). The cellular components include stromal cells and inflammatory cells, and inflammatory cells are predominant. It was reported that inflammatory cells in glioblastoma TME can constitute up to 30–50% of the tumor mass, which consists of resident microglia, infiltrated macrophages, and less abundant lymphoid T cells, NK cells, dendritic cells, and neutrophils. The inflammatory cells co-exist and interact with cancer cells, stromal cells, and non-cellular components, which shape the glioblastoma TME in both direct and indirect ways. The TME, especially neuroinflammation, has been regarded as the new therapeutic target for glioblastoma treatment [5,7–13,23–25].

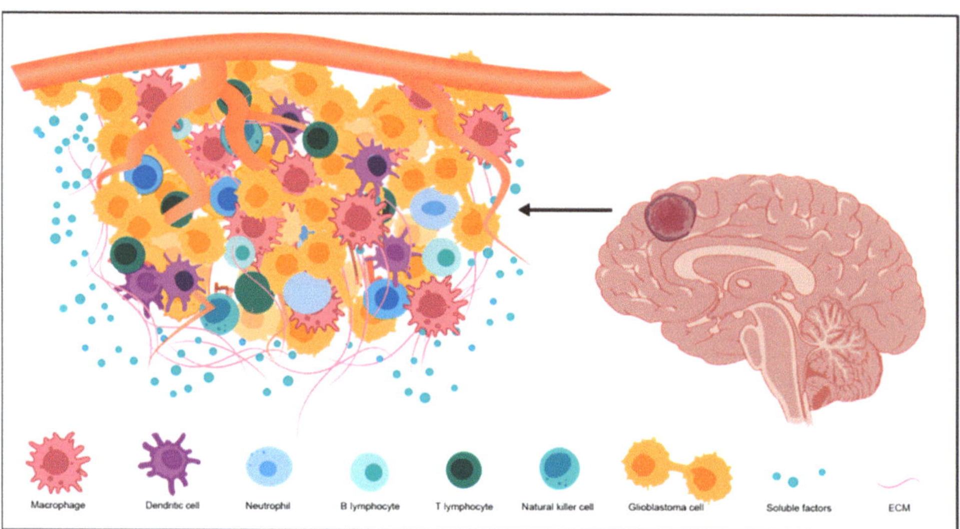

Figure 1. Components of the tumor microenvironment (TME) in glioblastoma.

3.1. Non-Cellular Components of TME in Glioblastoma

Extracellular matrix (ECM). The ECM is the non-cellular component that provides an important physical scaffold for all the cellular components embedded in it. However, the ECM in the brain is unique and different from the ECM normally found in many other tissues. The ECM of the brain is largely composed of hyaluronic acid (HA), proteoglycans, and glycoproteins and demonstrates a mesh-like appearance as compared with the fibrous ECM in other tissues [26]. In the case of glioblastoma, the components of the ECM change, which can be physically reflected in the increased stiffness of TME in glioblastoma [27]. Recent studies have shown that some ECM components (e.g., HA) increase in glioblastoma tissue as compared to non-tumor tissue and contribute to the increased mobility and invasiveness of glioblastoma cells [28].

Soluble molecular chemicals. For the soluble chemical components, a variety of inflammation mediators, including cytokines, chemokines, and growth factors, have been identified in the TME of glioblastoma and are involved in various signaling pathways [29]. Cytokines are signaling proteins secreted from specific immune cells, and the action modes of the cytokines include pro-inflammatory functions (e.g., IL-6, IL-8, and TNF-α) and anti-inflammatory functions (e.g., IL-4, IL-10, and TGF-β). Chemokines are small proteins that serve to mediate the migration of different cell types throughout the body. The chemokines that are highly expressed in the TME of glioblastoma have been characterized by CXCL2, IL-8, and CCL2, which can promote tumor invasion via facilitating cell proliferation, tumor growth, and angiogenesis. Additionally, tumor acidosis and low oxygen concentration are also important hallmarks of glioblastoma TME [30].

3.2. Cellular Components of TME in Glioblastoma
3.2.1. Stromal Cells

In glioblastoma TME, the stromal cells consist of astrocytes, neurons, and vascular endothelial cells. There is growing interest in the study of glioblastomas to communicate with astrocytes and neurons [31]. Among them, astrocytes have been found to undergo reactive astrogliosis during the growth of the tumor, which could further contribute to tumor cell infiltration. These tumor-associated reactive astrocytes have also been characterized as regulating the immune environment in glioblastoma [32,33]. The neurons have also been found in the pathologic process of glioblastoma, and they may interact with glioblastoma cells via paracrine stimulation, synaptic transmission, and some other indirect means [34]. For the endothelial cells, as mentioned above, glioblastoma is one of the most vascularized malignancies with extensive endothelial cell proliferation and even the formation of glomerular structures [22,35]. Various angiogenic factors have been reported to contribute to this hallmark, including the hypoxic tumor environment, which leads to highly elevated expressions of vascular endothelial growth factor (VEGF) and, therefore, proliferation of the endothelial cells. However, these newly formed microvasculature are abnormal blood vessels and cannot provide enough blood flow and oxygen to the tumor tissue, which will accelerate the necrosis of tumor tissues, another hallmark of glioblastoma.

3.2.2. Inflammatory Cells

Microglia/Macrophages. Microglia and macrophages are collectively referred to as tumor-associated macrophages (TAMs), which account for 30% of the total tumor volume and are the predominant inflammatory cell populations in glioblastoma [32]. Microglial cells are the resident myeloid cells in the brain, while macrophages are infiltrating monocytes derived from the peripheral blood due to the breakdown of the BBB under pathological conditions, e.g., tumors. Microglia and macrophages have similar functions and are difficult to differentiate in most cases. However, some studies reported that they have different localization sites within the tumor tissue: resident microglia are typically found at the tumor periphery, while infiltrating macrophages tend to be more enriched in the tumor core. Recent studies reported the employment of single-cell sequencing to precisely differentiate these two cell subpopulations [36].

According to their phenotype and function, TAMs are further classified into two subtypes: the pro-inflammatory subtype (M1 macrophages) and the anti-inflammatory subtype (M2 macrophages). M1 macrophages exhibit immune-supportive and anti-tumoral functions, while M2 macrophages have immune-suppressive and pro-tumoral functions [37]. The dual function of TAMs in glioblastoma pathogenesis and progression has been a robustly debated topic in the neuroinflammation field [37,38]. The acquisition of the M1 or M2 phenotype depends on the cytokines expressed in TME. It has been found that the M1 phenotype is acquired after being stimulated with pro-inflammatory factors, such as Toll-like receptor 4 (TLR-4) ligands and interferon-gamma (IFN-γ), and then eliminates tumor cells by producing inflammatory factors (e.g., TNF-a) [39,40]. On the other hand, the M2 phenotype is triggered after receiving stimulation with anti-inflammatory factors, for example, IL-4 and IL-10. M2-phenotype TAMs show less cytotoxicity in tumor cells by producing anti-inflammatory factors (e.g., TGF-β) and are associated with promoting tumor growth. However, it should be noted that the phenotypes of TAMs are dynamic, and as the tumor progresses, the M1 and M2 phenotypes can switch to each other [30,31]. Due to their dominant number, TAMs have been regarded as promising therapeutic targets for glioblastoma treatment. The cytokines that can contribute to TAM infiltration have been characterized, including colony-stimulating factor-1 (CSF-1), CCL2, and CCL5. Moreover, multiple immunotherapy approaches that could target TAMs have been tried [37], which is detailed in the Section 4 below.

T cells. T cells are also an important population of inflammatory cells in TME of glioblastoma, in spite of the fact that they constitute only a small proportion of the total cell numbers in TME (~0.25%) [41]. Of particular importance are the regulatory T cells (Tregs). Tregs are a unique population of CD4+ T cells that can regulate the overall immune homeostasis in an immunosuppressive manner [42,43]. In glioblastoma, Tregs inhibit the anti-tumor response and promote tumor-killing tolerance by secreting immunomodulatory cytokines (e.g., TGF-β and IL-10). This will, in turn, inhibit the production of anti-tumor cytokines, such as IL-2 and IFN-γ, leading to a decrease in effector cells necessary to control tumor growth. Tregs are reported to be recruited to the TME of glioblastoma by specific cytokines such as CXCR3 and CCR5, which can be secreted by glioblastoma cells and innate immune cells within the brain. In addition to Tregs, CD8+ cytotoxic T cells are another important subpopulation of T cells in glioblastoma, which can induce a tumor-killing effect and mediate tumor regression like natural killer cells. Various immunotherapy efforts have been tried to boost the cytotoxic CD8+ T cell function to treat glioblastoma, such as immune checkpoint inhibitors and CAR-T cell therapy [44–46]. However, these therapies are generally less efficacious in glioblastoma as compared to other malignancies due to the relatively low number of tumor-infiltrating T cells in the TME of glioblastoma.

Natural killer cells. Natural killer (NK) cells have also been characterized as an important part of inflammatory cells in glioblastoma TME. Although they account for a relatively small proportion (~2% of total infiltrating inflammatory cells) like T cells, NK cells are critical for the anti-tumor immune response in glioblastoma [47]. NK cells can not only provoke tumor cell apoptosis through their direct natural cytotoxicity (e.g., granzyme B and perforin); they can also control tumor growth via secreting cytokines or regulating the activity of other inflammatory cells. For example, NK cells have been demonstrated to be able to regulate T cell-mediated immune responses by maintaining the function of dendritic cells and promoting tumor antigen presentation. On the other hand, NK cells can also be regulated by the TME. For example, glioblastoma cells express transforming growth factor (TGF-β), which can inhibit the activation of NK cell function. Glioblastoma cells can also express unique MHC-I molecules to inhibit the function of NK cells by acting as inhibitory receptor ligands [48]. Therefore, although glioblastoma is often infiltrated by NK cells, these NK cells are functionally inhibited by glioblastoma cells and TME.

Dendritic cells. Dendritic cells (DCs) are a class of professional antigen-processing and presenting cells that play key roles in cancer immunity [49]. Similar to NK cells, DCs are recruited to glioblastoma via specific chemokines such as CXCL1 and CCL5. It has been shown that DCs can produce anti-tumor cytokines (e.g., IL-12), which in turn recruit more CD8+ T cells [50]. Preclinical studies have shown that the activation of DCs can improve long-term tumor survival in the mouse model of glioblastoma [50]. Clinical studies of DC vaccines in glioblastoma patients have also shown some efficacy in improving the median overall survival [50]. However, it remains to be elucidated for the standardization of DC vaccine therapy, e.g., the antigens used and the injection sites [48]. Therefore, future work on improving the efficacy of DC-based therapy in more clinical trials is needed [51].

Neutrophil cells. Neutrophil cells are the most abundant population of granulocytes in the human body, which account for approximately 70% of the total number of white blood cells. In glioblastoma, neutrophils are observed to be negatively correlated with the prognosis of glioblastoma patients [52–54]. Neutrophils are commonly found in the center area of the glioblastoma tumor bulk and aid in tumor progression and angiogenesis. Neutrophils are attracted to the TME core by specific chemokines, e.g., CXCL8 and IL-8. They can also promote tumor proliferation and angiogenesis by secreting elastase. Recently, it was found that neutrophils are involved in the proliferation and invasion of glioblastoma cells by activating the NF-κB signaling. Additionally, there was a positive feedback loop between IL-8 expression and neutrophil infiltration into tumor sites [55]. In glioma, it was also found that the neutrophil-to-lymphocyte ratio (NLR) in the peripheral blood was positively associated with tumor grading, in which an increase in NLR may indicate a higher tumor grade and poorer patient outcomes. Moreover, compared with traditional molecular prognostic markers, e.g., IDH1 mutations, NLR can better evaluate the prognosis of glioblastoma patients and guide the treatment regimen.

4. Anti-TME Intervention for the Therapy of Glioblastoma

The functional role of TME, especially neuroinflammation, in the pathogenesis and tumor progression of glioblastoma makes anti-TME intervention a major novel therapy strategy for glioblastoma treatment [12,56–59]. Currently, there are two main types of TME-based therapy for glioblastoma: anti-vasculature therapy and neuroinflammation-based therapy. The latter could be further classified into four strategies: immune checkpoint inhibitors (ICIs), chimeric antigen receptor T-cell (CAR-T) therapies, vaccines, and oncolytic viruses (OVs) [57–66]. Additionally, a combined multimodal therapy of these different strategies is also extensively studied. In the following, we will provide an overview of each of these therapeutic approaches in the clinical setting for glioblastoma treatment (Tables 1–4).

Table 1. Clinical trials of anti-vasculature therapy for glioblastoma treatment.

	ClinicalTrials.gov ID	Target	Brief Description of the Trial	Phase	Year of Start to Completion
1	NCT01753713	VEGF	Dovitinib in treating patients with recurrent or progressive glioblastoma	2	2012–2017
2	NCT01609790	VEGF	Bevacizumab with or without trebananib in treating patients with recurrent brain tumors	2	2012–2022
3	NCT02330562	VEGF	Marizomib alone or in combination with bevacizumab in patients with recurrent glioblastoma	1 and 2	2015–2021
4	NCT02511405	VEGF	A phase 3, pivotal trial of VB-111 plus bevacizumab vs. bevacizumab in patients with recurrent glioblastoma	3	2015–2018
5	NCT02342379	VEGF	TH-302 in combination with bevacizumab for glioblastoma	2	2015–2019

Note: The data are retrieved from clinicaltrials.gov.

Table 2. Clinical trials of ICIs for glioblastoma treatment.

	ClinicalTrials.gov ID	Target	Brief Description of the Trial	Phase	Year of Start to Completion
1	NCT02550249	PD-1	Neoadjuvant nivolumab in glioblastoma	2	2015–2017
2	NCT02336165	PD-1	Phase 2 study of durvalumab (MEDI4736) in patients with glioblastoma	2	2015–2021
3	NCT02337491	PD-1	Pembrolizumab +/− bevacizumab for recurrent glioblastoma	2	2015–2020
4	NCT02617589	PD-1	An investigational immuno-therapy study of nivolumab compared to temozolomide, each given with radiation therapy, for newly diagnosed patients with glioblastoma	3	2016–2022
5	NCT02798406	PD-1	Combination adenovirus + pembrolizumab to trigger immune virus effects	2	2016–2021
6	NCT03018288	PD-1	Radiation therapy plus temozolomide and pembrolizumab with and without HSPPC-96 in newly diagnosed glioblastoma	2	2017–2022
7	NCT02794883	CTLA-4	Tremelimumab and durvalumab in combination or alone in treating patients with recurrent malignant glioma	2	2016–2020
8	NCT03367715	CTLA-4	Nivolumab, ipilimumab, and short-course radiotherapy in adults with newly diagnosed MGMT unmethylated glioblastoma	2	2018–2022

Note: The data are retrieved from clinicaltrials.gov.

Table 3. Clinical trials of vaccine therapy for glioblastoma treatment.

	ClinicalTrials.gov ID	Vaccine Type	Title	Phase	Year of Start to Completion
1	NCT00293423	Peptide vaccine	GP96 heat shock protein-peptide complex vaccine in treating patients with recurrent or progressive glioma	1 and 2	2005–2013
2	NCT00458601	Peptide vaccine	Phase II study of rindopepimut (CDX-110) in patients with glioblastoma multiforme	2	2007–2016
3	NCT00643097	Peptide vaccine	Vaccine therapy in treating patients with newly diagnosed glioblastoma	2	2007–2016
4	NCT00905060	Peptide vaccine	HSPPC-96 vaccine with temozolomide in patients with newly diagnosed glioblastoma	2	2009–2014
5	NCT01480479	Peptide vaccine	Phase III study of rindopepimut/GM-CSF in patients with newly diagnosed glioblastoma	3	2011–2016

Table 3. *Cont.*

	ClinicalTrials.gov ID	Vaccine Type	Title	Phase	Year of Start to Completion
6	NCT01920191	Peptide vaccine	Phase I/II trial of IMA950 multi-peptide vaccine plus poly-ICLC in glioblastoma	1 and 2	2013–2016
7	NCT00639639	DC vaccine	Vaccine therapy in treating patients with newly diagnosed glioblastoma	1	2006–2022
8	NCT00323115	DC vaccine	Phase II feasibility study of dendritic cell vaccination for newly diagnosed glioblastoma	2	2006–2013
9	NCT00846456	DC vaccine	Safe study of dendritic cell (DC) based therapy targeting tumor stem cells in glioblastoma	1 and 2	2009–2013
10	NCT01006044	DC vaccine	Efficacy and safety of autologous dendritic cell vaccination in glioblastoma after complete surgical resection	2	2009–2014
11	NCT01213407	DC vaccine	Dendritic cell cancer vaccine for high-grade glioma	2	2010–2015
12	NCT02465268	DC vaccine	Vaccine therapy for the treatment of newly diagnosed glioblastoma	2	2016–2023

Note: The data are retrieved from clinicaltrials.gov.

Table 4. Clinical trials of CAR-T cell therapy for glioblastoma treatment.

	ClinicalTrials.gov ID	Target	Brief Description of the Trial	Phase	Year of Start to Completion
1	NCT01082926	GRm13Z40-2	Phase I study of cellular immunotherapy for recurrent/refractory malignant glioma using intratumoral infusions of GRm13Z40-2, an allogeneic CD8+ Cytolitic T-Cell line genetically modified to express the IL 13-Zetakine and HyTK and to be resistant to glucocorticoids, in combination with interleukin-2	1	2010–2013
2	NCT01109095	HER2	CMV-specific cytotoxic T lymphocytes expressing CAR targeting HER2 in patients with glioblastoma	1	2010–2018
3	NCT01454596	EGFRvIII	CAR T cell receptor immunotherapy targeting EGFRvIII for patients with malignant gliomas expressing EGFRvIII	1 and 2	2012–2019
4	NCT03726515	EGFRvIII	CART-EGFRvIII + pembrolizumab in glioblastoma	1	2019–2021

Note: The data are retrieved from clinicaltrials.gov.

4.1. Anti-Vasculature Therapy

Due to the hallmark of microvascular proliferation in glioblastoma, anti-vasculature has become one of the most studied therapy approaches. A series of clinical trials have been performed to test the effectiveness of anti-vasculature therapy in glioblastoma [22,35]. The majority of these studies focus on blocking the VEGF/VEGFR signaling pathway, either through a monoclonal antibody against VEGF or with small-molecule inhibitors against VEGFR. For example, VEGF inhibition with bevacizumab, a humanized monoclonal antibody targeting VEGF, has shown effects in improving glioblastoma patients' survival [67–69]. Moreover, it was found that anti-VEGF therapy can decrease vasogenic brain edema and improve blood perfusion and subsequent oxygenation, which creates conditions for better drug delivery and the efficacy of other treatments. It can also decrease the immune suppression in glioblastoma TME. Therefore, there are some strategies to combine anti-VEGF therapy with other treatment regimens, such as combining anti-vasculature therapy with immune-based approaches.

Overall, however, anti-VEGF therapy has benefitted only a subset of glioblastoma patients; the outcome in most anti-VEGF studies failed to demonstrate the benefit in patient survival [70–72]. There are several underlying reasons for this treatment failure. One of the major problems is the inefficient drug delivery to the tumor, which is frequent in almost all types of therapeutics for brain disease because of the BBB. Other reasons include the existence of VEGF-independent angiogenesis, such as neoangiogenesis through the CXCR4/CXCL12 axis [73]. Therefore, efforts that target angiogenesis through different action mechanisms will be helpful to increase treatment efficacy. Further analysis also revealed that the effects of anti-VEGF therapy may be dependent on the glioblastoma genetic subtypes, e.g., IDH1 mutation status, suggesting the necessity of patient subtype stratification before clinical trials [74]. Table 1 lists some of the completed phase II or III clinical trials of anti-vasculature therapy for the treatment of glioblastoma.

4.2. Neuroinflammation-Based Therapy
4.2.1. Immune Checkpoint Inhibitors

Immune checkpoint inhibitors (ICIs) primarily refer to the monoclonal antibodies that can target the cell immune checkpoints and allow for a more robust anti-tumor effect [75]. The majority of studies on ICIs have focused on programmed death-1 (PD-1), programmed death-ligand 1 (PD-L1), and cytotoxic T-lymphocyte antigen 4 (CTLA-4), which are all important proteins of immune checkpoint pathways [76]. PD-1 is a cell surface protein of T cells and normally acts as a T cell checkpoint that keeps T cells from attacking tumor cells by binding with PD-L1. The use of PD-1 inhibitors has led to increased survival in patients with various tumor types, including glioblastoma [75–77]. For example, in neoadjuvant anti-PD-1 immunotherapy (pembrolizumab) in recurrent glioblastoma, satisfactory results were observed in the overall survival of patients receiving neoadjuvant pembrolizumab compared to the control group [78]. Moreover, functional activation of tumor-infiltrating T lymphocytes was observed, and interferon responses were induced in TME.

In the majority of clinical trials applying the anti-PD-1 antibody to recurrent glioblastoma patients, however, very limited efficacy in improving patients' survival was observed. Generally, the anti-tumor effects of ICI treatment are less efficacious in glioblastoma than in other malignancies (e.g., melanoma). For example, in a phase 3 trial for a PD-1 inhibitor, although the safety of the treatment was found to be consistent with that in other tumor types, no clinical benefit was observed [79]. The scarcity of T cells in glioblastoma TME is a potential reason because the existence of infiltrating T cells in TME is fundamental to the success of ICI treatment [80]. With continuous efforts in this field, e.g., the characterization of new checkpoints and the combination of ICI treatment with other treatment regimens, the treatment efficacy of ICIs may be improved. Table 2 lists some of the completed phase II or III clinical trials of ICI therapy for the treatment of glioblastoma.

4.2.2. Vaccine Therapy

Vaccine therapy utilizes one or multiple tumor-associated antigens to stimulate anti-tumor effects and has been extensively studied in multiple malignancies, including glioblastoma [81–83]. There are several types of tumor vaccines that are being used in cancer treatment, and the peptide-based vaccine and the dendritic cell (DC)-based vaccine are two main strategies for glioblastoma [84,85]. Peptide-based vaccines are vaccines developed based on short peptides that have epitopes with cancer cells and can act as antigenic targets to induce effective anti-tumor responses. Peptide-based vaccines have simple structures and are relatively easy to manipulate. Moreover, they have relatively lower variability as compared to other vaccines. One of the most frequently used tumor-associated antigens for peptide vaccines in glioblastoma is EGFRvIII, a deletion mutation found in approximately 20% of glioblastoma patients [84]. Dendritic cell (DC)-based vaccines employ DCs primed with whole tumor cell lysates or tumor-associated antigens to stimulate the adaptive immune system and, therefore, to control the growth of tumors. Currently, both peptide- and DC-based vaccines have been investigated in clinical trials in glioblastoma patients and represent attractive approaches for the immunotherapy of glioblastoma [82–86]. For example, in a phase II clinical trial of the EGFRvIII peptide vaccine study, a substantial increase in patient survival was observed [87]. For the DC-based vaccines, preclinical studies have demonstrated promising results for vaccine treatment in combination with PD-1 inhibitors. Moreover, a phase I clinical study revealed that the combination of DC-based vaccine therapy with TMZ is safe and tolerable in glioblastoma patients. Additionally, in phase III clinical trials in newly diagnosed and recurrent glioblastoma patients, it was found that autologous tumor lysate-loaded DCs could extend the patients' survival [88].

While vaccine therapy presents an attractive method for glioblastoma therapy, disappointing results remain. In a phase III clinical trial that used the EGFRvIII peptide vaccine (rindopepimut), the results did not show a significant improvement in patient survival [89]. The potential reason is the heterogeneous and unstable expression of EGFRvIII in glioblastoma cells, which leads to the outgrowth of tumor cells that lack this antigen. Future studies are needed to reveal the effectiveness of this treatment and its impact on the overall survival of glioblastoma patients. Table 3 lists some of the completed phase II or III clinical trials of vaccine therapy for the treatment of glioblastoma.

4.2.3. CAR-T Cell Therapy

CAR-T cell therapy is a novel type of immunotherapy in which T cells are modified to bind chimeric antigen receptors (CARs) to increase their ability to recognize and target tumor cells [90–92]. Currently, clinical trials employing CAR-T cell therapy in cancer treatment have shown safety and encouraging results, especially in hematologic malignancies. The effectiveness of CAR-T cell therapy in treating solid malignancies, e.g., glioblastoma, has also been identified [90,92,93]. There are three commonly used antigens for CAR-T cell therapy in glioblastoma: EGFRvIII, HER2, and IL-13 receptor (IL-13R) [90–92]. For example, a phase I clinical trial targeting EGFRvIII CAR-T cell therapy demonstrated that the intravenous administration could transfer the CAR-T cells to the brain tumor site and that the EGFRvIII level on glioblastoma tumor cells was reduced by the CAR-T cell treatment [94]. Moreover, this study also demonstrated the effects of CAR-T cell therapy on improving the immunosuppressive tumor environment, indicating the promising perspective of combinational therapy of CAR-T cell therapy with other treatment approaches. In another phase I clinical trial employing IL13R-targeted CAR-T cell therapy, it was found that the intracranial injection of CAR-T cells can achieve an anti-tumor effect in glioblastoma treatment [95]. In a phase I study employing HER2-specific CAR-T cell therapy, of the 17 recruited glioblastoma patients, only 8 demonstrated clinical benefits in overall survival [96].

As for the confounding and limited effects of CAR-T cell therapy in glioblastoma, one major problem is the heterogeneity of target-antigen expressions in tumor cells, which finally leads to heterogeneous treatment effects. Another substantial issue is how to maximize and maintain the activity of the injected CAR-T cells. It has been reported that CAR-T cell administration can induce immunosuppressive responses in the brain [94]. Therefore, successful treatment needs the development of engineered CAR-T cells that are resistant to immunosuppression. Table 4 lists some of the completed phase II or III clinical trials of CAR-T cell therapy for the treatment of glioblastoma.

4.2.4. Oncolytic Virus Therapy

Oncolytic virus (OV) is able to infect cancer cells to present tumor-associated antigens and then lyse the tumor cells. Moreover, it was found that the cellular proteins released from the OV-lysed tumor cells can activate the anti-tumor immune response in multiple ways. For example, viruses can activate macrophages, and activated macrophages can enhance the infiltration of T cells into TME and, therefore, improve the immunosuppressive characteristic of glioblastoma. Therefore, OV therapy is becoming a very promising approach for the treatment of malignancies [97,98]. In glioblastoma, the effects of OV on tumor-killing have also been widely studied [98–102]. Multiple types of viruses are being tested for OV therapy, including retrovirus, adenovirus, herpes simplex virus, poliovirus, and measles virus [97,98]. In 2018, the recombinant oncolytic poliovirus PVSRIPO was tested in recurrent glioblastoma patients [103]. The study confirmed the potential of intratumor infusions of PVSRIPO for improving patients' clinical outcomes. It was observed that the survival rate among patients who received PVSRIPO therapy was higher than that of historical controls.

While OV therapy has become an important focus of anti-tumor therapy, its safety and efficacy need to be tested in future research. The initial studies usually used replication-incompetent viruses to avoid complications (e.g., encephalitis), and now, an increasing number of types of viruses have been utilized as aforementioned. However, no full safety or preliminary efficacy data are currently available in the public domain. While the main goal of the current work is not to discuss each of these studies in detail, a comprehensive discussion of oncolytic virus therapy for glioblastoma can be found in several other excellent review papers [94–97]. Table 5 lists some of the completed phase II/III clinical trials of OV therapy for glioblastoma treatment.

Table 5. Clinical trials of OV therapy for glioblastoma treatment.

	ClinicalTrials.gov ID	Virus Type	Brief Description of the Trial	Phase	Year of Start to Completion
1	NCT00028158	Herpes Simplex Virus	Safety and effectiveness study of G207, a tumor-killing virus, in patients with recurrent brain cancer	1 and 2	2002–2003
2	NCT00528684	Reovirus	Safety and efficacy study of REOLYSIN® in the treatment of recurrent malignant gliomas	1	2006–2010
3	NCT01301430	Parvovirus	Parvovirus H-1 (ParvOryx) in patients with progressive primary or recurrent glioblastoma multiforme.	1 and 2	2011–2015
4	NCT01956734	Adenovirus	Virus DNX2401 and temozolomide in recurrent glioblastoma	1	2013–2015
5	NCT02197169	Adenovirus	DNX-2401 with interferon-gamma (IFN-γ) for recurrent glioblastoma or gliosarcoma brain tumors	1	2014–2018
6	NCT02798406	Adenovirus	Combination adenovirus + pembrolizumab to trigger immune virus effects	2	2016–2021
7	NCT03072134	Adenovirus	Neural stem cell-based virotherapy of newly diagnosed malignant glioma	1	2017–2021

Note: The data are retrieved from clinicaltrials.gov.

5. Challenges and Perspectives

Despite encouraging results achieved so far, however, the anti-TME treatment for glioblastoma is still facing many challenges [5,58]. One of the primary impediments is the BBB. Although the BBB is compromised and more permeable during the tumor state, the anti-tumor drugs cannot cross the BBB inadequately to achieve sufficient drug accumulation in the tumor. For this reason, multiple efforts have been made to deliver pharmaceutical agents to the brain efficiently. One such important advance is focused ultrasound (FUS), which can increase the permeability of the BBB in a temporary way and enhance the delivery of drugs to the brain [104]. Currently, FUS-mediated BBB disruption has demonstrated robustness in non-invasive drug delivery to the brain and provides encouraging perspectives for the treatment of brain diseases, including glioblastoma. Another important impediment to achieving effective treatment responses, especially for immunotherapy-based approaches, is the immunosuppressive nature of glioblastoma TME. Therefore, strategies that could boost the immune response in glioblastoma TME will be helpful—for example, recruiting cytotoxic or tumor-killing inflammation cells, improving immunosuppressive properties through drugs, and transforming a 'cold' tumor into a 'hot' tumor. The third important impediment to developing effective treatment responses is the aforementioned complexity of the glioblastoma microenvironment. The enormous inter-tumor and intra-tumor heterogeneity of glioblastoma has made it one of the most difficult-to-treat malignancies in the world. Therefore, continued efforts are needed to fully understand the complex cellular and molecular components as well as their interactions involved in the TME of glioblastoma. At the same time, a synergic combination of different treatment strategies may lead to a promising curing regimen. Actually, there have been efforts to employ combinatorial therapies between immunotherapy and the current standard of care [105].

6. Conclusions

In summary, this review highlights the functions of neuroinflammation in glioblastoma at the cellular, molecular, and therapeutic levels. While increasing and promising results have been achieved in the anti-neuroinflammation therapy of glioblastoma, there are still many challenges. The immunosuppressive and heterogeneous characteristics of the glioblastoma microenvironment ultimately lead to resistance to anti-inflammatory therapies. Continued efforts into the tumor microenvironment will help our understanding of how these components interact with one another and contribute to the therapeutic response. This will lead to the development of more efficient and targeted therapy strategies for the treatment of glioblastoma in the future.

Funding: This work was supported by the National Natural Science Foundation of China (No. 81702945).

Acknowledgments: We thank Zhong Lan for helpful discussions and insightful comments.

Conflicts of Interest: The authors declare no conflicts of interest.

References

1. Wen, P.Y.; Weller, M.; Lee, E.Q.; Alexander, B.M.; Barnholtz-Sloan, J.S.; Barthel, F.P.; Batchelor, T.T.; Bindra, R.S.; Chang, S.M.; Chiocca, E.A.; et al. Glioblastoma in adults: A Society for Neuro-Oncology (SNO) and European Society of Neuro-Oncology (EANO) consensus review on current management and future directions. *Neuro-Oncology* **2020**, *22*, 1073–1113. [CrossRef] [PubMed]
2. Schaff, L.R.; Mellinghoff, I.K. Glioblastoma and Other Primary Brain Malignancies in Adults: A Review. *J. Am. Med. Assoc.* **2023**, *329*, 574–587. [CrossRef] [PubMed]
3. Tan, A.C.; Ashley, D.M.; López, G.Y.; Malinzak, M.; Friedman, H.S.; Khasraw, M. Management of glioblastoma: State of the art and future directions. *CA—Cancer J. Clin.* **2020**, *70*, 299–312. [CrossRef] [PubMed]
4. Aldape, K.; Brindle, K.M.; Chesler, L.; Chopra, R.; Gajjar, A.; Gilbert, M.R.; Gottardo, N.; Gutmann, D.H.; Hargrave, D.; Holland, E.C.; et al. Challenges to curing primary brain tumours. *Nat. Rev. Clin. Oncol.* **2019**, *16*, 509–520. [CrossRef] [PubMed]
5. Bikfalvi, A.; da Costa, C.A.; Avril, T.; Barnier, J.V.; Bauchet, L.; Brisson, L.; Cartron, P.F.; Castel, H.; Chevet, E.; Chneiweiss, H.; et al. Challenges in glioblastoma research: Focus on the tumor microenvironment. *Trends Cancer* **2023**, *9*, 9–27. [CrossRef] [PubMed]
6. Becker, A.P.; Sells, B.E.; Haque, S.J.; Chakravarti, A. Tumor Heterogeneity in Glioblastomas: From Light Microscopy to Molecular Pathology. *Cancers* **2021**, *13*, 761. [CrossRef] [PubMed]
7. Gieryng, A.; Pszczolkowska, D.; Walentynowicz, K.A.; Rajan, W.D.; Kaminska, B. Immune microenvironment of gliomas. *Lab. Investig.* **2017**, *97*, 498–518. [CrossRef] [PubMed]
8. Virtuoso, A.; Giovannoni, R.; De Luca, C.; Gargano, F.; Cerasuolo, M.; Maggio, N.; Lavitrano, M.; Papa, M. The Glioblastoma Microenvironment: Morphology, Metabolism, and Molecular Signature of Glial Dynamics to Discover Metabolic Rewiring Sequence. *Int. J. Mol. Sci.* **2021**, *22*, 3301. [CrossRef]
9. Di Nunno, V.; Franceschi, E.; Tosoni, A.; Gatto, L.; Bartolini, S.; Brandes, A.A. Tumor-Associated Microenvironment of Adult Gliomas: A Review. *Front. Oncol.* **2022**, *12*, 891543. [CrossRef]
10. Sharma, P.; Aaroe, A.; Liang, J.Y.; Puduvalli, V.K. Tumor microenvironment in glioblastoma: Current and emerging concepts. *Neuro-Oncol. Adv.* **2023**, *5*, vdad009. [CrossRef]
11. Nóbrega, A.H.L.; Pimentel, R.S.; Prado, A.P.; Garcia, J.; Frozza, R.L.; Bernardi, A. Neuroinflammation in Glioblastoma: The Role of the Microenvironment in Tumour Progression. *Curr. Cancer Drug Targets* **2024**, *24*, 579–594. [CrossRef]
12. Alghamri, M.S.; McClellan, B.L.; Hartlage, M.S.; Haase, S.; Faisal, S.M.; Thalla, R.; Dabaja, A.; Banerjee, K.; Carney, S.V.; Mujeeb, A.A.; et al. Targeting Neuroinflammation in Brain Cancer: Uncovering Mechanisms, Pharmacological Targets, and Neuropharmaceutical Developments. *Front. Pharmacol.* **2021**, *12*, 680021. [CrossRef] [PubMed]
13. Catalano, M. Editorial: Brain Tumors and Neuroinflammation. *Front. Cell. Neurosci.* **2022**, *16*, 941263. [CrossRef]
14. Bianconi, A.; Aruta, G.; Rizzo, F.; Salvati, L.F.; Zeppa, P.; Garbossa, D.; Cofano, F. Systematic Review on Tumor Microenvironment in Glial Neoplasm: From Understanding Pathogenesis to Future Therapeutic Perspectives. *Int. J. Mol. Sci.* **2022**, *23*, 4166. [CrossRef]
15. Bianconi, A.; Palmieri, G.; Aruta, G.; Monticelli, M.; Zeppa, P.; Tartara, F.; Melcarne, A.; Garbossa, D.; Cofano, F. Updates in Glioblastoma Immunotherapy: An Overview of the Current Clinical and Translational Scenario. *Biomedicines* **2023**, *11*, 1520. [CrossRef]
16. Obermeier, B.; Daneman, R.; Ransohoff, R.M. Development, maintenance and disruption of the blood-brain barrier. *Nat. Med.* **2013**, *19*, 1584–1596. [CrossRef] [PubMed]
17. Alahmari, A. Blood-Brain Barrier Overview: Structural and Functional Correlation. *Neural Plast.* **2021**, *2021*, 6564585. [CrossRef] [PubMed]
18. Louveau, A.; Smirnov, I.; Keyes, T.J.; Eccles, J.D.; Rouhani, S.J.; Peske, J.D.; Derecki, N.C.; Castle, D.; Mandell, J.W.; Lee, K.S.; et al. Structural and functional features of central nervous system lymphatic vessels. *Nature* **2015**, *523*, 337–341. [CrossRef]

19. Cugurra, A.; Mamuladze, T.; Rustenhoven, J.; Dykstra, T.; Beroshvili, G.; Greenberg, Z.J.; Baker, W.; Papadopoulos, Z.; Drieu, A.; Blackburn, S.; et al. Skull and vertebral bone marrow are myeloid cell reservoirs for the meninges and CNS parenchyma. *Science* **2021**, *373*, eabf7844. [CrossRef] [PubMed]
20. Steeg, P.S. The blood-tumour barrier in cancer biology and therapy. *Nat. Rev. Clin. Oncol.* **2021**, *18*, 696–714. [CrossRef]
21. Arvanitis, C.D.; Ferraro, G.B.; Jain, R.K. The blood-brain barrier and blood-tumour barrier in brain tumours and metastases. *Nat. Rev. Cancer* **2020**, *20*, 26–41. [CrossRef]
22. Mosteiro, A.; Pedrosa, L.; Ferrés, A.; Diao, D.; Sierra, A.; González, J.J. The Vascular Microenvironment in Glioblastoma: A Comprehensive Review. *Biomedicines* **2022**, *10*, 1285. [CrossRef]
23. Baghba, R.; Roshangar, L.; Jahanban-Esfahlan, R.; Seidi, K.; Ebrahimi-Kalan, A.; Jaymand, M.; Kolahian, S.; Javaheri, T.; Zare, P. Tumor microenvironment complexity and therapeutic implications at a glance. *Cell Commun. Signal.* **2020**, *18*, 59. [CrossRef]
24. Dapash, M.; Hou, D.; Castro, B.; Lee-Chang, C.; Lesniak, M.S. The Interplay between Glioblastoma and Its Microenvironment. *Cells* **2021**, *10*, 2257. [CrossRef] [PubMed]
25. Crivii, C.B.; Bosca, A.B.; Melincovici, C.S.; Constantin, A.M.; Marginean, M.; Dronca, E.; Sufletel, R.; Gonciar, D.; Bungardean, M.; Sovrea, A. Glioblastoma Microenvironment and Cellular Interactions. *Cancers* **2022**, *14*, 1092. [CrossRef]
26. Lam, D.; Enright, H.A.; Cadena, J.; Peters, S.K.G.; Sales, A.P.; Osburn, J.J.; Soscia, D.A.; Kulp, K.S.; Wheeler, E.K.; Fischer, N.O. Tissue-specific extracellular matrix accelerates the formation of neural networks and communities in a neuron-glia co-culture on a multi-electrode array. *Sci. Rep.* **2019**, *9*, 4159. [CrossRef]
27. Belousov, A.; Titov, S.; Shved, N.; Garbuz, M.; Malykin, G.; Gulaia, V.; Kagansky, A.; Kumeiko, V. The Extracellular Matrix and Biocompatible Materials in Glioblastoma Treatment. *Front. Bioeng. Biotech.* **2019**, *7*, 341. [CrossRef] [PubMed]
28. Wang, C.; Sinha, S.; Jiang, X.Y.; Murphy, L.; Fitch, S.; Wilson, C.; Grant, G.; Yang, F. Matrix Stiffness Modulates Patient-Derived Glioblastoma Cell Fates in Three-Dimensional Hydrogels. *Tissue Eng. Part A* **2021**, *27*, 390–401. [CrossRef]
29. Yeo, E.C.F.; Brown, M.P.; Gargett, T.; Ebert, L.M. The Role of Cytokines and Chemokines in Shaping the Immune Microenvironment of Glioblastoma: Implications for Immunotherapy. *Cells* **2021**, *10*, 607. [CrossRef]
30. Corbet, C.; Feron, O. Tumour acidosis: From the passenger to the driver's seat. *Nat. Rev. Cancer* **2017**, *17*, 577–593. [CrossRef] [PubMed]
31. Uyar, R. Glioblastoma microenvironment: The stromal interactions. *Pathol. Res. Pract.* **2022**, *232*, 153813. [CrossRef]
32. Lin, C.C.J.; Yu, K.; Hatcher, A.; Huang, T.W.; Lee, H.K.; Carlson, J.; Weston, M.C.; Chen, F.J.; Zhang, Y.Q.; Zhu, W.Y.; et al. Identification of diverse astrocyte populations and their malignant analogs. *Nat. Neurosci.* **2017**, *20*, 396–405. [CrossRef]
33. Heiland, D.H.; Ravi, V.M.; Behringer, S.P.; Frenking, J.H.; Wurm, J.; Joseph, K.; Garrelfs, N.W.C.; Strähle, J.; Heynckes, S.; Grauvogel, J.; et al. Tumor-associated reactive astrocytes aid the evolution of immunosuppressive environment in glioblastoma. *Nat. Commun.* **2019**, *10*, 2541. [CrossRef] [PubMed]
34. Quail, D.F.; Joyce, J.A. The Microenvironmental Landscape of Brain Tumors. *Cancer Cell* **2017**, *31*, 326–341. [CrossRef] [PubMed]
35. Ahir, B.K.; Engelhard, H.H.; Lakka, S.S. Tumor Development and Angiogenesis in Adult Brain Tumor: Glioblastoma. *Mol. Neurobiol.* **2020**, *57*, 2461–2478. [CrossRef] [PubMed]
36. Kaminska, B.; Ochocka, N.; Segit, P. Single-Cell Omics in Dissecting Immune Microenvironment of Malignant Gliomas-Challenges and Perspectives. *Cells* **2021**, *10*, 2264. [CrossRef] [PubMed]
37. Khan, F.; Pang, L.Z.; Dunterman, M.; Lesniak, M.S.; Heimberger, A.B.; Chen, P.W. Macrophages and microglia in glioblastoma: Heterogeneity, plasticity, and therapy. *J. Clin. Investig.* **2023**, *133*, e163446. [CrossRef] [PubMed]
38. Hu, X.M. Microglia/macrophage polarization: Fantasy or evidence of functional diversity? *J. Cereb. Blood Flow Metab.* **2020**, *40*, S134–S136. [CrossRef]
39. Mantovani, A.; Sica, A.; Sozzani, S.; Allavena, P.; Vecchi, A.; Locati, M. The chemokine system in diverse forms of macrophage activation and polarization. *Trends Immunol.* **2004**, *25*, 677–686. [CrossRef]
40. Shapouri-Moghaddam, A.; Mohammadian, S.; Vazini, H.; Taghadosi, M.; Esmaeili, S.A.; Mardani, F.; Seifi, B.; Mohammadi, A.; Afshari, J.T.; Sahebkar, A. Macrophage plasticity, polarization, and function in health and disease. *J. Cell. Physiol.* **2018**, *233*, 6425–6440. [CrossRef]
41. Woroniecka, K.I.; Rhodin, K.E.; Chongsathidkiet, P.; Keith, K.A.; Fecci, P.E. T-cell Dysfunction in Glioblastoma: Applying a New Framework. *Clin. Cancer Res.* **2018**, *24*, 3792–3802. [CrossRef]
42. Humphries, W.; Wei, J.; Sampson, J.H.; Heimberger, A.B. The Role of Tregs in Glioma-Mediated Immunosuppression: Potential Target for Intervention. *Neurosurg. Clin. N. Am.* **2010**, *21*, 125–137. [CrossRef] [PubMed]
43. Li, C.X.; Jiang, P.; Wei, S.H.; Xu, X.F.; Wang, J.J. Regulatory T cells in tumor microenvironment: New mechanisms, potential therapeutic strategies and future prospects. *Mol. Cancer* **2020**, *19*, 116. [CrossRef] [PubMed]
44. Kelly, W.J.; Giles, A.J.; Gilbert, M. T lymphocyte-targeted immune checkpoint modulation in glioma. *J. Immunother. Cancer* **2020**, *8*, e000379. [CrossRef] [PubMed]
45. Chongsathidkiet, P.; Jackson, C.; Koyama, S.; Loebel, F.; Cui, X.Y.; Farber, S.H.; Woroniecka, K.; Elsamadicy, A.A.; Dechant, C.A.; Kemeny, H.R.; et al. Sequestration of T cells in bone marrow in the setting of glioblastoma and other intracranial tumors. *Nat. Med.* **2018**, *24*, 1459–1468. [CrossRef]
46. Jackson, C.M.; Choi, J.; Lim, M. Mechanisms of immunotherapy resistance: Lessons from glioblastoma. *Nat. Immunol.* **2019**, *20*, 1100–1109. [CrossRef]

47. Sedgwick, A.J.; Ghazanfari, N.; Constantinescu, P.; Mantamadiotis, T.; Barrow, A.D. The Role of NK Cells and Innate Lymphoid Cells in Brain Cancer. *Front. Immunol.* **2020**, *11*, 1549. [CrossRef]

48. Burster, T.; Gärtner, F.; Bulach, C.; Zhanapiya, A.; Gihring, A.; Knippschild, U. Regulation of MHC I Molecules in Glioblastoma Cells and the Sensitizing of NK Cells. *Pharmaceuticals* **2021**, *14*, 236. [CrossRef] [PubMed]

49. Böttcher, J.P.; Sousa, C.R.E. The Role of Type 1 Conventional Dendritic Cells in Cancer Immunity. *Trends Cancer* **2018**, *4*, 784–792. [CrossRef] [PubMed]

50. Srivastava, S.; Jackson, C.; Kim, T.; Choi, J.; Lim, M. A Characterization of Dendritic Cells and Their Role in Immunotherapy in Glioblastoma: From Preclinical Studies to Clinical Trials. *Cancers* **2019**, *11*, 537. [CrossRef]

51. Peng, Q.; Qiu, X.Y.; Zhang, Z.H.; Zhang, S.L.; Zhang, Y.Y.; Liang, Y.; Guo, J.Y.; Peng, H.; Chen, M.Y.; Fu, Y.X.; et al. PD-L1 on dendritic cells attenuates T cell activation and regulates response to immune checkpoint blockade. *Nat. Commun.* **2020**, *11*, 4835. [CrossRef]

52. Massara, M.; Persico, P.; Bonavita, O.; Poeta, V.M.; Locati, M.; Simonelli, M.; Bonecchi, R. Neutrophils in Gliomas. *Front. Immunol.* **2017**, *8*, 1349. [CrossRef] [PubMed]

53. Masucci, M.T.; Minopoli, M.; Carriero, M.V. Tumor Associated Neutrophils. Their Role in Tumorigenesis, Metastasis, Prognosis and Therapy. *Front. Oncol.* **2019**, *9*, 1146. [CrossRef] [PubMed]

54. Wu, L.; Zhang, X.H.F. Tumor-Associated Neutrophils and Macrophages-Heterogenous but Not Chaotic. *Front. Immunol.* **2020**, *11*, 553967. [CrossRef] [PubMed]

55. Khan, S.; Mittal, S.; McGee, K.; Alfaro-Munoz, K.D.; Majd, N.; Balasubramaniyan, V.; de Groot, J.F. Role of Neutrophils and Myeloid-Derived Suppressor Cells in Glioma Progression and Treatment Resistance. *Int. J. Mol. Sci.* **2020**, *21*, 1954. [CrossRef] [PubMed]

56. Di Nunno, V.; Franceschi, E.; Tosoni, A.; Gatto, L.; Bartolini, S.; Brandes, A.A. Glioblastoma Microenvironment: From an Inviolable Defense to a Therapeutic Chance. *Front. Oncol.* **2022**, *12*, 852950. [CrossRef] [PubMed]

57. Zhang, X.H.; Zhao, L.L.; Zhang, H.; Zhang, Y.R.; Ju, H.Y.; Wang, X.Y.; Ren, H.; Zhu, X.; Dong, Y.C. The immunosuppressive microenvironment and immunotherapy in human glioblastoma. *Front. Immunol.* **2022**, *13*, 1003651. [CrossRef] [PubMed]

58. Yasinjan, F.; Xing, Y.; Geng, H.Y.; Guo, R.; Yang, L.; Liu, Z.L.; Wang, H. Immunotherapy: A promising approach for glioma treatment. *Front. Immunol.* **2023**, *14*, 1255611. [CrossRef] [PubMed]

59. Sampson, J.H.; Maus, M.V.; June, C.H. Immunotherapy for Brain Tumors. *J. Clin. Oncol.* **2017**, *35*, 2450–2456. [CrossRef] [PubMed]

60. Lim, M.; Xia, Y.X.; Bettegowda, C.; Weller, M. Current state of immunotherapy for glioblastoma. *Nat. Rev. Clin. Oncol.* **2018**, *15*, 422–442. [CrossRef]

61. Sampson, J.H.; Gunn, M.D.; Fecci, P.E.; Ashley, D.M. Brain immunology and immunotherapy in brain tumours. *Nat. Rev. Cancer* **2020**, *20*, 12–25. [CrossRef]

62. Xu, S.C.; Tang, L.; Li, X.Z.; Fan, F.; Liu, Z.X. Immunotherapy for glioma: Current management and future application. *Cancer Lett.* **2020**, *476*, 1–12. [CrossRef] [PubMed]

63. Chokshi, C.R.; Brakel, B.A.; Tatari, N.; Savage, N.; Salim, S.K.; Venugopal, C.; Singh, S.K. Advances in Immunotherapy for Adult Glioblastoma. *Cancers* **2021**, *13*, 3400. [CrossRef] [PubMed]

64. Ma, Y.F.; Yang, Z.G.; Huntoon, K.; Jiang, W.; Kim, B.Y.S. Advanced Immunotherapy Approaches for Glioblastoma. *Adv. Ther.* **2021**, *4*, 2100046. [CrossRef]

65. Mende, A.L.; Schulte, J.D.; Okada, H.; Clarke, J.L. Current Advances in Immunotherapy for Glioblastoma. *Curr. Oncol. Rep.* **2021**, *23*, 21. [CrossRef] [PubMed]

66. Mahmoud, A.B.; Ajina, R.; Aref, S.; Darwish, M.; Alsayb, M.; Taher, M.; AlSharif, S.A.; Hashem, A.M.; Alkayyal, A.A. Advances in immunotherapy for glioblastoma multiforme. *Front. Immunol.* **2022**, *13*, 944452. [CrossRef] [PubMed]

67. Gilbert, M.R.; Dignam, J.J.; Armstrong, T.S.; Wefel, J.S.; Blumenthal, D.T.; Vogelbaum, M.A.; Colman, H.; Chakravarti, A.; Pugh, S.; Won, M.; et al. A Randomized Trial of Bevacizumab for Newly Diagnosed Glioblastoma. *N. Engl. J. Med.* **2014**, *370*, 699–708. [CrossRef] [PubMed]

68. Ferrara, N.; Hillan, K.J.; Gerber, H.P.; Novotny, W. Discovery and development of bevacizumab, an anti-VEGF antibody for treating cancer. *Nat. Rev. Drug Discov.* **2004**, *3*, 391–400. [CrossRef] [PubMed]

69. Friedman, H.S.; Prados, M.D.; Wen, P.Y.; Mikkelsen, T.; Schiff, D.; Abrey, L.E.; Yung, W.K.A.; Paleologos, N.; Nicholas, M.K.; Jensen, R.; et al. Bevacizumab Alone and in Combination with Irinotecan in Recurrent Glioblastoma. *J. Clin. Oncol.* **2009**, *27*, 4733–4740. [CrossRef] [PubMed]

70. Clavreul, A.; Pourbaghi-Masouleh, M.; Roger, E.; Menei, P. Nanocarriers and nonviral methods for delivering antiangiogenic factors for glioblastoma therapy: The story so far. *Int. J. Nanomed.* **2019**, *14*, 2497–2513. [CrossRef]

71. Martin, J.D.; Seano, G.; Jain, R.K. Normalizing Function of Tumor Vessels: Progress, Opportunities, and Challenges. *Annu. Rev. Physiol.* **2019**, *81*, 505–534. [CrossRef]

72. Lombardi, G.; De Salvo, G.L.; Brandes, A.A.; Eoli, M.; Rudà, R.; Faedi, M.; Lolli, I.; Pace, A.; Daniele, B.; Pasqualetti, F.; et al. Regorafenib compared with lomustine in patients with relapsed glioblastoma (REGOMA): A multicentre, open-label, randomised, controlled, phase 2 trial. *Lancet Oncol.* **2019**, *20*, 110–119. [CrossRef] [PubMed]

73. Belotti, D.; Pinessi, D.; Taraboletti, G. Alternative Vascularization Mechanisms in Tumor Resistance to Therapy. *Cancers* **2021**, *13*, 1912. [CrossRef]

74. Thomas Sandmann, R.B.; Garcia, J.; Li, C.; Cloughesy, T.; Chinot, O.L.; Wick, W.; Nishikawa, R.; Mason, W.; Henriksson, R.; Saran, F.; et al. Patients with Proneural Glioblastoma May Derive Overall Survival Benefit from the Addition of Bevacizumab to First-Line Radiotherapy and Temozolomide: Retrospective Analysis of the AVAglio Trial. *J. Clin. Oncol.* **2015**, *33*, 10. [CrossRef] [PubMed]

75. Korman, A.J.; Garrett-Thomson, S.C.; Lonberg, N. The foundations of immune checkpoint blockade and the ipilimumab approval decennial. *Nat. Rev. Drug Discov.* **2022**, *21*, 509–528. [CrossRef]

76. Berger, K.N.; Pu, J.J. PD-1 pathway and its clinical application: A 20 year journey after discovery of the complete human-gene. *Gene* **2018**, *638*, 20–25. [CrossRef] [PubMed]

77. Han, Y.Y.; Liu, D.D.; Li, L.H. PD-1/PD-L1 pathway: Current researches in cancer. *Am. J. Cancer Res.* **2020**, *10*, 727–742. [PubMed]

78. Cloughesy, T.F.; Mochizuki, A.Y.; Orpilla, J.R.; Hugo, W.; Lee, A.H.; Davidson, T.B.; Wang, A.C.; Ellingson, B.M.; Rytlewski, J.A.; Sanders, C.M.; et al. Neoadjuvant anti-PD-1 immunotherapy promotes a survival benefit with intratumoral and systemic immune responses in recurrent glioblastoma. *Nat. Med.* **2019**, *25*, 477–486. [CrossRef] [PubMed]

79. Lim, M.; Weller, M.; Idbaih, A.; Steinbach, J.; Finocchiaro, G.; Raval, R.R.; Ansstas, G.; Baehring, J.; Taylor, J.W.; Honnorat, J.; et al. Phase III trial of chemoradiotherapy with temozolomide plus nivolumab or placebo for newly diagnosed glioblastoma with methylated promoter. *Neuro-Oncology* **2022**, *24*, 1935–1949. [CrossRef] [PubMed]

80. Han, S.; Ma, E.L.; Wang, X.N.; Yu, C.Y.; Dong, T.; Zhan, W.; Wei, X.Z.; Liang, G.B.; Feng, S.Z. Rescuing defective tumor-infiltrating T-cell proliferation in glioblastoma patients. *Oncol. Lett.* **2016**, *12*, 2924–2929. [CrossRef]

81. Sayegh, E.T.; Oh, T.; Fakurnejad, S.; Bloch, O.; Parsa, A.T. Vaccine therapies for patients with glioblastoma. *J. Neuro-Oncol.* **2014**, *119*, 531–546. [CrossRef]

82. Zhao, B.H.; Wu, J.M.; Li, H.Z.; Wang, Y.K.; Wang, Y.N.; Xing, H.; Wang, Y.; Ma, W.B. Recent advances and future challenges of tumor vaccination therapy for recurrent glioblastoma. *Cell Commun. Signal* **2023**, *21*, 74. [CrossRef] [PubMed]

83. Platten, M.; Bunse, L.; Riehl, D.; Bunse, T.; Ochs, K.; Wick, W. Vaccine Strategies in Gliomas. *Curr. Treat. Options Neurol.* **2018**, *20*, 11. [CrossRef] [PubMed]

84. Swartz, A.M.; Batich, K.A.; Fecci, P.E.; Sampson, J.H. Peptide vaccines for the treatment of glioblastoma. *J. Neuro-Oncol.* **2015**, *123*, 433–440. [CrossRef] [PubMed]

85. Li, L.H.; Zhou, J.; Dong, X.T.; Liao, Q.J.; Zhou, D.B.; Zhou, Y.H. Dendritic cell vaccines for glioblastoma fail to complete clinical translation: Bottlenecks and potential countermeasures. *Int. Immunopharmacol.* **2022**, *109*, 108929. [CrossRef] [PubMed]

86. Zhang, M.; Choi, J.; Lim, M. Advances in Immunotherapies for Gliomas. *Curr. Neurol. Neurosci.* **2022**, *22*, 1–10. [CrossRef] [PubMed]

87. Schuster, J.; Lai, R.K.; Recht, L.D.; Reardon, D.A.; Paleologos, N.A.; Groves, M.D.; Mrugala, M.M.; Jensen, R.; Baehring, J.M.; Sloan, A.; et al. A phase II, multicenter trial of rindopepimut (CDX-110) in newly diagnosed glioblastoma: The ACT III study. *Neuro-Oncology* **2015**, *17*, 854–861. [CrossRef] [PubMed]

88. Weller, M.; Butowski, N.; Tran, D.D.; Recht, L.D.; Lim, M.; Hirte, H.; Ashby, L.; Mechtler, L.; Goldlust, S.A.; Iwamoto, F.; et al. Rindopepimut with temozolomide for patients with newly diagnosed, EGFRvIII-expressing glioblastoma (ACT IV): A randomised, double-blind, international phase 3 trial. *Lancet Oncol.* **2017**, *18*, 1373–1385. [CrossRef] [PubMed]

89. Liau, L.M.; Ashkan, K.; Brem, S.; Campian, J.L.; Trusheim, J.E.; Iwamoto, F.M.; Tran, D.D.; Ansstas, G.; Cobbs, C.S.; Heth, J.A.; et al. Association of Autologous Tumor Lysate-Loaded Dendritic Cell Vaccination with Extension of Survival Among Patients with Newly Diagnosed and Recurrent Glioblastoma A Phase 3 Prospective Externally Controlled Cohort Trial. *JAMA Oncol.* **2023**, *9*, 112–121. [CrossRef] [PubMed]

90. Chen, D.; Yang, J. Development of novel antigen receptors for CAR T-cell therapy directed toward solid malignancies. *Transl. Res.* **2017**, *187*, 11–21. [CrossRef]

91. Bagley, S.J.; Desai, A.S.; Linette, G.P.; June, C.H.; O'Rourke, D.M. CAR T-cell therapy for glioblastoma: Recent clinical advances and future challenges. *Neuro-Oncology* **2018**, *20*, 1429–1438. [CrossRef]

92. Li, L.; Zhu, X.Q.; Qian, Y.; Yuan, X.L.; Ding, Y.; Hu, D.S.; He, X.; Wu, Y. Chimeric Antigen Receptor T-Cell Therapy in Glioblastoma: Current and Future. *Front. Immunol.* **2020**, *11*, 594271. [CrossRef] [PubMed]

93. Jenkins, M.R.; Drummond, K.J. CAR T-Cell Therapy for Glioblastoma. *N. Engl. J. Med.* **2024**, *390*, 1329–1332. [CrossRef] [PubMed]

94. O'Rourke, D.M.; Nasrallah, M.P.; Desai, A.; Melenhorst, J.J.; Mansfield, K.; Morrissette, J.J.D.; Martinez-Lage, M.; Brem, S.; Maloney, E.; Shen, A.; et al. A single dose of peripherally infused EGFRvIII-directed CAR T cells mediates antigen loss and induces adaptive resistance in patients with recurrent glioblastoma. *Sci. Transl. Med.* **2017**, *9*, aaa0984. [CrossRef]

95. Brown, C.E.; Alizadeh, D.; Starr, R.; Weng, L.H.; Wagner, J.R.; Naranjo, A.; Ostberg, J.R.; Blanchard, M.S.; Kilpatrick, J.; Simpson, J.; et al. Regression of Glioblastoma after Chimeric Antigen Receptor T-Cell Therapy. *N. Engl. J. Med.* **2016**, *375*, 2561–2569. [CrossRef] [PubMed]

96. Ahmed, N.; Brawley, V.; Hegde, M.; Bielamowicz, K.; Kalra, M.; Landi, D.; Robertson, C.; Gray, T.L.; Diouf, O.; Wakefield, A.; et al. HER2-Specific Chimeric Antigen Receptor-Modified Virus-Specific T Cells for Progressive Glioblastoma A Phase 1 Dose-Escalation Trial. *JAMA Oncol.* **2017**, *3*, 1094–1101. [CrossRef] [PubMed]

97. Aurelian, L. Oncolytic viruses as immunotherapy: Progress and remaining challenges. *Oncotargets Ther.* **2016**, *9*, 2627–2637. [CrossRef] [PubMed]

98. Lawler, S.E.; Speranza, M.C.; Cho, C.F.; Chiocca, E.A. Oncolytic Viruses in Cancer Treatment A Review. *JAMA Oncol.* **2017**, *3*, 841–849. [CrossRef] [PubMed]

99. Wollmann, G.; Ozduman, K.; van den Pol, A.N. Oncolytic Virus Therapy for Glioblastoma Multiforme. *Cancer J.* **2012**, *18*, 69–81. [CrossRef] [PubMed]
100. Martikainen, M.; Essand, M. Virus-Based Immunotherapy of Glioblastoma. *Cancers* **2019**, *11*, 186. [CrossRef]
101. Liu, P.H.; Wang, Y.N.; Wang, Y.K.; Kong, Z.R.; Chen, W.Q.; Li, J.T.; Chen, W.L.; Tong, Y.R.; Ma, W.B.; Wang, Y. Effects of oncolytic viruses and viral vectors on immunity in glioblastoma. *Gene Ther.* **2022**, *29*, 115–126. [CrossRef]
102. Qi, Z.B.; Long, X.Y.; Liu, J.Y.; Cheng, P. Glioblastoma microenvironment and its reprogramming by oncolytic virotherapy. *Front. Cell. Neurosci.* **2022**, *16*, 819363. [CrossRef] [PubMed]
103. Desjardins, A.; Gromeier, M.; Herndon, J.E.; Beaubier, N.; Bolognesi, D.P.; Friedman, A.H.; Friedman, H.S.; McSherry, F.; Muscat, A.M.; Nair, S.; et al. Recurrent Glioblastoma Treated with Recombinant Poliovirus. *N. Engl. J. Med.* **2018**, *379*, 150–161. [CrossRef] [PubMed]
104. Durham, P.G.; Butnariu, A.; Alghorazi, R.; Pinton, G.; Krishna, V.; Dayton, P.A. Current clinical investigations of focused ultrasound blood-brain barrier disruption: A review. *Neurotherapeutics* **2024**, *21*, e00352. [CrossRef] [PubMed]
105. Niedbala, M.; Malarz, K.; Sharma, G.; Kramer-Marek, G.; Kaspera, W. Glioblastoma: Pitfalls and Opportunities of Immunotherapeutic Combinations. *Oncotargets Ther.* **2022**, *15*, 437–468. [CrossRef]

Disclaimer/Publisher's Note: The statements, opinions and data contained in all publications are solely those of the individual author(s) and contributor(s) and not of MDPI and/or the editor(s). MDPI and/or the editor(s) disclaim responsibility for any injury to people or property resulting from any ideas, methods, instructions or products referred to in the content.

Article

Genetically Predicted Association of 91 Circulating Inflammatory Proteins with Multiple Sclerosis: A Mendelian Randomization Study

Xin'ai Li [1], Zhiguo Ding [1,2,3], Shuo Qi [1,2,3], Peng Wang [4], Junhui Wang [5,*] and Jingwei Zhou [6,*]

1   Department of Thyropathy, Dongzhimen Hospital, Beijing University of Chinese Medicine, Beijing 100013, China; xinaili@bucm.edu.cn (X.L.); dingzhiguo_1@163.com (Z.D.); shuoqi@bucm.edu.cn (S.Q.)

2   Sun Simiao Institute, Beijing University of Chinese Medicine, Tongchuan 727000, China

3   Thyropathy Hospital, Sun Simiao Hospital, Beijing University of Chinese Medicine, Tongchuan 727000, China

4   The Key Laboratory of Cardiovascular Remodelling and Function Research, Chinese Ministry of Education and Chinese Ministry of Public Health, Department of Cardiology, Qilu Hospital of Shandong University, Jinan 250012, China; 17866806032@163.com

5   Lunenfeld-Tanenbaum Research Institute, Mount Sinai Hospital, Toronto, ON M5G 1X5, Canada

6   The 1st Ward, Department of Nephrology and Endocrinology, Dongzhimen Hospital, Beijing University of Chinese Medicine, Beijing 100010, China

*   Correspondence: junhui@lunenfeld.ca (J.W.); 13910634708@163.com (J.Z.)

Abstract: Previous studies have validated a close association between inflammatory factors and multiple sclerosis (MS), but their causal relationship is not fully profiled yet. This study used Mendelian randomization (MR) to investigate the causal effect of circulating inflammatory proteins on MS. Data from a large-scale genome-wide association study (GWAS) were analyzed using a two-sample MR method to explore the relationship between 91 circulating inflammatory proteins and MS. The inverse-variance-weighted (IVW) analysis was employed as the main method for evaluating exposures and outcomes. Furthermore, series of the methods of MR Egger, weighted median, simple mode, and weighted mode were used to fortify the final results. The results of the IVW method were corrected with Bonferroni (bon) and false discovery rate (fdr) for validating the robustness of results and ensuring the absence of heterogeneity and horizontal pleiotropy. The sensitivity analysis was also performed. The results of the forward MR analysis showed that higher levels of CCL25 were found to be associated with an increased risk of MS according to IVW results, OR: 1.085, 95% CI (1.011, 1.165), $p = 2.42 \times 10^{-2}$, adjusted p_adj_bon = 1, p_adj_fdr = 0.307. Similarly, higher levels of CXCL10 were found to be associated with an increased risk of MS, OR: 1.231, 95% CI (1.057, 1.433), $p = 7.49 \times 10^{-3}$, adjusted p_adj_bon = 0.682, p_adj_fdr = 0.227. In contrast, elevated levels of neurturin (NRTN) were associated with a decreased risk of MS, OR: 0.815, 95% CI (0.689, 0.964), $p = 1.68 \times 10^{-2}$, adjusted p_adj_bon = 1, p_adj_fdr = 0.307. Reverse MR analysis showed no causal relationship between MS and the identified circulating inflammatory cytokines. The effects of heterogeneity and level pleiotropy were further excluded by sensitivity analysis. This study provides new insights into the relationship between circulating inflammatory proteins and MS and brings up a new possibility of using these cytokines as potential biomarkers and therapeutic targets. The data in this study show that there are only weak associations between inflammatory molecules and MS risk, which did not survive bon and fdr correction, and the obtained p-values are quite low. Therefore, further studies on larger samples are needed.

Keywords: multiple sclerosis; circulating inflammatory protein; mendelian randomization; biomarkers

Citation: Li, X.; Ding, Z.; Qi, S.; Wang, P.; Wang, J.; Zhou, J. Genetically Predicted Association of 91 Circulating Inflammatory Proteins with Multiple Sclerosis: A Mendelian Randomization Study. *Brain Sci.* **2024**, *14*, 833. https://doi.org/10.3390/brainsci14080833

Academic Editors: Andrew Clarkson and Allan Bieber

Received: 14 July 2024
Revised: 7 August 2024
Accepted: 15 August 2024
Published: 19 August 2024

Copyright: © 2024 by the authors. Licensee MDPI, Basel, Switzerland. This article is an open access article distributed under the terms and conditions of the Creative Commons Attribution (CC BY) license (https://creativecommons.org/licenses/by/4.0/).

1. Introduction

Characterized by reactive gliosis, axonal damage, neuronal degeneration, and inflammatory cell infiltration, MS is an autoimmune disorder affecting the central nervous system (CNS) [1]. The occurrence of this disease is mainly associated with genetic factors [2], environmental factors [3], lifestyle [4], viral infection [5], and immunologic factors [6]. Its pathology is featured by a loss of myelin sheath in the CNS with infiltration of a large number of inflammatory cells [7]. It leads to a heterogeneous set of symptoms and signs due to involvement of different motor, sensory, and autonomic nervous systems [8]. Patients usually present with neurological symptoms such as cognitive impairment, motor ataxia, blindness, and loss of coordination [9]. The disease poses a significant burden to patients and families [10]. The prevalence of MS has been increasing globally [11]. Therefore, identifying modifiable risk factors is an imperative task of the medical community for the purpose of developing novel strategies to manage the disease.

Chronic inflammatory response in MS arises from the activation of innate and adaptive immune responses in the CNS [12]. Although the CNS is considered an immune-privileged organ with highly controlled adaptive immunity and inflammation [13], recent findings have shed light on the fact that neuroinflammation or neuroimmune response play an essential role on the development of neurodegenerative conditions [14,15]. Among these, several cytokines with elevated expression levels in MS patients have been considered to be important biomarkers for this disease [16,17]. However, comprehensive preclinical and clinical studies are still not in place, and the clinical significance of these cytokines is not yet fully understood. The study of identifying certain circulating inflammatory proteins that are closely associated with MS may provide a new perspective in terms of the diagnosis and treatment of MS.

Genetic variation is the foundation for causal inference in MR. In order to deduce the influence of biological factors on disease, the fundamental idea involves utilizing the influence of randomly assigned genotypes on phenotype [18,19]. This method is effective in diminishing the effects of biases and confounders caused by behavioral or environmental influences while relying on the random distribution of genetic variation during meiosis [20]. When dealing with rare diseases, it proves to be highly effective in tackling the drawbacks of conventional randomized controlled trials and observational studies [21]. The GWAS database was utilized for data mining in order to facilitate a two-sample MR analysis, which aimed to uncover the causal association between circulating inflammatory proteins and MS. Our primary aim is to investigate how genetic proxy inflammatory protein levels influence the likelihood of developing MS.

2. Methods

2.1. Study Design

Using the two-sample MR study, this research was performed to examine the causal link between circulating inflammatory proteins and MS. Achieving valid results in MR analysis hinges on satisfying three key assumptions. Single nucleotide polymorphisms (SNPs) were used as instrumental variables (IVs) in this study, and it is vital for IVs acting as risk factors to meet three conditions, as illustrated in Figure 1. First, a reliable connection to the risk factor being evaluated must be established (correlation assumption). Second, it is essential to ensure independence from any recognized or unrecognized confounding variables (independence assumption). Third, only the risk factor should influence the outcome, while any other direct causal pathway will be excluded (exclusionary restriction assumption) [20]. Through the utilization of openly accessible data derived from extensive GWAS and consortia, ethical clearance was not a prerequisite for the conduct of this study. Visual summary of the analysis is shown in Figure 2.

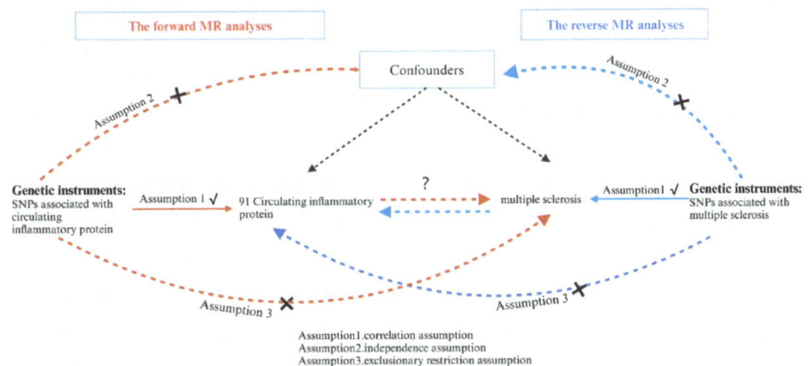

Figure 1. Flowchart of bidirectional MR analysis. Red arrows indicate the flowchart for forward MR and blue arrows indicate the flowchart for reverse MR, black arrows indicate forward MR and reverse MR sharing.

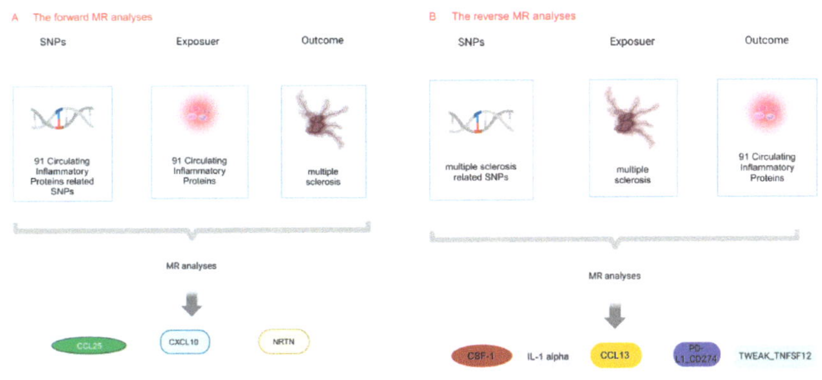

Figure 2. Visual summary of the analysis. The MR analysis unveils the relationships between circulating inflammatory proteins and MS. (**A**) The forward MR analyses; (**B**) the reverse MR analyses.

2.2. Selection of IVs for MR Analyses

Genetic variants achieving genome-wide significance were recognized as IVs. We adjusted the criteria by increasing the threshold to $p < 5 \times 10^{-6}$ to incorporate more inflammatory proteins for all 91 inflammatory proteins, and this allowed us the opportunity to retrieve IVs. Similarly, in the reverse analysis, SNPs that reached the threshold $p < 5 \times 10^{-6}$ were used as IVs for MS. To maintain variant independence, SNPs with high linkage disequilibrium (specified as $r^2 > 0.001$ and kb < 10,000 kb) were omitted. Variants with conflicting allelic frequencies underwent harmonization or elimination to align with the estimated effects. Palindromic SNPs were adjusted based on a maximum minor allele frequency (MAF) criterion set at 0.01. In instances where the results lacked directly related SNPs for the exposure, we opted for proxy SNPs exhibiting high correlation ($r^2 > 0.8$) with the desired variant. These rigorously selected SNPs served as the definitive genetic IVs for the ensuing MR evaluations. For each SNP, the F-statistic was derived using Beta^2/se^2, where Beta symbolizes the estimated effect allele for the exposure and SE indicated the standard error. The formula $2 \times \text{Beta}^2 \times \text{MAF} \times (1 - \text{MAF})$ [22] was employed to compute the variance fraction attributed to each SNP. An F-statistic exceeding 10 indicated a robust correlation between the IV and exposure, ensuring that the MR outcomes remained unaffected by weak instrument bias [23].

2.3. Data Sources

We obtained summary data related to MS from GWAS, including genetic data from 47,429 cases and 68,374 subjects from controls of European ancestry [24]. In 11 cohorts, 91 circulating inflammatory proteins were identified from a population of 14,824 individuals of European origin. The original paper describes the procedures utilized for assessing inflammatory proteins [25]. The complete per-protein GWAS summary statistics can be downloaded at https://www.phpc.cam.ac.uk/ceu/proteins (accessed on 1 June 2024) and from the EBI GWAS catalog (accession number GCST90274758-GCST90274848). Between the exposure and outcome groups, specific information on the 91 circulating inflammatory proteins is shown in Table 1. There will be no overlap in population selection.

Table 1. Ninety-one circulating inflammatory proteins.

Number	Abbreviations	Full Name	ID
1	CXCL9	Chemokine (C-X-C motif) ligand 9	GCST90274784
2	TWEAK_TNFSF12	Tumor Necrosis Factor-like Weak Inducer of Apoptosis—Tumor Necrosis Factor Superfamily Member 12	GCST90274846
3	CCL23	Chemokine (C-C motif) ligand 23	GCST90274767
4	ADA	Adenosine Deaminase	GCST90274759
5	CASP-8	Caspase-8	GCST90274763
6	CXCL6	Chemokine (C-X-C motif) ligand 6	GCST90274783
7	PD-L1_CD274	Programmed Cell Death 1 Ligand 1-Cluster of Differentiation 274	GCST90274832
8	IL-15RA	Interleukin-15 Receptor Alpha	GCST90274800
9	IL-1 alpha	Interleukin-1 alpha	GCST90274805
10	CST5	Cystatin 5	GCST90274777
11	IL-10	Interleukin-10	GCST90274795
12	IL-10RA	Interleukin-10 Receptor Alpha	GCST90274796
13	NTF3	Neurotrophin 3	GCST90274829
14	IL-12B	Interleukin-12B	GCST90274798
15	IL-33	Interleukin-33	GCST90274812
16	CCL13	Chemokine (C-C motif) ligand 13	GCST90274824
17	OPG_TNFRSF11B	Osteoprotegerin—Tumor Necrosis Factor Receptor Superfamily Member 11B	GCST90274830
18	IL-4	Interleukin-4	GCST90274813
19	LIF	Leukemia Inhibitory Factor	GCST90274819
20	FIt3L	Fms-related tyrosine kinase 3 ligand	GCST90274791
21	TNFRSF9	Tumor Necrosis Factor Receptor Superfamily Member 9	GCST90274841
22	IL-5	Interleukin-5	GCST90274814
23	DNER	Delta/Notch-like EGF repeat-containing receptor	GCST90274785
24	CCL20	Chemokine (C-C motif) ligand 20	GCST90274766
25	TNFSF14	Tumor Necrosis Factor Superfamily Member 14	GCST90274842
26	IL-6	Interleukin-6	GCST90274815
27	CCL19	Chemokine (C-C motif) ligand 19	GCST90274765
28	TNFB_LTA	Tumor Necrosis Factor B/Lymphotoxin-alpha	GCST90274840
29	SIRT2	Sirtuin 2	GCST90274834
30	STAMPB	Six transmembrane proteins of prostate B	GCST90274837
31	4EBP1_EIF4EBP1	Eukaryotic Translation Initiation Factor 4E-Binding Protein 1	GCST90274758
32	IL-18R1	Interleukin-18 Receptor 1	GCST90274804
33	LAP TGF-beta-1	Latency Associated Peptide Transforming Growth Factor-beta 1	GCST90274818
34	IL-7	Interleukin-7	GCST90274816
35	CXCL5	Chemokine (C-X-C motif) ligand 5	GCST90274782
36	NRTN	Neurturin	GCST90274828
37	IL-13	Interleukin-13	GCST90274799
38	CDCP1	CUB Domain Containing Protein 1	GCST90274775
39	TGF-alpha	Transforming Growth Factor-alpha	GCST90274838
40	FGF-21	Fibroblast Growth Factor 21	GCST90274788
41	SLAMF1	Signaling Lymphocytic Activation Molecule Family Member 1	GCST90274835
42	CXCL1	Chemokine (C-X-C motif) ligand 1	GCST90274779
43	TRAIL	Tumor Necrosis Factor-related Apoptosis-inducing Ligand	GCST90274843
44	IL-17C	Interleukin-17C	GCST90274802
45	MMP-1	Matrix Metalloproteinase-1	GCST90274826

Table 1. *Cont.*

Number	Abbreviations	Full Name	ID
46	CXCL11	Chemokine (C-X-C motif) ligand 11	GCST90274781
47	FGF-23	Fibroblast Growth Factor 23	GCST90274789
48	uPA_PLAU	Urokinase- type Plasminogen Activator (uPA)/Plasminogen Activator, Urokinase (PLAU)	GCST90274847
49	FGF-19	Fibroblast Growth Factor 19	GCST90274787
50	CX3CL1	Chemokine (C-X3-C motif) ligand 1	GCST90274778
51	CXCL10	Chemokine (C-X-C motif) ligand 10	GCST90274780
52	FGF-5	Fibroblast Growth Factor 5	GCST90274790
53	CCL25	Chemokine (C-C motif) ligand 25	GCST90274768
54	ARTN	Artemin	GCST90274760
55	VEGF_A	Vascular Endothelial Growth Factor A	GCST90274848
56	SCF_KITLG	Stem Cell Factor (SCF)/KIT Ligand (KITLG)	GCST90274833
57	ST1A1_SULT1A1	Sulfotransferase Family 1A Member 1 (SULT1A1)	GCST90274836
58	CD244	Cluster of Differentiation 244	GCST90274771
59	CCL11	Chemokine (C-C motif) ligand 11	GCST90274764
60	MMP-10	Matrix Metalloproteinase-10	GCST90274827
61	TRANCE	Tumor Necrosis Factor (TNF)-related Activation-induced Cytokine	GCST90274844
62	IL-2	Interleukin-2	GCST90274806
63	Beta-NGF_NGF	Beta-Nerve Growth Factor (NGF)	GCST90274762
64	IL-2RB	Interleukin-2 Receptor Subunit Beta	GCST90274811
65	OSM	Oncostatin M	GCST90274831
66	AXIN1	Axis Inhibition Protein 1	GCST90274761
67	CCL2	Chemokine (C-C motif) ligand 2	GCST90274821
68	CCL8	Chemokine (C-C motif) ligand 8	GCST90274822
69	CD6	Cluster of Differentiation 6	GCST90274774
70	MIP-1 alpha_CCL3	Macrophage Inflammatory Protein-1 alpha (MIP-1 alpha)/Chemokine (C-C motif) ligand 3 (CCL3)	GCST90274825
71	hGDNF_GDNF	Human Glial Cell Line-Derived Neurotrophic Factor (GDNF)	GCST90274792
72	CCL7	Chemokine (C-C motif) ligand 7	GCST90274823
73	LIF-R	Leukemia Inhibitory Factor Receptor	GCST90274820
74	EN-RAGE_S100A12	Extracellular Newly Identified RAGE Binding Protein (EN-RAGE)/S100 Calcium Binding Protein A12 (S100A12)	GCST90274786
75	CCL4	Chemokine (C-C motif) ligand 4	GCST90274770
76	HGF	Hepatocyte Growth Factor	GCST90274793
77	IL-17A	Interleukin-17A	GCST90274801
78	CD5	Cluster of Differentiation 5	GCST90274773
79	CSF-1	Colony-Stimulating Factor 1	GCST90274776
80	CCL28	Chemokine (C-C motif) ligand 28	GCST90274769
81	CD40	Cluster of Differentiation 40	GCST90274772
82	IL-22RA1	Interleukin-22 Receptor Subunit Alpha 1	GCST90274809
83	TNF	Tumor Necrosis Factor	GCST90274839
84	IL-18	Interleukin-18	GCST90274803
85	IL-20RA	Interleukin-20 Receptor Subunit Alpha	GCST90274808
86	IL-24	Interleukin-24	GCST90274810
87	IFNG	Interferon gamma	GCST90274794
88	TSLP	Thymic Stromal Lymphopoietin	GCST90274845
89	IL-20	Interleukin-20	GCST90274807
90	IL10RB	Interleukin 10 Receptor Subunit Beta	GCST90274797
91	IL-8	Interleukin-8	GCST90274817

2.4. MR and Sensitivity Analysis

The results of five Mendelian methods, MR Egger [26], weighted median [27], IVW [28,29], simple mode, and weighted mode [30], were used, and the results of the IVW method were corrected with bon and fdr [31]. When there were no multiple validities in IV [32], the IVW method demonstrated the greatest statistical validity and effectiveness [32]. Hence, IVW was employed as the primary research methodology in this study [33–35]. Furthermore, the techniques we incorporated, specifically weighted mode, simple mode, weighted median, and MR Egger, enhanced the conclusive findings [26,27]. By applying Cochran's Q test, the heterogeneity of SNPs in IVW and MR Egger was evaluated, consequently bolstering the robustness of the results [36]. The intercept of MR-Egger [26] was deployed to examine horizontal pleiotropy [26]. In order to ascertain whether a single SNP was the sole factor influencing the causal effect, we conducted a leave-one-out analysis [37]. The detection of pleiotropic residuals and outliers was executed using MR-Presso [26]. MR-Steiger was employed to establish the correct direction of causality. In instances where the exposure was anticipated to result in the outcome, it was classified as TRUE; if not, it was marked as FALSE [38]. All analyses were two-sided and executed through the Two Sample MR and MRPRESSO packages in R software version 4.3.2.

3. Results

3.1. Effect of 91 Circulating Inflammatory Proteins on MS

In the forward MR analysis, details of the genetic tools used to assess the effects of 91 plasma proteins on MS were recorded separately (Table S1). All MS-associated SNPs used as IVs had F-statistics higher than 10, suggesting a strong prediction of MS, whereas there was less evidence of weak IV bias in our study. According to the IVW results, higher levels of CCL25 were found to be associated with an increased risk of MS, OR: 1.085, 95% CI (1.011, 1.165) $p = 2.42 \times 10^{-2}$, adjusted p_adj_bon = 1, p_adj_fdr = 0.307. Similarly, there was an association between higher levels of CXCL10 and increased risk of MS, OR: 1.231, 95% CI (1.057, 1.433), $p = 7.49 \times 10^{-3}$, adjusted p_adj_bon = 0.682, p_adj_fdr = 0.227. On the contrary, elevated NRTN levels were associated with a reduced risk of MS, OR: 0.815, 95% CI (0.689, 0.964), $p = 1.68 \times 10^{-2}$, adjusted p_adj_bon = 1, p_adj_fdr = 0.307, as shown in Table 2.

Table 2. MR analysis of the causal association between circulating inflammatory proteins and risk of MS.

Exposure	Outcome	Method	Nsnp	B	Se	Pval	OR (95%CI)	P_adj_bon	P_adj_fdr
CCL25		MR Egger		0.156	0.053	7.49×10^{-3}	1.169 (1.054, 1.297)		
		Weighted median		0.113	0.044	9.72×10^{-3}	1.120 (1.028, 1.220)		
		IVW	24	0.082	0.036	2.42×10^{-2}	1.085 (1.011, 1.165)	1	0.307
		Simple mode		−0.037	0.100	7.17×10^{-1}	0.964 (0.793, 1.172)		
		Weighted mode		0.122	0.039	5.17×10^{-3}	1.130 (1.046, 1.221)		
CXCL10	MS	MR Egger		0.188	0.188	3.30×10^{-1}	1.207 (0.835, 1.746)		
		Weighted median		0.149	0.101	1.39×10^{-1}	1.161 (0.953, 1.414)		
		IVW	20	0.208	0.078	7.49×10^{-3}	1.231 (1.057, 1.433)	0.682	0.227
		Simple mode		0.063	0.181	7.31×10^{-1}	1.065 (0.747, 1.518)		
		Weighted mode		0.105	0.215	6.32×10^{-1}	1.110 (0.729, 1.692)		
NRTN		MR Egger		0.160	0.282	5.85×10^{-1}	1.173 (0.674, 2.040)		
		Weighted median		−0.304	0.118	1.02×10^{-2}	0.738 (0.585, 0.931)		
		IVW	12	−0.205	0.086	1.68×10^{-2}	0.815 (0.689, 0.964)	1	0.307
		Simple mode		−0.411	0.187	5.00×10^{-2}	0.663 (0.459, 0.956)		
		Weighted mode		−0.374	0.189	7.39×10^{-2}	0.688 (0.475, 0.997)		

3.2. Effect of MS on 91 Circulating Inflammatory Proteins

In the inverse MR analysis, we found that MS was associated with five circulating inflammatory proteins, among which IL-1 alpha was positively causally associated with MS: $p = 3.30 \times 10^{-4}$, OR:1.035, 95% CI (1.016, 1.054), adjusted p_adj_bon = 0.026, p_adj_fdr = 0.026; and the blood inflammatory factor CSF-1: $p = 2.35 \times 10^{-2}$, OR: 0.983, 95% CI (0.969, 0.998), adjusted p_adj_bon = 1, p_adj_fdr = 0.376; CCL13: $p = 1.45 \times 10^{-2}$, OR: 0.980, 95% CI (0.964, 0.996), adjusted p_ adj_bon = 1, p_adj_fdr = 0.300; PD-L1_CD274:

$p = 1.50 \times 10^{-2}$, OR:0.982, 95% CI (0.968, 0.996), adjusted p_adj_bon = 1, p_adj_fdr = 0.300; TWEAK_TNFSF12: $p = 1.43 \times 10^{-2}$, OR: 0.981, 95% CI (0.966, 0.996), adjusted p_adj_bon = 1, p_adj_fdr = 0.300 were negatively causally associated with MS, as shown in Table 3. However, no bi-directional genetic causality was found. Genetic tools used to assess the association of MS and 91 plasma proteins were documented in Table S2.

Table 3. MR analysis of the causal association between MS and risk of circulating inflammatory proteins.

Exposure	Outcome	Method	Nsnp	B	Se	Pval	OR (95%CI)	P_adj_bon	P_adj_fdr
MS	CSF-1	MR Egger		−0.008	0.011	4.75×10^{-1}	0.992 (0.971, 1.014)		
		Weighted median		−0.006	0.012	6.06×10^{-1}	0.994 (0.972, 1.017)		
		IVW	122	−0.017	0.007	2.35×10^{-2}	0.983 (0.969, 0.998)	1	0.376
		Simple mode		−0.022	0.026	4.05×10^{-1}	0.979 (0.930, 1.029)		
		Weighted mode		−0.010	0.011	3.60×10^{-1}	0.990 (0.970, 1.011)		
	IL-1 alpha	MR Egger		0.048	0.015	1.46×10^{-3}	1.049 (1.019, 1.080)		
		Weighted median		0.017	0.016	2.70×10^{-1}	1.018 (0.987, 1.050)		
		IVW	122	0.034	0.009	3.30×10^{-4}	1.035 (1.016, 1.054)	0.026	0.026
		Simple mode		0.017	0.0345	6.33×10^{-1}	1.017 (0.950, 1.088)		
		Weighted mode		0.023	0.018	1.95×10^{-1}	1.023 (0.989, 1.059)		
	CCL13	MR Egger		−0.027	0.013	3.10×10^{-2}	0.973 (0.949, 0.997)		
		Weighted median		−0.027	0.012	3.12×10^{-2}	0.974 (0.950, 0.998)		
		IVW	122	−0.020	0.008	1.45×10^{-2}	0.980 (0.964, 0.996)	1	0.300
		Simple mode		−0.015	0.029	5.89×10^{-1}	0.985 (0.932, 1.041)		
		Weighted mode		−0.027	0.011	1.81×10^{-2}	0.973 (0.952, 0.995)		
	PD-L1_CD274	MR Egger		−0.017	0.0112	1.32×10^{-1}	0.983 (0.962, 1.005)		
		Weighted median		−0.015	0.013	2.51×10^{-1}	0.985 (0.961, 1.011)		
		IVW	123	−0.018	0.007	1.50×10^{-2}	0.982 (0.968, 0.996)	1	0.300
		Simple mode		−0.003	0.026	9.10×10^{-1}	0.997 (0.948, 1.049)		
		Weighted mode		−0.015	0.011	1.64×10^{-1}	0.985 (0.964, 1.006)		
	TWEAK_TNFSF12	MR Egger		−0.0135	0.0116	2.49×10^{-1}	0.987 (0.964, 1.009)		
		Weighted median		−0.0123	0.011	2.76×10^{-1}	0.988 (0.9663, 1.010)		
		IVW	123	−0.019	0.008	1.43×10^{-2}	0.981 (0.966, 0.996)	1	0.300
		Simple mode		0.008	0.027	7.68×10^{-1}	1.008 (0.955, 1.064)		
		Weighted mode		−0.010	0.011	3.97×10^{-1}	0.990 (0.969, 1.013)		

3.3. Sensitivity Analysis

As shown in Tables 4 and 5, in the IVW and MR-Egger analysis based on Cochran's Q test, the results indicated no heterogeneity of SNPs. No signs of horizontal pleiotropy were found in the MR-Egger intercept. The MR-Presso method did not identify any outliers. In addition, scatter plots ruled out potential outliers and horizontal pleiotropy (Figures 3 and 4). In addition, no SNPs with large effect sizes were tested for bias estimation by the leave-one-out test (Figures 5 and 6). The MR-Steiger analysis results validated the accuracy of the directionality and ruled out any indication of reverse causality. Sensitivity analysis eliminated the impacts of horizontal pleiotropy and heterogeneity, confirming the reliability of the outcomes. Presently, there is evidence from MS indicating a connection between MS and inflammatory proteins [39–42]. Nevertheless, the precise cause and effect association is still unclear at the genetic level as a result of constraints in research. In the context of the potential causal link of 91 circulating inflammatory proteins with MS in this exploratory study, we completed a comprehensive two-sample MR analysis.

Table 4. Sensitive analysis of the causal association between circulating inflammatory proteins and risk of MS.

Inflammatory Proteins	Outcomes	SNPs	Cochran's Q Test		MR-Egger Intercept		MR-Presso		MR-Steiger
			IVW	MR Egger	Egger Intercept	*p* Value	Global Test RSSobs	*p* Value	Causal Direction
CCL25		24	0.165	0.273	−0.013	0.078	0.187	0.034 (Outlier-corrected, 0 Outlier)	TRUE
CXCL10	MS	20	0.098	0.074	0.002	0.911	0.064	0.015 (Outlier-corrected, 0 Outlier)	TRUE
NRTN		12	0.501	0.581	−0.030	0.206	0.534	0.031 (Outlier-corrected, 0 Outlier)	TRUE

Table 5. Sensitive analysis of the causal association between MS and risk of circulating inflammatory proteins.

Exposure	Outcomes	Nsnp	Cochran's Q Test		MR-Egger Intercept		MR-Presso		MR-Steiger
			IVW	MR Egger	Egger Intercept	*p* Value	Global Test RSSobs	*p* Value	Causal Direction
	CSF-1	122	0.59	0.595	−0.002	0.282	0.622	0.023 (Outlier-corrected, 0 Outlier)	TRUE
	IL-1 alpha	122	0.092	0.099	−0.003	0.222	0.081	0.0005 (Outlier-corrected, 0 Outlier)	TRUE
MS	CCL13	122	0.050	0.050	0.002	0.459	0.053	0.016 (Outlier-corrected, 0 Outlier)	TRUE
	PD-L1_CD274	123	0.639	0.615	−0.0003	0.885	0.616	0.014 (Outlier-corrected, 0 Outlier)	TRUE
	TWEAK_TNFSF12	123	0.269	0.257	−0.001	0.526	0.308	0.016 (Outlier-corrected, 0 Outlier)	TRUE

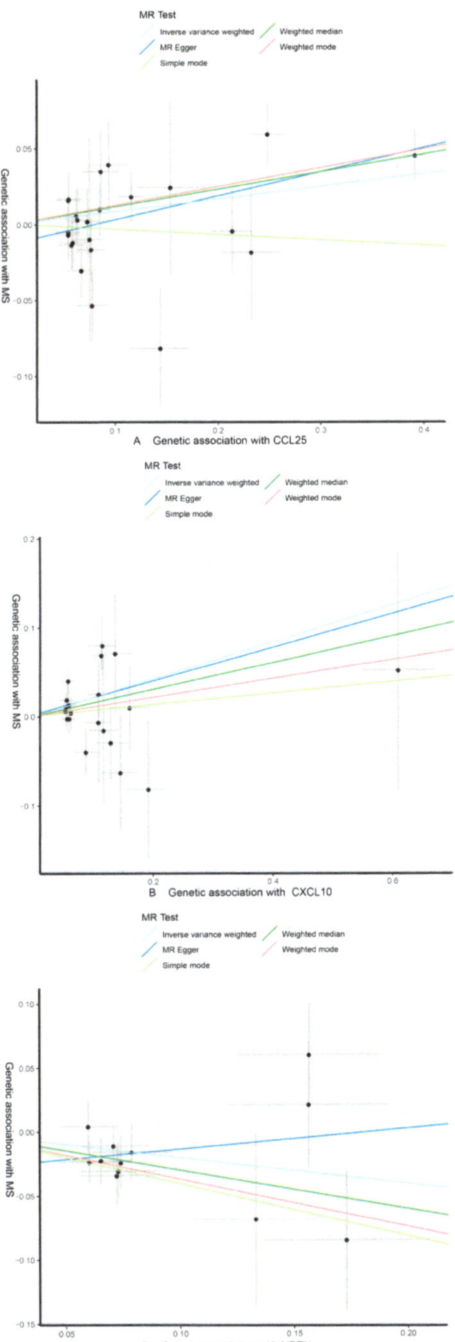

Figure 3. CCL25, CXCL10, and NRTN of MS with scatter plots, respectively. (**A**) MS as the outcome, with CCL25 as the exposure; (**B**) MS as the outcome, with CXCL10 as the exposure; (**C**) MS as the outcome, with NRTN as the exposure.

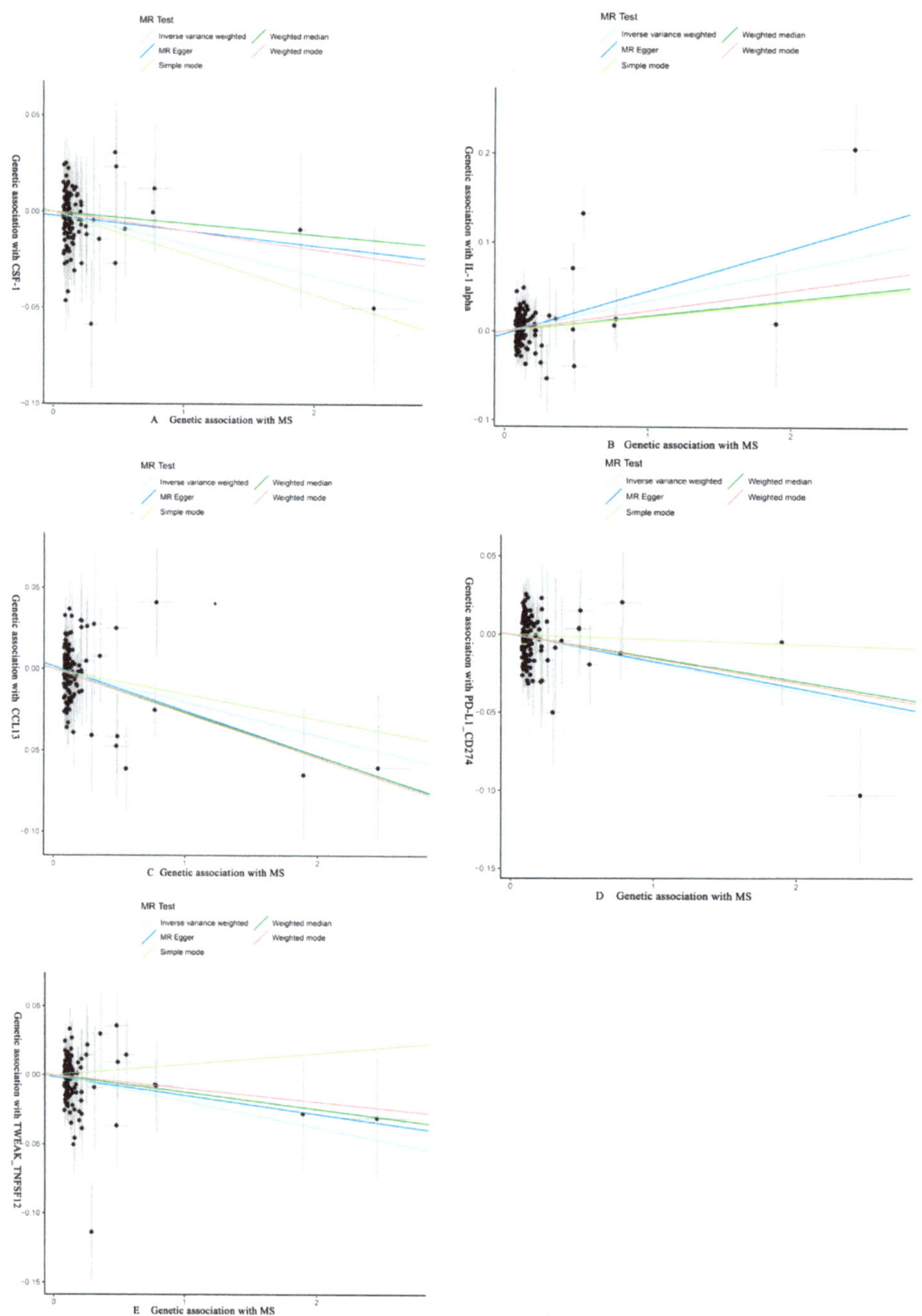

Figure 4. MS of CSF-1, IL-1 alpha, CCL13, PD-L1_CD274, TWEAK_TNFSF12 with scatter plots, respectively. (**A**) CSF-1 as the outcome, with MS as the exposure; (**B**) IL-1 alpha as the outcome, with MS as the exposure; (**C**) CCL13 as the outcome, with MS as the exposure; (**D**) PD-L1_CD274 as the outcome, with MS as the exposure; (**E**) TWEAK_TNFSF12 as the outcome, with MS as the exposure.

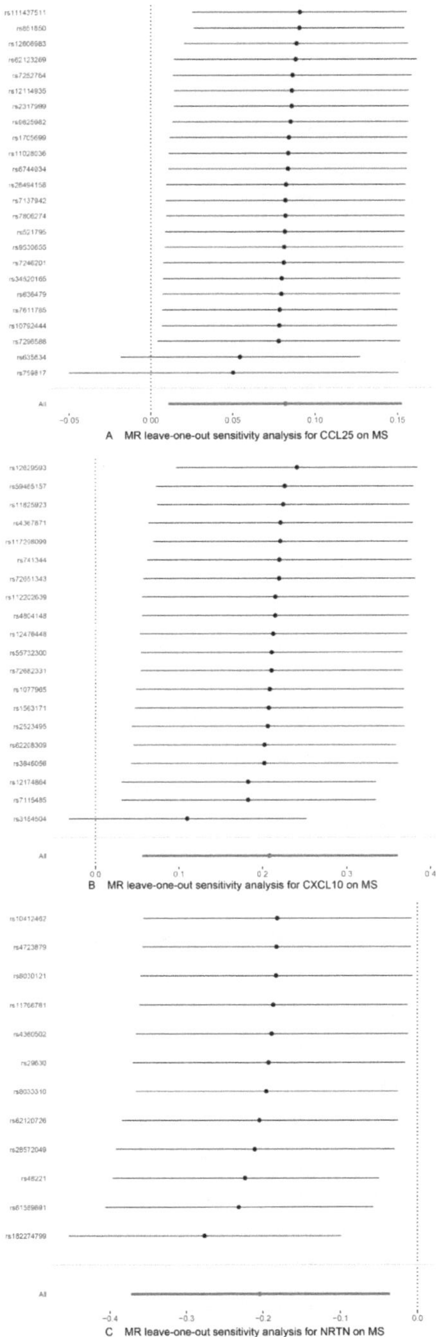

Figure 5. Use of the IVW method to display the results of leave-one-out analyses and assess the impact of individual SNPs on the overall MR results by excluding each SNP in turn. (**A**) MS as the outcome, with CCL25 as the exposure; (**B**) MS as the outcome, with CXCL10 as the exposure; (**C**) MS as the outcome, with NRTN as the exposure.

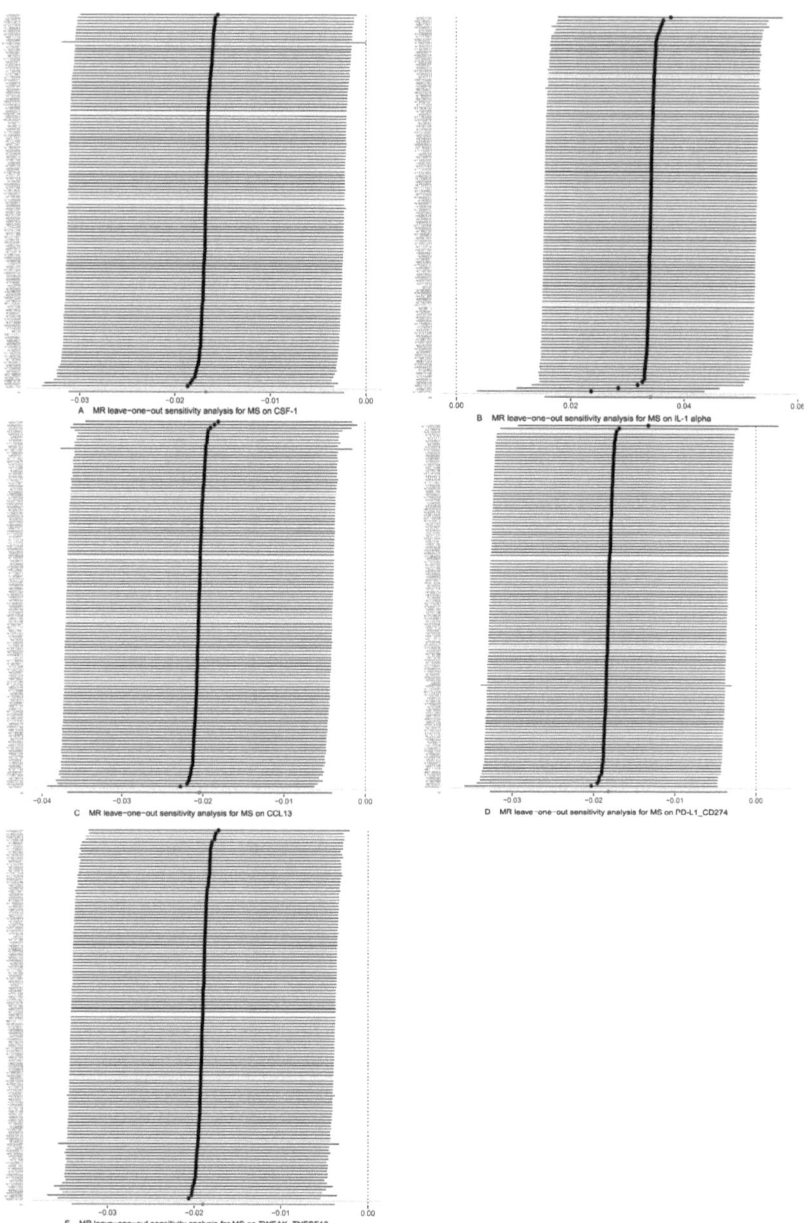

Figure 6. Use of the IVW method to display the results of leave-one-out analyses and assess the impact of individual SNPs on the overall MR results by excluding each SNP in turn. (**A**) CSF-1 as the outcome, with MS as the exposure; (**B**) IL-1 alpha as the outcome, with MS as the exposure; (**C**) CCL13 as the outcome, with MS as the exposure; (**D**) PD-L1_CD274 as the outcome, with MS as the exposure; (**E**) TWEAK_TNFSF12 as the outcome, with MS as the exposure.

4. Discussion

In this study, bidirectional MR analysis was performed to explore the association of 91 circulating inflammatory cytokine proteins with MS. The aim was to explore genetic evidence for a potential causal relationship between circulating inflammatory cytokines and MS risk. Our study showed that CCL25, CXCL10, and NRTN levels were associated with the likelihood of developing MS according to forward analysis. In addition, reverse MR analysis showed that CSF-1, IL-1 alpha, CCL13, PD-L1_CD274, and TWEAK_TNFSF12 were genetically causally associated with MS. No circulating inflammatory proteins were found to be bi-directionally causally associated with the disease. This study is the first to explore the interrelationship between inflammatory proteins and MS through bidirectional MR analysis. This study provides some evidence to use medications targeting inflammatory factors to treat MS in the future.

Focal cerebral white matter lesions characterized by inflammation and demyelination are the most obvious hallmark of MS histopathology. The inflammatory infiltrate consists mainly of phagocytes, T cells, and B cells originating from the blood [43]. Cortical lesions present in early MS are associated with significant inflammation [44]. It has been shown that Th1 and Th17 responses are the main cause of MS progression [45]. It has also been shown [46] that the neutrophil–lymphocyte ratio (NLR) is significantly increased in MS patients compared to controls. A study has shown that impairment of CD200-CD200R-mediated macrophage silencing exacerbates CNS inflammation and neuronal degeneration [47]. Our study showed that high levels of CCL25 and CXCL10 were associated and positively correlated with the development of MS, while NRTN levels were negatively correlated with MS risk. This suggests that elevated levels of CCL25 and CXCL10 could promote the development of MS, whereas elevated NRTN could reduce the risk of developing MS. A notable feature of this study was the use of MR analysis to assess the pathogenic impact of circulating inflammatory proteins on MS risk. An MR approach could skillfully handle confounding factors, reverse causality, and increase the confidence in causal inferences. These results could position these proteins as potential biomarkers for MS diagnosis and provide a new way to extensively understand the pathogenesis of the disease.

CCL25 is classified as a chemokine that is expressed in the thymus [48]. CCL25 is predominantly found in the intestinal epithelium and thymus. However, other parenchymal cells, such vascular endothelial cells, can produce it as well. These CCL25 expression cells can guide immature T cells to migrate into the thymus, where they turn mature and release [49], and subsequently are able to engage in numerous inflammatory responses. In recent years, research has brought more evidence of how CCR9/CCL25 contributes to inflammation, which are associated with several diseases, including cardiovascular disease (CVD), hepatitis, arthritis [50], inflammatory bowel disease [51], and asthma. Toll-like receptor 4 (TLR4) plays a role in the pathogenesis of experimental autoimmune encephalomyelitis (EAE) by regulating CCL25/CCR9 expression in response to Th17 infiltration [52]. Recent data suggest that CCR9 blockade or inhibition leads to a reduction in lymphocyte infiltration and amelioration of clinical symptoms in many clinical inflammatory disorders [53,54]. The fact that CCR9 mediates effector T-cell infiltration into the CNS suggests that CCL25/CCR9 is a potential new biologic target for the inhibiting of pathologic lymphocyte recruitment in MS therapy [55]. This indicates CCL25 as a risk factor for MS, which is revealed in this study as well (OR: 1.085, 95% CI: 1.011–1.165, $p = 2.42 \times 10^{-2}$, adjusted p_adj_bon = 1, p_adj_fdr = 0.307) and suggests its potential as a biomarker and therapeutic target of MS.

As a tiny protein, CXCL10 is an "inflammatory" chemokine that attaches to CXCR3 and enables immune response via leukocyte activation and recruitment, including eosinophils, T cells, NK cells, and monocytes [56]. Specimens of cerebrospinal fluid (CSF) were obtained from active MS patients and CXCL10 exhibited a higher level than those from the patients with non-inflammatory symptoms, according to the report of Sørensen et al. [57]. In these MS patients, CXCR3 was found to be expressed in over 90% of T cells from CSF, a substantially larger percentage than those T cells from peripheral blood. Previous studies have confirmed that the CXCL10/CXCR3 axis plays a critical role in MS patients [57–59]. In another study, Sørensen et al. [60] discovered that the CSF from MS patients contained significant CXCL10 levels, in line with the presence of more leukocytes. Furthermore, Comini-Frota et al. [61] discovered that the levels of serum CXCL10 were elevated among MS patients in comparison to the normal group. Here, our results also demonstrated that elevated CXCL10 levels were a risk factor of MS (OR: 1.231, 95% CI: 1.057–1.433, $p = 7.49 \times 10^{-3}$, adjusted p_adj_bon = 0.682, p_adj_fdr = 0.227), which is in line with the previous results.

There are a number of studies that have shown that administration of neurotrophic factors improves the survival of injured neurons in models of neuronal injury [62,63]. Neurotrophic factors show promise in promoting functional recovery following demyelination or nerve injury, making them good candidates in the study of MS to unveil pathogenesis and explore new treatment. Transplanted fibroblasts expressing BDNF or NT3 in adult rats with spinal cord injury could lead to improved myelin formation, OPC (oligodendrocyte progenitor cell) proliferation, and axonal growth [64]. When Schwann cells expressing BDNF or NT3 were transplanted into the spinal cord of demyelinated mice, they showed similar recovery of motor function [65]. The neuroglial cell-lineage-derived neurotrophic factor (GDNF) families, including NRTN, have been reported to play key roles in the maturation of neuromuscular synapses during development and post-nerve injury regeneration [66]. The results of this study suggest that NRTN has a potential protective effect on MS (OR: 0.815, 95% CI: 0.689–0.964, 1.68×10^{-2}, adjusted p_adj_bon = 1, p_adj_fdr = 0.307), which is consistent with the above-mentioned results from the research of GDNF families. However, the underlying mechanism by which NRTN influences MS needs to be further investigated in future studies.

This study employed MR analysis to evaluate the causal relationship between circulating inflammatory proteins and MS. This approach was selected to minimize confounding factors and potential reverse causation in causal inference. Genetic variants associated with these proteins were derived from recent GWAS meta-analyses, ensuring robust instrumental strength in MR analysis. The regression intercept tests of MR-PRESSO and MR-Egger were completed to determine multiplicity levels. To reduce the potential for bias, we applied a two-sample MR framework with outcome pooled data and exposure that does not overlap.

However, this study is subject to a few limitations. First, the exclusion of horizontal pleiotropy and IV assumptions were the specific assumptions integrated by the MR analysis. Sensitivity analyses were conducted to tackle these concerns. However, we cannot completely eliminate the possibility of unmeasured pleiotropy or confounders. Second, our research only involved individuals of European descent, potentially restricting the applicability of our conclusions to other demographic groups. Third, the obtained p-values are quite low, and no causal connection of circulating inflammatory proteins with MS had statistical significance in the wake of applying bon correction and fdr correction. The stringent parameters utilized in our analysis may have contributed to the false negative outcomes. Further studies on larger samples to confirm these findings are needed.

5. Conclusions

This study evaluated the potential causal relationship between 91 circulating inflammatory proteins and the risk of MS. We identified the plasma proteins CCL25 and CCL10 as being associated with an increased risk of MS, whereas NRTN was associated with a reduction in MS risk. However, only weak associations of inflammatory molecules and MS risk were found in our data, which did not survive bon and fdr correction. Therefore, further studies on larger samples are needed. The findings highlight that these inflammatory proteins in circulation are closely associated with MS to a certain extent, although they may not be the direct cause of MS. More research is needed to further substantiate these findings and investigate additional possible mechanisms for the association between inflammatory proteins and the risk of MS. However, the implications of these results are still significant for future studies in providing a research direction of deciphering the involvement of inflammation in MS and could help the development of new therapies of MS by targeting specific inflammatory pathways.

Supplementary Materials: The following supporting information can be downloaded from: https://www.mdpi.com/article/10.3390/brainsci14080833/s1. Table S1: Detailed information of genetic instruments of 91 plasma proteins for effect on multiple sclerosis. Table S2: Detailed information of genetic instruments of multiple sclerosis for effect on 91 plasma proteins.

Author Contributions: X.L. wrote the draft; Z.D. and S.Q. helped review and edit the manuscript; P.W. prepared tables and pictures; J.W. and J.Z. proposed the idea and designed the structure. All authors have read and agreed to the published version of the manuscript.

Funding: This study is financially supported by the Science and Technology Department of Shaanxi Province (2022KXJ-019): Scientist + engineer team construction for Research and development and application of Integrated Chinese and Western Medicine in diagnosis and treatment of difficult thyroid diseases, China; Shanxi Central Administration Bureau (2022-SLRH-LG-005), "Double Chain Integration" Research and Innovation Team of Chinese and Western Integrated Thyroid Disease Diagnosis and Treatment, China; Science and Technology Department of Shanxi Province (2023-ZLSF-56), Joint Research and Development of hospital preparations for Hashimoto thyroiditis, Exploration of therapeutic mechanism and Formulation of TCM diagnosis and treatment Plan, China.

Institutional Review Board Statement: This research has been conducted using published studies and consortia providing publicly available summary statistics. All original studies have been approved by the corresponding ethical review board. In addition, no individual-level data were used in this study. Therefore, no new ethical review board approval was required.

Informed Consent Statement: The participants have provided informed consent.

Data Availability Statement: Data are available in a publicly accessible repository. The original data used in the study are openly available in GWAS Catalog (ebi.ac.uk): https://www.ebi.ac.uk/gwas/ (accessed on 1 June 2024) and https://gwas.mrcieu.ac.uk/ (accessed on 1 June 2024).

Acknowledgments: We would like to express our gratitude to UK biobank dataset and Finngen study for their generosity of openly sharing their data.

Conflicts of Interest: The authors declare no conflicts of interest.

References

1. Coclitu, C.; Constantinescu, C.S.; Tanasescu, R. The future of multiple sclerosis treatments. *Expert Rev. Neurother.* **2016**, *16*, 1341–1356. [CrossRef]
2. Harirchian, M.H.; Fatehi, F.; Sarraf, P.; Honarvar, N.M.; Bitarafan, S. Worldwide prevalence of familial multiple sclerosis: A systematic review and meta-analysis. *Mult. Scler. Relat. Disord.* **2018**, *20*, 43–47. [CrossRef] [PubMed]
3. Browne, P.; Chandraratna, D.; Angood, C.; Tremlett, H.; Baker, C.; Taylor, B.V.; Thompson, A.J. Atlas of Multiple Sclerosis 2013: A growing global problem with widespread inequity. *Neurology* **2014**, *83*, 1022–1024. [CrossRef] [PubMed]
4. Mandoj, C.; Renna, R.; Plantone, D.; Sperduti, I.; Cigliana, G.; Conti, L.; Koudriavtseva, T. Anti-annexin antibodies, cholesterol levels and disability in multiple sclerosis. *Neurosci. Lett.* **2015**, *606*, 156–160. [CrossRef] [PubMed]
5. Voumvourakis, K.I.; Fragkou, P.C.; Kitsos, D.K.; Foska, K.; Chondrogianni, M.; Tsiodras, S. Human herpesvirus 6 infection as a trigger of multiple sclerosis: An update of recent literature. *BMC Neurol.* **2022**, *22*, 57. [CrossRef]
6. Kamphuis, W.W.; Derada Troletti, C.; Reijerkerk, A.; Romero, I.A.; de Vries, H.E. The blood-brain barrier in multiple sclerosis: microRNAs as key regulators. *CNS Neurol. Disord. Drug Targets* **2015**, *14*, 157–167. [CrossRef] [PubMed]
7. Garg, N.; Smith, T.W. An update on immunopathogenesis, diagnosis, and treatment of multiple sclerosis. *Brain Behav.* **2015**, *5*, e00362. [CrossRef]
8. Ward, M.; Goldman, M.D. Epidemiology and Pathophysiology of Multiple Sclerosis. *Continuum* **2022**, *28*, 988–1005. [CrossRef] [PubMed]
9. Dutra, R.C.; Moreira, E.L.; Alberti, T.B.; Marcon, R.; Prediger, R.D.; Calixto, J.B. Spatial reference memory deficits precede motor dysfunction in an experimental autoimmune encephalomyelitis model: The role of kallikrein-kinin system. *Brain Behav. Immun.* **2013**, *33*, 90–101. [CrossRef]
10. GBD 2016 Multiple Sclerosis Collaborators. Global, regional, and national burden of multiple sclerosis 1990–2016: A systematic analysis for the Global Burden of Disease Study 2016. *Lancet Neurol.* **2019**, *18*, 269–285. [CrossRef]

11. Walton, C.; King, R.; Rechtman, L.; Kaye, W.; Leray, E.; Marrie, R.A.; Robertson, N.; La Rocca, N.; Uitdehaag, B.; van der Mei, I.; et al. Rising prevalence of multiple sclerosis worldwide: Insights from the Atlas of MS, third edition. *Mult. Scler.* **2020**, *26*, 1816–1821. [CrossRef] [PubMed]

12. Sun, Y.; Yu, H.; Guan, Y. Glia Connect Inflammation and Neurodegeneration in Multiple Sclerosis. *Neurosci. Bull.* **2023**, *39*, 466–478. [CrossRef] [PubMed]

13. Rauf, A.; Badoni, H.; Abu-Izneid, T.; Olatunde, A.; Rahman, M.M.; Painuli, S.; Semwal, P.; Wilairatana, P.; Mubarak, M.S. Neuroinflammatory Markers: Key Indicators in the Pathology of Neurodegenerative Diseases. *Molecules* **2022**, *27*, 3194. [CrossRef] [PubMed]

14. Schain, M.; Kreisl, W.C. Neuroinflammation in Neurodegenerative Disorders-a Review. *Curr. Neurol. Neurosci. Rep.* **2017**, *17*, 25. [CrossRef] [PubMed]

15. Liu, Z.; Cheng, X.; Zhong, S.; Zhang, X.; Liu, C.; Liu, F.; Zhao, C. Peripheral and Central Nervous System Immune Response Crosstalk in Amyotrophic Lateral Sclerosis. *Front. Neurosci.* **2020**, *14*, 575. [CrossRef] [PubMed]

16. Tehrani, A.R.; Gholipour, S.; Sharifi, R.; Yadegari, S.; Abbasi-Kolli, M.; Masoudian, N. Plasma levels of CTRP-3, CTRP-9 and apelin in women with multiple sclerosis. *J. Neuroimmunol.* **2019**, *333*, 576968. [CrossRef] [PubMed]

17. Grzegorski, T.; Iwanowski, P.; Kozubski, W.; Losy, J. The alterations of cerebrospinal fluid TNF-alpha and TGF-beta2 levels in early relapsing-remitting multiple sclerosis. *Immunol. Res.* **2022**, *70*, 708–713. [CrossRef] [PubMed]

18. Evans, D.M.; Davey Smith, G. Mendelian Randomization: New Applications in the Coming Age of Hypothesis-Free Causality. *Annu. Rev. Genom. Hum. Genet.* **2015**, *16*, 327–350. [CrossRef]

19. Davey Smith, G.; Hemani, G. Mendelian randomization: Genetic anchors for causal inference in epidemiological studies. *Hum. Mol. Genet.* **2014**, *23*, R89–R98. [CrossRef]

20. Emdin, C.A.; Khera, A.V.; Kathiresan, S. Mendelian Randomization. *Jama* **2017**, *318*, 1925–1926. [CrossRef] [PubMed]

21. Larsson, S.C.; Butterworth, A.S.; Burgess, S. Mendelian randomization for cardiovascular diseases: Principles and applications. *Eur. Heart J.* **2023**, *44*, 4913–4924. [CrossRef]

22. Teumer, A.; Chaker, L.; Groeneweg, S.; Li, Y.; Di Munno, C.; Barbieri, C.; Schultheiss, U.T.; Traglia, M.; Ahluwalia, T.S.; Akiyama, M.; et al. Genome-wide analyses identify a role for SLC17A4 and AADAT in thyroid hormone regulation. *Nat. Commun.* **2018**, *9*, 4455. [CrossRef]

23. Shungin, D.; Winkler, T.W.; Croteau-Chonka, D.C.; Ferreira, T.; Locke, A.E.; Mägi, R.; Strawbridge, R.J.; Pers, T.H.; Fischer, K.; Justice, A.E.; et al. New genetic loci link adipose and insulin biology to body fat distribution. *Nature* **2015**, *518*, 187–196. [CrossRef] [PubMed]

24. International Multiple Sclerosis Genetics Consortium. Multiple sclerosis genomic map implicates peripheral immune cells and microglia in susceptibility. *Science* **2019**, *365*, eaav7188. [CrossRef]

25. Zhao, J.H.; Stacey, D.; Eriksson, N.; Macdonald-Dunlop, E.; Hedman, Å.K.; Kalnapenkis, A.; Enroth, S.; Cozzetto, D.; Digby-Bell, J.; Marten, J.; et al. Genetics of circulating inflammatory proteins identifies drivers of immune-mediated disease risk and therapeutic targets. *Nat. Immunol.* **2023**, *24*, 1540–1551. [CrossRef]

26. Bowden, J.; Davey Smith, G.; Burgess, S. Mendelian randomization with invalid instruments: Effect estimation and bias detection through Egger regression. *Int. J. Epidemiol.* **2015**, *44*, 512–525. [CrossRef]

27. Bowden, J.; Davey Smith, G.; Haycock, P.C.; Burgess, S. Consistent Estimation in Mendelian Randomization with Some Invalid Instruments Using a Weighted Median Estimator. *Genet. Epidemiol.* **2016**, *40*, 304–314. [CrossRef] [PubMed]

28. Burgess, S.; Scott, R.A.; Timpson, N.J.; Davey Smith, G.; Thompson, S.G. Using published data in Mendelian randomization: A blueprint for efficient identification of causal risk factors. *Eur. J. Epidemiol.* **2015**, *30*, 543–552. [CrossRef] [PubMed]

29. Burgess, S.; Butterworth, A.; Thompson, S.G. Mendelian randomization analysis with multiple genetic variants using summarized data. *Genet. Epidemiol.* **2013**, *37*, 658–665. [CrossRef]

30. Hartwig, F.P.; Davey Smith, G.; Bowden, J. Robust inference in summary data Mendelian randomization via the zero modal pleiotropy assumption. *Int. J. Epidemiol.* **2017**, *46*, 1985–1998. [CrossRef] [PubMed]

31. Glickman, M.E.; Rao, S.R.; Schultz, M.R. False discovery rate control is a recommended alternative to Bonferroni-type adjustments in health studies. *J. Clin. Epidemiol.* **2014**, *67*, 850–857. [CrossRef] [PubMed]

32. Burgess, S.; Davey Smith, G.; Davies, N.M.; Dudbridge, F.; Gill, D.; Glymour, M.M.; Hartwig, F.P.; Kutalik, Z.; Holmes, M.V.; Minelli, C.; et al. Guidelines for performing Mendelian randomization investigations: Update for summer 2023. *Wellcome Open Res.* **2019**, *4*, 186. [CrossRef]

33. Burgess, S.; Dudbridge, F.; Thompson, S.G. Combining information on multiple instrumental variables in Mendelian randomization: Comparison of allele score and summarized data methods. *Stat. Med.* **2016**, *35*, 1880–1906. [CrossRef]

34. Yavorska, O.O.; Burgess, S. MendelianRandomization: An R package for performing Mendelian randomization analyses using summarized data. *Int. J. Epidemiol.* **2017**, *46*, 1734–1739. [CrossRef]

35. Slob, E.A.W.; Burgess, S. A comparison of robust Mendelian randomization methods using summary data. *Genet. Epidemiol.* **2020**, *44*, 313–329. [CrossRef]

36. Greco, M.F.; Minelli, C.; Sheehan, N.A.; Thompson, J.R. Detecting pleiotropy in Mendelian randomisation studies with summary data and a continuous outcome. *Stat. Med.* **2015**, *34*, 2926–2940. [CrossRef]

37. Burgess, S.; Bowden, J.; Fall, T.; Ingelsson, E.; Thompson, S.G. Sensitivity Analyses for Robust Causal Inference from Mendelian Randomization Analyses with Multiple Genetic Variants. *Epidemiology* **2017**, *28*, 30–42. [CrossRef]

38. Hemani, G.; Tilling, K.; Davey Smith, G. Orienting the causal relationship between imprecisely measured traits using GWAS summary data. *PLoS Genet.* **2017**, *13*, e1007081.
39. Huang, J.; Khademi, M.; Fugger, L.; Lindhe, Ö.; Novakova, L.; Axelsson, M.; Malmeström, C.; Constantinescu, C.; Lycke, J.; Piehl, F.; et al. Inflammation-related plasma and CSF biomarkers for multiple sclerosis. *Proc. Natl. Acad. Sci. USA* **2020**, *117*, 12952–12960. [CrossRef]
40. Ruiz, F.; Vigne, S.; Pot, C. Resolution of inflammation during multiple sclerosis. *Semin. Immunopathol.* **2019**, *41*, 711–726. [CrossRef] [PubMed]
41. Pegoretti, V.; Swanson, K.A.; Bethea, J.R.; Probert, L.; Eisel, U.L.M.; Fischer, R. Inflammation and Oxidative Stress in Multiple Sclerosis: Consequences for Therapy Development. *Oxidative Med. Cell. Longev.* **2020**, *2020*, 7191080. [CrossRef]
42. Jank, L.; Bhargava, P. Relationship Between Multiple Sclerosis, Gut Dysbiosis, and Inflammation: Considerations for Treatment. *Neurol. Clin.* **2024**, *42*, 55–76. [CrossRef]
43. Kuhlmann, T.; Ludwin, S.; Prat, A.; Antel, J.; Brück, W.; Lassmann, H. An updated histological classification system for multiple sclerosis lesions. *Acta Neuropathol.* **2017**, *133*, 13–24. [CrossRef] [PubMed]
44. Popescu, B.F.; Lucchinetti, C.F. Meningeal and cortical grey matter pathology in multiple sclerosis. *BMC Neurol.* **2012**, *12*, 11. [CrossRef]
45. Hedegaard, C.J.; Krakauer, M.; Bendtzen, K.; Lund, H.; Sellebjerg, F.; Nielsen, C.H. T helper cell type 1 (Th1), Th2 and Th17 responses to myelin basic protein and disease activity in multiple sclerosis. *Immunology* **2008**, *125*, 161–169. [CrossRef]
46. Fahmi, R.M.; Ramadan, B.M.; Salah, H.; Elsaid, A.F.; Shehta, N. Neutrophil-lymphocyte ratio as a marker for disability and activity in multiple sclerosis. *Mult. Scler. Relat. Disord.* **2021**, *51*, 102921. [CrossRef]
47. Meuth, S.G.; Simon, O.J.; Grimm, A.; Melzer, N.; Herrmann, A.M.; Spitzer, P.; Landgraf, P.; Wiendl, H. CNS inflammation and neuronal degeneration is aggravated by impaired CD200-CD200R-mediated macrophage silencing. *J. Neuroimmunol.* **2008**, *194*, 62–69. [CrossRef]
48. Qiuping, Z.; Jei, X.; Youxin, J.; Wei, J.; Chun, L.; Jin, W.; Qun, W.; Yan, L.; Chunsong, H.; Mingzhen, Y.; et al. CC chemokine ligand 25 enhances resistance to apoptosis in CD4+ T cells from patients with T-cell lineage acute and chronic lymphocytic leukemia by means of livin activation. *Cancer Res.* **2004**, *64*, 7579–7587. [CrossRef]
49. Wu, X.; Sun, M.; Yang, Z.; Lu, C.; Wang, Q.; Wang, H.; Deng, C.; Liu, Y.; Yang, Y. The Roles of CCR9/CCL25 in Inflammation and Inflammation-Associated Diseases. *Front. Cell Dev. Biol.* **2021**, *9*, 686548. [CrossRef] [PubMed]
50. Wu, W.; Doan, N.; Said, J.; Karunasiri, D.; Pullarkat, S.T. Strong expression of chemokine receptor CCR9 in diffuse large B-cell lymphoma and follicular lymphoma strongly correlates with gastrointestinal involvement. *Hum. Pathol.* **2014**, *45*, 1451–1458. [CrossRef] [PubMed]
51. Kalindjian, S.B.; Kadnur, S.V.; Hewson, C.A.; Venkateshappa, C.; Juluri, S.; Kristam, R.; Kulkarni, B.; Mohammed, Z.; Saxena, R.; Viswanadhan, V.N.; et al. A New Series of Orally Bioavailable Chemokine Receptor 9 (CCR9) Antagonists; Possible Agents for the Treatment of Inflammatory Bowel Disease. *J. Med. Chem.* **2016**, *59*, 3098–3111. [CrossRef]
52. Zhang, Y.; Han, J.; Wu, M.; Xu, L.; Wang, Y.; Yuan, W.; Hua, F.; Fan, H.; Dong, F.; Qu, X.; et al. Toll-Like Receptor 4 Promotes Th17 Lymphocyte Infiltration Via CCL25/CCR9 in Pathogenesis of Experimental Autoimmune Encephalomyelitis. *J. Neuroimmune Pharmacol.* **2019**, *14*, 493–502. [CrossRef] [PubMed]
53. Braitch, M.; Constantinescu, C.S. The role of osteopontin in experimental autoimmune encephalomyelitis (EAE) and multiple sclerosis (MS). *Inflamm. Allergy Drug Targets* **2010**, *9*, 249–256. [CrossRef]
54. Tubo, N.J.; Wurbel, M.A.; Charvat, T.T.; Schall, T.J.; Walters, M.J.; Campbell, J.J. A systemically-administered small molecule antagonist of CCR9 acts as a tissue-selective inhibitor of lymphocyte trafficking. *PLoS ONE* **2012**, *7*, e50498. [CrossRef]
55. Trivedi, P.J.; Schmidt, C.; Bruns, T. Letter: The therapeutic potential of targeting CCL25/CCR9 in colonic inflammatory bowel disease-reading between the lines. *Aliment. Pharmacol. Ther.* **2016**, *44*, 307–308. [CrossRef] [PubMed]
56. Vazirinejad, R.; Ahmadi, Z.; Kazemi Arababadi, M.; Hassanshahi, G.; Kennedy, D. The biological functions, structure and sources of CXCL10 and its outstanding part in the pathophysiology of multiple sclerosis. *Neuroimmunomodulation* **2014**, *21*, 322–330. [CrossRef]
57. Sørensen, T.L.; Tani, M.; Jensen, J.; Pierce, V.; Lucchinetti, C.; Folcik, V.A.; Qin, S.; Rottman, J.; Sellebjerg, F.; Strieter, R.M.; et al. Expression of specific chemokines and chemokine receptors in the central nervous system of multiple sclerosis patients. *J. Clin. Investig.* **1999**, *103*, 807–815. [CrossRef] [PubMed]
58. Simpson, J.E.; Newcombe, J.; Cuzner, M.L.; Woodroofe, M.N. Expression of the interferon-gamma-inducible chemokines IP-10 and Mig and their receptor, CXCR3, in multiple sclerosis lesions. *Neuropathol. Appl. Neurobiol.* **2000**, *26*, 133–142. [CrossRef]
59. Balashov, K.E.; Rottman, J.B.; Weiner, H.L.; Hancock, W.W. CCR5(+) and CXCR3(+) T cells are increased in multiple sclerosis and their ligands MIP-1alpha and IP-10 are expressed in demyelinating brain lesions. *Proc. Natl. Acad. Sci. USA* **1999**, *96*, 6873–6878. [CrossRef]
60. Sørensen, T.L.; Sellebjerg, F.; Jensen, C.V.; Strieter, R.M.; Ransohoff, R.M. Chemokines CXCL10 and CCL2: Differential involvement in intrathecal inflammation in multiple sclerosis. *Eur. J. Neurol.* **2001**, *8*, 665–672. [CrossRef]
61. Comini-Frota, E.R.; Teixeira, A.L.; Angelo, J.P.; Andrade, M.V.; Brum, D.G.; Kaimen-Maciel, D.R.; Foss, N.T.; Donadi, E.A. Evaluation of serum levels of chemokines during interferon-β treatment in multiple sclerosis patients: A 1-year, observational cohort study. *CNS Drugs* **2011**, *25*, 971–981. [CrossRef] [PubMed]

62. Yan, Q.; Elliott, J.; Snider, W.D. Brain-derived neurotrophic factor rescues spinal motor neurons from axotomy-induced cell death. *Nature* **1992**, *360*, 753–755. [CrossRef] [PubMed]
63. Mo, L.; Yang, Z.; Zhang, A.; Li, X. The repair of the injured adult rat hippocampus with NT-3-chitosan carriers. *Biomaterials* **2010**, *31*, 2184–2192. [CrossRef] [PubMed]
64. Rosenberg, S.S.; Ng, B.K.; Chan, J.R. The quest for remyelination: A new role for neurotrophins and their receptors. *Brain Pathol.* **2006**, *16*, 288–294. [CrossRef] [PubMed]
65. McTigue, D.M.; Horner, P.J.; Stokes, B.T.; Gage, F.H. Neurotrophin-3 and brain-derived neurotrophic factor induce oligodendrocyte proliferation and myelination of regenerating axons in the contused adult rat spinal cord. *J. Neurosci.* **1998**, *18*, 5354–5365. [CrossRef]
66. Baudet, C.; Pozas, E.; Adameyko, I.; Andersson, E.; Ericson, J.; Ernfors, P. Retrograde signaling onto Ret during motor nerve terminal maturation. *J. Neurosci.* **2008**, *28*, 963–975. [CrossRef]

Disclaimer/Publisher's Note: The statements, opinions and data contained in all publications are solely those of the individual author(s) and contributor(s) and not of MDPI and/or the editor(s). MDPI and/or the editor(s) disclaim responsibility for any injury to people or property resulting from any ideas, methods, instructions or products referred to in the content.

MDPI AG
Grosspeteranlage 5
4052 Basel
Switzerland
Tel.: +41 61 683 77 34

Brain Sciences Editorial Office
E-mail: brainsci@mdpi.com
www.mdpi.com/journal/brainsci

Disclaimer/Publisher's Note: The statements, opinions and data contained in all publications are solely those of the individual author(s) and contributor(s) and not of MDPI and/or the editor(s). MDPI and/or the editor(s) disclaim responsibility for any injury to people or property resulting from any ideas, methods, instructions or products referred to in the content.

www.ingramcontent.com/pod-product-compliance
Lightning Source LLC
LaVergne TN
LVHW071939160726
843515LV00011B/2647